Rare Earth Elements

Special Publications 79

RARE EARTH ELEMENTS

Sustainable Recovery, Processing, and Purification

Athanasios K. Karamalidis
Roderick Eggert

Editors

This work is a co-publication of
the American Geophysical Union and John Wiley and Sons, Inc.

ADVANCING EARTH AND SPACE SCIENCES

WILEY

This edition first published 2025

Published under the aegis of the AGU Publications Committee

Matthew Giampoala, Vice President, Publications
Steven A. Hauck, II, Chair, Publications Committee
For details about the American Geophysical Union, visit us at www.agu.org.

Registered Office
John Wiley & Sons, Inc., 111 River Street, Hoboken, NJ 07030, USA

Editorial Office
111 River Street, Hoboken, NJ 07030, USA

For details of our global editorial offices, customer services, and more information about Wiley products, visit us at www.wiley.com.

Wiley also publishes its books in a variety of electronic formats and by print-on-demand. Some content that appears in standard print versions of this book may not be available in other formats.

Library of Congress Cataloging-in-Publication Data
Names: Karamalidis, Athanasios K., editor. | Eggert, Roderick G., editor.
Title: Rare earth elements : sustainable recovery, processing, and purification / Athanasios K. Karamalidis, Roderick Eggert, editors.
Other titles: Rare earth elements (John Wiley & Sons)
Description: Hoboken, NJ, USA : Wiley, 2025. | Includes bibliographical references and index.
Identifiers: LCCN 2024035432 (print) | LCCN 2024035433 (ebook) | ISBN 9781119515036 (hardback) | ISBN 9781119515067 (adobe pdf) | ISBN 9781119515043 (epub)
Subjects: LCSH: Rare earths. | Ore-dressing. | Strategic materials. | Chemicals–Purification.
Classification: LCC TN490.A2 R368 2024 (print) | LCC TN490.A2 (ebook) | DDC 661/.041–dc23/eng/20240813
LC record available at https://lccn.loc.gov/2024035432
LC ebook record available at https://lccn.loc.gov/2024035433

Cover design by Wiley
Cover image: © Phawat/Adobe Stock Photos

Set in 10/12pt TimesNewRomanMTStd by Straive, Pondicherry, India

CONTENTS

LIST OF CONTRIBUTORS

Andre Anderko
Critical Materials Innovation Hub
OLI Systems Inc.
Parsippany, New Jersey, USA

Darwin Argumedo
Brightmark Energy
Illinois Institute of Technology
Fort Wayne, Indiana, USA

Efthymios Balomenos
Laboratory of Metallurgy
National Technical University of Athens
Athens, Greece

Raquel Crossman
Freeport-McMoRan
Phoenix, Arizona, USA

Gaurav Das
Critical Materials Innovation Hub
OLI Systems Inc.
Parsippany, New Jersey, USA

Panagiotis Davris
Laboratory of Metallurgy
National Technical University of Athens
Athens, Greece

David DePaoli
Oak Ridge National Laboratory
Oak Ridge, Tennessee, USA

Luis A. Diaz
Critical Materials Innovation Hub
Idaho National Laboratory
Idaho Falls, Idaho, USA

Roderick Eggert
Critical Materials Innovation Hub
Colorado School of Mines
Golden, Colorado, USA

Eugene Engmann
Critical Materials Innovation Hub
Idaho National Laboratory
Idaho Falls, Idaho, USA
and
Department of Nuclear Engineering and Industrial Management
University of Idaho
Idaho Falls, Idaho, USA

Ali Eslamimanesh
Critical Materials Innovation Hub
OLI Systems Inc.
Parsippany, New Jersey, USA

Kerstin Forsberg
Department of Chemical Engineering
KTH Royal Institute of Technology
Stockholm, Sweden

Yoshiko Fujita
Critical Materials Innovation Hub
Idaho National Laboratory
Idaho Falls, Idaho, USA

Joseph Gambogi
National Minerals Information Center
U.S. Geological Survey
Reston, Virginia, USA

Michael Heinrichs
Battelle Memorial Institute
Columbus, Ohio, USA

James C. Hower
Center for Applied Energy Research
University of Kentucky
Lexington, Kentucky, USA

Heileen Hsu-Kim
Department of Civil & Environmental Engineering
Duke University
Durham, North Carolina, USA

Yongqin Jiao
Critical Materials Innovation Hub
Lawrence Livermore National Laboratory
Livermore, California, USA

Zhen Jin
Florida Industrial and Phosphate Research Institute
Florida Polytechnic University
Lakeland, Florida, USA

Athanasios K. Karamalidis
Department of Energy and Mineral Engineering
Pennsylvania State University
State College, Pennsylvania, USA

Allan Kolker
Geology, Energy & Minerals Science Center
U.S. Geological Survey
Reston, Virginia, USA

Margaret Lencka
Critical Materials Innovation Hub
OLI Systems Inc.
Parsippany, New Jersey, USA

Haijun Liang
Florida Industrial and Phosphate Research Institute
Florida Polytechnic University
Lakeland, Florida, USA

Chen-Luh Lin
Department of Metallurgical Engineering
University of Utah
Salt Lake City, Utah, USA

Tedd E. Lister
Critical Materials Innovation Hub
Idaho National Laboratory
Idaho Falls, Idaho, USA

Jan Miller
Department of Metallurgical Engineering
University of Utah
Salt Lake City, Utah, USA

Dimitrios Panias
Laboratory of Metallurgy
National Technical University of Athens
Athens, Greece

Dan Park
Critical Materials Innovation Hub
Lawrence Livermore National Laboratory
Livermore, California, USA

Ioannis Paspaliaris
Laboratory of Metallurgy
National Technical University of Athens
Athens, Greece

Madhav Patel
Department of Energy and Mineral Engineering
Pennsylvania State University
State College, Pennsylvania, USA

Rick Peterson
Brightmark Energy
San Francisco, California, USA

Desirée L. Plata
Department of Civil & Environmental Engineering
Massachusetts Institute of Technology
Cambridge, Massachusetts, USA

David Reed
Critical Materials Innovation Hub
Idaho National Laboratory
Idaho Falls, Idaho, USA

Michael Svärd
Department of Chemical Engineering
KTH Royal Institute of Technology
Stockholm, Sweden

Vicki Thompson
Critical Materials Innovation Hub
Idaho National Laboratory
Idaho Falls, Idaho, USA

Patrick Zhang
Florida Industrial and Phosphate Research Institute
Florida Polytechnic University
Lakeland, Florida, USA

Haiyan Zhao
Department of Chemical and Biological Engineering
University of Idaho
Moscow, Idaho, USA
and
Department of Nuclear Engineering and Industrial Management
University of Idaho
Idaho Falls, Idaho, USA

PREFACE

Rare earths are a family of 17 chemical elements—the 15 lanthanides, plus scandium and yttrium. Given their similar chemical characteristics, rare earths are typically found together in nature. These elements can be considered the vitamins and spices of many modern engineering materials. In small quantities, they provide essential properties to permanent magnets in electric motors, phosphors in LED lights, lasers, catalysts, and ceramic pigments, to name but a few important applications.

Demand for rare earths will increase significantly in the coming decades. In part, growth in demand will reflect improvements in living standards as the incidence of poverty falls and as low-income countries become middle-income countries and middle-income countries become high-income countries. Moreover, much of the growth in demand will reflect structural or systemic changes in economic activity associated with decarbonization and development and deployment of technologies that have lower environmental footprints than current technologies.

Society requires that these increased demands be met in more sustainable ways. Although sustainability means different things to different people, we define recovery, processing, and purification as "sustainable" if this production meets or exceeds society's requirements for environmental performance, social performance (social sustainability, in the sense that they are done with the support and participation of the local communities in which they occur), and economic performance (delivering sufficient, affordable, and secure supplies of rare earths from the perspective of rare earth users).

This book describes sources of rare earths and methods of production that have the potential to make rare earth recovery, processing, and purification more sustainable. The chapters consider innovations that would enhance environmental and economic performance.

Some chapters focus on improving the process of rare earth recovery—reducing both the costs and environmental impacts of production and increasing the likelihood of both community acceptance and profitable production. These processes include enhanced beneficiation using biotechnology, separations via crystallization and improved solvent extraction, and aqueous electrochemical processing.

Other chapters focus on unlocking new sources of rare earths that are unconventional, secondary, or both. Unconventional means recovering rare earths from sources from which, to date, there has not been technically and commercially feasible production at scale. Secondary means sources from which rare earths would not be recovered as the principal or primary product but rather as a byproduct of other activities. These include coal fly ash, bauxite residues, and various waste streams associated with phosphate rock and phosphoric acid production.

Contributions to this book come from more than 35 authors from 15 different organizations in different parts of the world. They carry out their work at universities, national laboratories, government agencies, institutes, nonprofit organizations, and companies.

The book should be of interest to a range of readers including academic and government researchers, industry professionals over the entire supply chain who rely on rare earths, market analysts, and graduate students.

We are grateful for the support and guidance we have received from many individuals and groups, especially the chapter authors and staff members at AGU and Wiley. We are also thankful for an outstanding set of colleagues, at both universities and other organizations in the United States and internationally, who were willing to take time from their busy schedules to read and provide insightful comments and feedback on different chapters.

Summing up, the book's purpose is to shine a light on new processes and sources of rare earths that have the potential to enhance the availability of rare earths, allowing all of us to benefit from the special properties rare earths provide to modern technologies.

Athanasios K. Karamalidis
Pennsylvania State University, USA

Roderick Eggert
Colorado School of Mines, USA

LIST OF ACRONYMS AND ABBREVIATIONS

AC	Activated carbon
ADP	Acid digestion process
AEM	Anion exchange membrane
AER	Anion-exchange resin
APTES	3-aminopropyl triethoxysilane
APTMS	3-aminopropyl trimethoxysilane
APTS	3-[2-(2-aminoethylamino) ethylamino] propyl-trimethoxysilane
ATMP	Amino tris(methylene phosphonic acid)
ATS	Aluminum silicotitanates
BHPA	N-Benzoyl-N-phenylhydroxylamine
BPG	Bis(phosphonomethyl) glycine
BR	Bauxite residue
CBMM	Companhia Brasileira de Metalurgia e Mineração
CER	Cation-exchange resin
CMI	Critical Materials Innovation (Hub)
CN	Coordination number
CNT	Carbon nanotube
COK	Centre for Research Chemistry and Catalysis
COMEX	Commodity Exchange
CT	Computed tomography
Cyanex 272	Bis(2,4,4-trimethylpentyl) phosphinic acid
DEHPA	Di(2-ethylhexyl) phosphoric acid
DETA	Diethylenetriamine
DGA	Diglycolamide
DMF	N,N-dimethylformamide
DOE	United States Department of Energy
DOODA	3,6-dioxaoctanediamidopropyl
DSP	Desilication product
DTPA	Diethylenetriaminepentaacetic acid
DTPADA	Diethylenetriaminepentaacetic dianhydride
EAF	Electric arc furnace
EDTA	Ethylenediaminetetraacetic acid
EELS	Electron energy-loss spectroscopy
EPA	Environmental Protection Agency
EXAFS	Extended X-ray absorption fine structure
FCC	Fluid catalytic cracking

FDGA	Furan-2,4- diamidopropyltriethoxysilane
FIB	Focused ion beam
FIPR	Florida Industrial and Phosphate Research (Institute)
GO	Graphene oxide
GONS	Graphene oxide nanosheet
HDEHP	Bis(2-ethylhexyl) hydrogen phosphate
HER	Hydrogen evolution reaction
HMS	Heavy-mineral sands
HREE	Heavy rare earth element
HRXMT	High-resolution X-ray microtomography
HS	Harmonized system
HTS	Harmonized tariff schedule
ICP-AES	Inductively coupled plasma atomic emission spectroscopy
ICP-MS	Inductively coupled plasma mass spectroscopy
ICP-OES	Inductively coupled plasma optical emission spectroscopy
IER	Ion-exchange resin
IIP	Ion-imprinted polymer
IL	Ionic liquids
IUPAC	International Union of Pure and Applied Chemistry
KAIST	Korean Advanced Institute of Science and Technology
KMt	Kilo metric tons
LA-ICP-MS	Laser-ablation - inductively coupled plasma - mass spectroscopy
LBT	Lanthanide-binding tag
LMC	Liquid-metal cathode
LME	London Metal Exchange
LREE	Light rare earth element
MCM	Mobil Composition of Matter
MIP	Molecular imprinted polymer
MISA	Moscow Institute of Steel and Alloys
MLR	Ministry of Land and Resources (China)
MMt	Million metric tons
MOF	Metal-organic framework
MRC	Metallothermic reduction and comproportionation
MSE	Molten salt electrolysis
MWh	Megawatt hour
NIST	National Institute of Standards and Technology
ORNL	Oak Ridge National Laboratory
ORP	Oxidation-reduction potential
PCC	Pulverized coal combustion
PEGDA	Polyethylene glycol diacrylate
QC	Quality control

REE	Rare earth elements
REO	Rare earth oxides
REY	Rare earth elements + Yttrium
SAED	Selected area electron diffraction
SAIL	Steel Authority of India Limited
SEM-EDS	Scanning electron micrography energy dispersive x-ray spectroscopy
SEM-EDX	Scanning electron micrography energy dispersive x-ray spectroscopy
SX	Solvent extraction
TBP	Tributyl phosphate
TEA	Techno-Economic Analysis
TEM	Transmission electron microscopy
TOPO	Trioctylphosphine oxide
TREO	Total rare earth oxides
UCC	Upper continental crust
USBM	United States Bureau of Mines
USGS	United States Geological Survey
XANES	X-ray absorption near-edge structure
XRD	X-ray diffraction
μXRF	Microprobe X-ray fluorescence spectroscopy

1

RARE EARTH INDUSTRY OVERVIEW

Joseph Gambogi

National Minerals Information Center, U.S. Geological Survey, Reston, Virginia, USA

Any use of trade, firm, or product names is for descriptive purposes only and does not imply endorsement by the U.S. Government.

Abstract

The rare earth elements (REEs) are a group of moderately abundant elements comprising the 15 lanthanides, scandium (Sc), and yttrium (Y). Most REE production has been from deposits where specific minerals (usually bastnaesite, loparite, monazite, and xenotime) or ores such as ion-adsorption clays are amenable to known industrial-scale physical and chemical separation techniques. The rare earth industry is complex and evolving with changing geographic sources of supply and shifting demand for specific REEs. Global mine production tripled from 2010 to 2022. New sources of supply have changed trade patterns for mineral concentrates, metals, and compounds. While China has remained the largest global producer, its imports have increased from 12 thousand metric tons (KMt) in 2010 to over 126 KMt in 2021. A few mine producers have emerged as established sources of supply, and interests in developing new alternative sources of

Rare Earth Elements: Sustainable Recovery, Processing, and Purification, Special Publications 79, First Edition. Edited by Athanasios K. Karamalidis and Roderick Eggert.

DOI: 10.1002/9781119515005.ch1

supply are ongoing. Several factors have contributed to price volatility for rare earth materials including political tensions, shifts in demand patterns, and supply disruptions. Prices for rare earth materials peaked in 2011 following changes in China's export quotas and political tensions between China and Japan. In subsequent years, prices for most rare earths decreased significantly but were influenced by rising or falling demand for specific applications such as permanent magnets and phosphors.

1.1. Introduction

Despite their name, the rare earth elements (REEs) are a group of moderately abundant elements comprising the 15 lanthanides, scandium (Sc), and yttrium (Y). The lanthanides are the elements with atomic numbers 57 through 71, in the following order: lanthanum (La), cerium (Ce), praseodymium (Pr), neodymium (Nd), promethium (Pm), samarium (Sm), europium (Eu), gadolinium (Gd), terbium (Tb), dysprosium (Dy), holmium (Ho), erbium (Er), thulium (Tm), ytterbium (Yb), and lutetium (Lu).

Excluding scandium, REEs can be classified as either light rare earth elements (LREEs) or heavy rare earth elements (HREEs). The LREEs include the lanthanide elements from atomic number 57 (La) through atomic number 64 (Gd), and the HREEs include the lanthanide elements from atomic number 65 (Tb) through atomic number 71 (Lu). The division is based on the LREEs having unpaired electrons in the 4f electron shell and HREEs having paired electrons in the 4f electron shell. Yttrium is chemically similar to the HREE lanthanides and commonly occurs in the same minerals as a result of its similar ionic radius. Consequently, yttrium is considered an HREE even though it is not part of the lanthanide series (Rare Earth Industry Association, n.d.).

In rock-forming minerals, rare earths typically occur in compounds as trivalent cations in carbonates, oxides, phosphates, and silicates (Mason and Moore, 1982, p. 46). Because REEs often share a divalent or trivalent charge and similar ionic radii, they are found together in a complex and diverse suite of minerals. Hundreds of REE-bearing minerals have been identified in a variety of depositional environments (Van Gosen, et al., 2017).

1.2. Production

Most REE production has been from deposits where specific minerals (usually bastnaesite, loparite, monazite, xenotime) or ores such as ion-adsorption clays are amenable to known industrial-scale physical and chemical separation techniques.

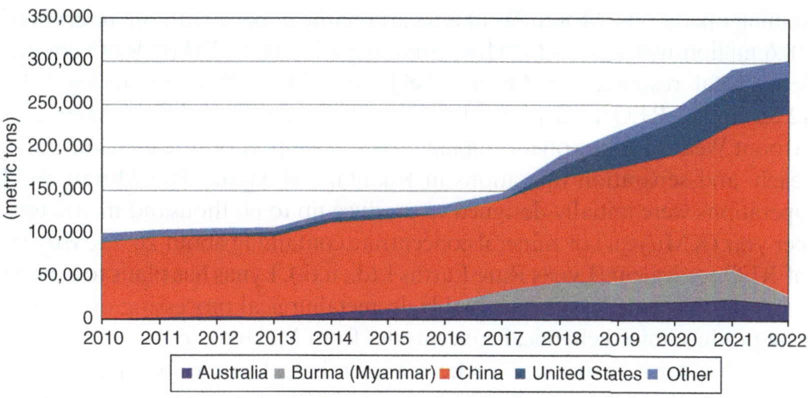

Figure 1.1 Global estimated rare earth mine production (metric tons (Mt) of rare-earth-oxide equivalent). Source: U.S. Geological Survey National Minerals Information Center, https://www.usgs.gov/centers/national-minerals-information-center/rare-earths-statistics-and-information./United States Global Change /Public Domain.

LREEs have been produced predominantly from bastnaesite and monazite found in carbonatite and placer deposits, while production of HREEs has largely been from ion-adsorption clays in Southern China and Burma (Myanmar). Sources of rare earths are evolving with changes in end uses, mining and processing technologies, and environmental concerns. Extensive research is ongoing to recycle REEs and recover REEs as byproducts of mine tailings and intermediate process streams.

Global production tripled during the period from 2010 to 2022 exceeding 300 thousand metric tons (KMt) of rare-earth-oxide (REO) equivalent in 2022. Since the 1980s, China has consistently led global production. Owing to increased production from Australia, Burma (Myanmar), and the United States, China's contribution to the total has fallen from 90% in 2010 to about 70% in 2022. Concurrent with increased production outside of China, China increased its mine production quota to 210 KMt in 2022 from 89 KMt in 2010 (Figure 1.1).

1.2.1. Australia

Australia has significant production, reserves, and resources of REEs. Prior to the commissioning of the Mount Weld mining operations in 2011 by Lynas Corp., Australia's REE mine production was as monazite–xenotime mineral concentrates, a byproduct of heavy-mineral sands (HMS) mining and processing. HMS operations typically produce titanium, zirconium, and a variety of other mineral concentrates. In 2022, Australia's REE production was based primarily on production of concentrates from the Mount Weld mining operations. Mount Weld is a carbonatite deposit with ore reserves of 18.6 million metric tons (MMt) containing 8.2% (1.53 MMt) REO equivalent. Total resources at Mount Weld were 54.7 MMt containing 5.3% (2.88 MMt) REO equivalent (Lynas Rare Earths Ltd., 2023). Mining at Mount Weld has been on a campaign basis in support of downstream crack, leach, and separation operations in Kuantan, Malaysia. The Mount Weld operations were initially designed to produce up to 66 thousand metric tons per year (KMt/year) of mineral concentrate containing about 26.5 KMt/year of REO equivalent (Lynas Rare Earths Ltd., n.d.). Lynas has plans to expand mine capacity and move its initial hydrometallurgical processing operations that produce mixed chemical concentrates from Malaysia to Australia.

In Western Australia, Northern Minerals Ltd. commissioned its pilot production at the Browns Range project in 2018. Hard rock ore containing xenotime was used to produce a mixed REE carbonate. In the Australian financial year 2022 (1st July 2021 to 30th June 2022), the pilot plant processed 8.5 KMt of ore and produced 70 Mt of carbonate. Indicated resources for the Browns Range project were estimated to be 4.54 MMt containing about 0.7% (31.7 KMt) REO equivalent. Probable reserves were 3.29 MMt containing about 0.7% (22.3 KMt) REO equivalent (Northern Minerals Ltd., 2022).

In 2020, heavy-mineral sands producer, Iluka Resources Ltd., began processing a stockpile of mine tailings to produce monazite-rich mineral concentrates at its Eneabba operations in Western Australia. The first phase of the project commenced in 2021, and the company produced 44 KMt of concentrate in 2020 and 58 KMt in 2021. In 2022, a second phase was completed to produce a 90% monazite concentrate from stockpiled tailings supplemented with ongoing process tailings. In the third phase of the project, the company planned to use monazite concentrates from potentially multiple sources to produce mixed and separated REE compounds. Reserves of the stockpile were estimated to be close to 1 MMt grading 84% heavy minerals. Monazite and xenotime comprised about 19% of the heavy minerals fraction (Iluka Resources Ltd., 2023a, p. 39; 2023b, p. 14).

Australia has been an active region for mineral exploration and development. Publicly listed companies developing projects with Joint Ore Reserve Committee (JORC)-compliant rare earth reserves in Australia included Alkane Resources Ltd. (Dubbo Zirconia), Arafura Resources Ltd. (Nolans Bore), Australian Mines Ltd. (Sconi), Clean TeQ Holdings Ltd. (Sunrise), Hastings Technology Metals Ltd. (Yangibana), Lynas (Mount Weld), Northern Minerals Ltd. (Browns Range), Platina Resources Ltd. (Owendale), and Scandium International Mining Corp. (Nyngan). According to Geoscience Australia, economic demonstrated resources in Australia in 2021 were 4.26 MMt of REO equivalent (Hughes et al., 2023, p. 12).

1.2.2. Brazil

Brazil was an early producer of monazite mineral concentrates from heavy mineral sands deposits. In addition to mining, some chemical processing of monazite took place from 1949 to 1992 until environmental concerns closed the operations. In 2021, Brazil exported about 833 Mt of mineral concentrates that were produced from existing stockpiles of mixed heavy-mineral concentrates. According to the Agência Nacional de Mineração, Brazil's prior exports were derived from Indústrias Nucleares do Brasil inventories in Sao Francisco do Itabapoana (Andrade, 2019, p. 168).

In 2018, Companhia Brasileira de Metalurgia e Mineração (CBMM) conducted trial processing of monazite-bearing tailings from its Araxá mine and processing operations in the State of Minas Gerais. CBMM Araxá operations produced niobium metal, ferroniobium, and compounds from pyrochlore ore containing about 2.3% niobium oxide. The plant capacity was 1 KMt/year of REO in the form of mixed compounds and pilot capacity for separated compounds was less than 20 Mt/year of REO equivalent. No production has been reported since 2018 (Ferreira Neto, 2021).

1.2.3. Burma (Myanmar)

Mine production in Burma (Myanmar) is estimated based on China's reported imports of REE compounds from the country. Mining of ion-adsorption clays has been reported in Kachin State, which borders China's Yunnan Province (Htoi Awng, 2022). Although much of the production is undocumented, exports of REE compounds from Burma (Myanmar) to China were first reported in 2014 and peaked at an estimated 35 KMt of REO equivalent in 2021 (Zen Innovations AG, 2023).

1.2.4. Burundi

Rainbow Rare Earths Ltd. began trial mine production at its Gakara project in 2017. Although the company was targeting 100 Mt/month of concentrate production, the project was placed on care and maintenance in June 2021 at the request of the Government of Burundi. Measured resources are yet to be defined. The Gakara project includes numerous occurrences where REEs are hosted in bastnaesite-rich veins and breccia pipes (Rainbow Rare Earths Ltd., 2022, p.56).

1.2.5. Canada

Canada has significant REE reserves and resources. According to Natural Resources Canada, Canada's measured and indicated resources were over 15.1 MMt of REO in 2022. Numerous other projects in British Columbia, Saskatchewan, Ontario, Quebec, Newfoundland and Labrador, and Northwest Territories were reported to be active in 2021. Active REE mineral projects included Alces Lake, Crater Lake, Eldor (Ashram), Falcon Point, Port Hope Simpson (Foxtrot), Kipawa (Zeus), Kwyjibo, Montviel, Nechalacho, Red Wine, and Strange Lake. The Kipawa (Zeus) and Nechalacho–Basal Zone were at the feasibility stage of development. Active processing facilities included the Nechalacho–T Zone and Tardif Zone, SRC Processing Facility, Vital Metals Processing Facility, and the St-Bruno Recycling Demonstration Plant (Natural Resources Canada, 2023).

In 2021, Vital Metals was conducting a mining demonstration project at the Nechalacho project in the Northwest Territories. Mineral concentrates were shipped to Saskatchewan, for further processing. In the North T deposit where the company was sourcing ore for the demonstration, total resources were 94 KMt containing about 9.1% REO. Using a 0.1% Nd oxide cutoff, total resources for the Tardiff Upper Zone were 95 MMt containing about 1.5% REO (Vital Metals, 2022, p. 18).

1.2.6. China

China has dominated the supply of mine production and downstream products for decades. The Government of China sets production quotas for the mining and separation of REEs. However, actual Chinese production is estimated to have exceeded the quotas, as illegal production has been a concern since 2010. The illegal production has been difficult to quantify; however, it was estimated by one news source to have been 60 KMt in 2017 and was expected to fall to 8 KMt in 2021 (Argus Media Group, 2020).

Owing in part to illegal mining activities and environmental concerns, China has been consolidating its rare earth industry and enforcing more restrictive environmental policies. In 2022, China's Ministry of Land and Resources (MLR) allotted most of the mining production quota to four entities (China Rare Earth Group Co., Northern Rare Earth, Xiamen Tungsten Co., and Guangdong Rare Earth Industry Co.) In 2021, the China Rare Earth Group was established to consolidate production from Chinalco, Minerals Rare Earth, and Southern Rare Earth. The mining quota for 2022 was set at a record high of 210 KMt, a 25% increase from 168 KMt in 2021 and more than triple the 89 KMt set in 2010. China's reserves were estimated to be 44 MMt of REO and the identified resources were estimated to be 1,400 MMt of REO (Argus Media Group, 2022a).

1.2.7. India

India is one of the earliest sources of monazite concentrates with production beginning in the early 1900s. Monazite concentrates containing REEs are produced in India as a byproduct of heavy mineral sands mining operations that also produce titanium and zirconium concentrates. Production capacity of monazite concentrates is reported to be 6 KMt/year at the Manavalakurichi operations and 240 Mt/year at the Chavara operations. A processing plant was commissioned in 2015 with more than 11 KMt/year of capacity to produce mixed rare earth compounds from monazite concentrates.

India's monazite resources in heavy mineral sands deposits were estimated to be 12.5 MMt of monazite. The distribution by state was Andhra Pradesh (3.69 MMt); Tamil Nadu (2.46 MMt); Odisha (3.06 MMt); Kerala (1.84 MMt); West Bengal (1.2 MMt); and Jharkhand (21 KMt) (Indian Bureau of Mines, 2020).

1.2.8. Madagascar

In Madagascar, monazite concentrates are a byproduct of heavy mineral sands mining operations near Fort-Dauphin that produce titanium and zirconium mineral concentrates (Rio Tinto plc, 2022). In 2018, initial shipments of mixed heavy mineral concentrates containing monazite were produced from mine tailings and exported to China. Based on the unit value of China's imports of these concentrates from Madagascar, the monazite content was estimated to be significantly less than those from other sources (Zen Innovations AG, 2023). Resource estimates for heavy mineral sands were not available at a national level. At the Tantalus ionic clay project in northern Madagascar, measured and indicated resources were 198 MMt containing 897 parts per million (177 KMt) of REO equivalent. In 2019, China Nonferrous agreed to provide engineering procurement and construction services to the project, purchase up to 3 KMt of output within 3 years of commissioning, and make future investments into the project (ISR Capital, 2019; Tantalus Rare Earths, 2014).

1.2.9. Malaysia

Historically, Malaysia produced monazite and xenotime concentrates that were recovered as a byproduct of tin and titanium–zirconium heavy mineral sands operations. In 2021, Malaysia produced and exported less than 100 Mt of REO equivalent in mineral concentrates. In downstream processing, an additional 15.6 KMt of REO equivalent in compounds was produced from Australian mineral concentrates (Lynas Rare Earths Corp., 2022b, p. 8).

1.2.10. Russia

Russia's Lovozersky GOK produces loparite concentrates from its Karnasurt Mine in Murmansk. Loparite concentrates are further processed into mixed rare earth compounds by Solikamsk Magnesium Works. Russia's mine production of REO was estimated to have been 2.6 KMt in 2021 (U.S. Geological Survey, 2023). Russia's exports of REE compounds totaled 5.6 KMt (gross weight) in 2021 and Estonia (80%) and China (16%) were the leading destinations (Zen Innovations AG, 2023).

1.2.11. South Africa

Although South Africa has no reported mine production of rare earths, monazite concentrates are a potential byproduct from heavy-mineral sands production at the Richards Bay and Exxaro Resources mining operations. In addition to recovery from byproduct sources, feasibility studies have been completed on the Steenkampskraal project in the Western Cape province where reserves were estimated to be 605 KMt containing 14.4% (87 KMt) of REO equivalent (Liedtke, 2019). Prefeasibility studies were also started at the Phalaborwa, Glenover, and Zandkopsdrift REE projects (Frontier Rare Earths Ltd., n.d.; Galileo Resources Plc, n.d.; Rainbow Rare Earths Ltd., 2022).

1.2.12. Tanzania

No production of REE mineral concentrates have been reported for Tanzania; however, feasibility studies have been completed on the Ngualla project in southwest Tanzania. Reserves were reported to be 18.5 MMt containing 4.8% (887 KMt) REO equivalent using a 1%-REO cutoff grade. Resources were reported to be 210 MMt containing 2.15% (4.62 MMt) REO equivalent, and 93% of the resources were classified as measured and indicated (Peak Resources Ltd., 2022, p. 8, 92, 94).

1.2.13. Thailand

Thailand is known to be an intermediate processer of tin tailings and mixed HMS concentrates and produces monazite–xenotime concentrates that are then exported to China. During the period from 2010 to 2022, production fluctuated between a low of less than 1 KMt REO equivalent in 2012, 2013, and 2015, and a high of 8.1 KMt REO equivalent in 2022. Estimates of Thailand's reserves and resources were not available (U.S. Geological Survey, 2023).

1.2.14. Vietnam

Despite limited commercial-scale mine production, Vietnam has significant reserves and resources of rare earths. Ore reserves have been estimated to be 22 MMt of REO equivalent. Notable REE deposits include Dong Pao (Lai Chau Province), Nam Xe (Lai Chau Province), and Yen Phu (Yen Bai Province). Historically, mine production from Vietnam was largely based on monazite–xenotime concentrates recovered from HMS. More recently, REEs have been recovered from ion-adsorption clay deposits. Vietnam has commercial-scale downstream capacity to produce REE metals and compounds. In 2022, Vietnam's exports of REE compounds and metals were 2.5 KMt and 4.3 KMt, respectively (Zen Innovations AG, 2023).

1.2.15. United States

Mine production of rare earths in the United States was almost entirely from bastnaesite mineral concentrates produced at Mountain Pass, CA, where mining operations were most recently recommissioned in 2018. Monazite concentrates were also produced as a separated concentrate or included as an accessory mineral in mixed HMS produced in the southeastern United States. In 2022, mine production at Mountain Pass was reported to be 42.5 KMt of REO. Reconfigured separation operations were scheduled to be recommissioned in 2023 (MP Materials Corp., 2023).

Government agencies in the United States have been partnering with industry and academia to foster domestic production and identify new byproduct recovery sources of rare earths such as coal fly ash. In 2020 and 2021, respectively, Lynas announced plans to establish light and heavy separation operations in the United States with support from the U.S. Department of Defense (Lynas Rare Earths Ltd., 2021, 2022a).

1.3. Trade

Rare earths are traded on a global basis and are contained in mineral concentrates, compounds, and metals. Global trade flows of rare earths are usually reported on a gross weight basis under selected harmonized system (HS) and country-specific trade codes. The HS codes are common between countries to the 6-digit level. Country specific codes like the U.S. Harmonized Tariff Schedule may have up to 10 digits and add specificity for categorizing either imports or exports or both. Rare earths are also contained in numerous downstream products such as catalysts, batteries, and magnets. Downstream codes are often not specific to rare earths and it is difficult to estimate their rare earth content.

1.3.1. Mineral Concentrates

For mineral concentrates, there is not a 6-digit HS code dedicated to rare earths. However, monazite and xenotime concentrates are usually traded under the HS code 2612.20 (defined as thorium ore and concentrates). At the 8-digit code level, China Customs has a dedicated code (2530.90.20) for rare earths (defined as ores of rare earth metals).

China is the largest and one of only a few countries that has the capability to process rare earth mineral concentrates at an industrial scale. In addition to being the largest mine producer, China is the largest importer of mineral concentrates. The volume of concentrates imported into China has increased significantly in recent years rising from 12 KMt in 2010 to over 126 KMt in 2021. The unit value of concentrate imports into China has varied significantly by country reflecting the differences in grade and quality of source materials. In 2010, China imported 12 KMt of concentrates, and import levels remained below 10 KMt/year until 2018 when imports from the United States and Madagascar increased the volume significantly. In the 5-year period between 2018 and 2022, the United States was the leading supplier of rare earth mineral concentrates to China, accounting for more than 60% of China's imports. Madagascar supplied 23% of imports during this period and China's total volume of imports peaked in 2021 at 127 KMt (Figure 1.2).

Malaysia is the second largest importer of rare earth mineral concentrates. Malaysia's production of REO compounds in Kuantan is sourced entirely from imports of mineral concentrates from Australia. The Mount Weld mineral concentrate operations were commissioned in 2011 and exports to Malaysia began in 2012. Although these exports are mineral concentrates, they have been included under the HS code for rare earth metals (2805.30). Imports from Australia to Malaysia of 50 KMt in 2014 may include misclassified material, as this import level does not align well with the 7.2 KMt of REO production of REO compounds in Malaysia (Figure 1.3).

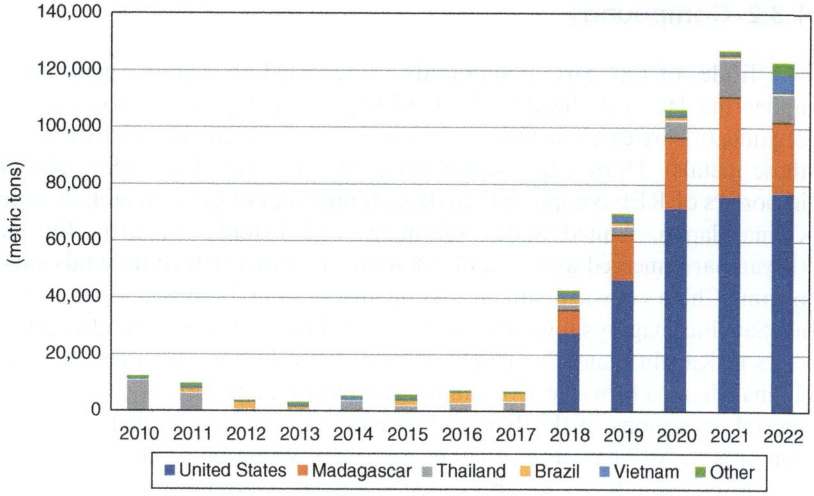

Figure 1.2 China's imports of rare earth mineral concentrates by country of origin (metric tons, gross weight). Source: Global Trade Tracker, https://www.globaltradetracker.com/.

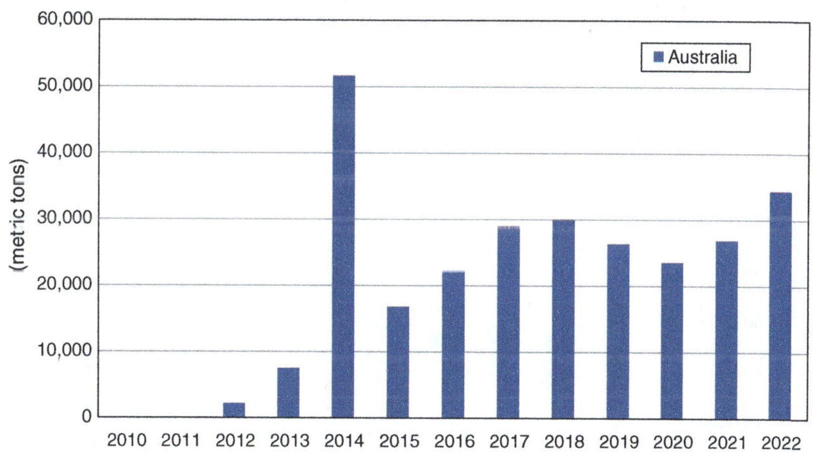

Figure 1.3 Malaysia's imports of rare earth metals from Australia (metric tons, gross weight). Source: Global Trade Tracker, https://www.globaltradetracker.com/.

1.3.2. Compounds

Trades of rare earth compounds are reported on a gross weight basis under the HS code heading 2846 (defined as compounds, inorganic or organic, of rare earth metals, of yttrium or of scandium, or of mixtures of these metals). During the period from 2010 to 2021, the leading global importers of REE compounds, in descending order of gross weight, include China, Japan, United States, Germany, and Estonia. In 2016, Burma (Myanmar) emerged as a significant source of rare earth compounds supporting China's downstream processing into separated compounds and metals. Satellite imagery shows significant growth in the mining of ion-adsorption clays in Kachin State, Burma (Myanmar), similar to mining operations in China's Jiangxi Province. Coinciding with increased imports of compounds, several operations in the Jiangxi Province were idled over environmental concerns in 2017 (Argus Media Group, 2020). With limited ability to separate mixed compounds, Japan, United States, and Germany imported mixed and separated compounds (primarily from China) for downstream processing into catalysts, polishing compounds, etc. Estonia imported mixed compounds primarily from Russia and the United States to support its separation and metallurgical operations in Sillamäe (Figure 1.4).

1.3.3. Metals

Imports and exports of rare earth metals are reported on a gross weight basis primarily under the HS code heading 2805.30 (defined as

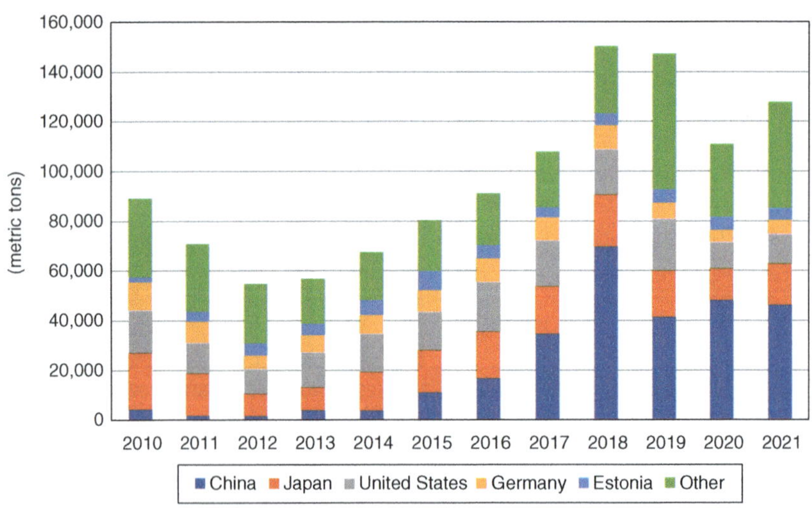

Figure 1.4 Global importers of rare earth compounds (metric tons, gross weight). Source: Global Trade Tracker, https://www.globaltradetracker.com/.

rare earth metals, scandium and yttrium, whether or not intermixed or interalloyed). On a gross weight basis, global trade in metals is significantly less compared with trade in compounds or concentrates. Japan was the leading importer of REE metals with about 60% of global imports. The next three countries (Norway, India, and the United States) sum to about 15% of global imports during the period from 2010 to 2021. Global trade in metals reached a high of 14.2 KMt in 2021 compared with 10.7 KMt in 2020.

In addition to trade under 2805.30, significant volumes of REE metals are contained in other ferrous and nonferrous alloys and are traded under different base metal trade codes. Tracking these can be problematic. For example, ferrocerium is traded under the HS code 3606.90 (defined as ferrocerium and other pyrophoric alloys in all forms). The REE content within this HS code varies significantly and other pyrophoric materials that do not contain REEs are included within the data. Another significant volume of rare earths is contained in finished and semifinished materials like permanent magnets (HS 8505.11 defined as permanent magnets and articles intended to become permanent magnets after magnetization). Specific HS trade codes for rare earth permanent magnets (NdFeB and SmCo) are not defined at the 6-digit level but are available with certain country-specific codes like the HTS codes for NdFeB-sintered magnets (8505.11.0070) and SmCo-sintered magnets (8505.11.0050) (Figure 1.5).

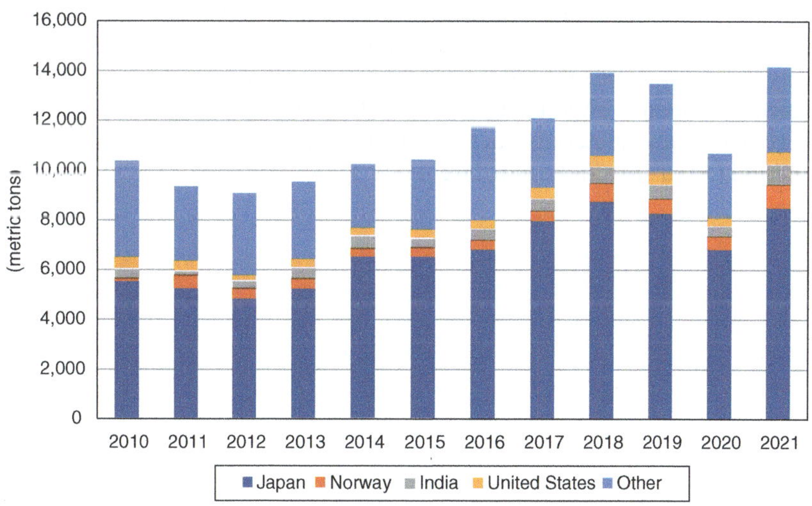

Figure 1.5 Global importers of rare earth metals. Source: Global Trade Tracker, https://www.globaltradetracker.com/.

1.4. Prices

Unlike base metals, rare earths are not traded on the Commodity Exchange (COMEX) or the London Metal Exchange (LME), but price data for rare earths are available from a variety of sources. For LREEs, the data are available; however, for certain HREEs that are used in niche markets, the data are unavailable or have a limited price history. For purposes of discussion, it is useful to group certain elements that are relatively close on the periodic table, are treated together during the separation processing steps, or have overlapping uses. For this review, prices are reviewed for the period 2010–2022. Thulium prices were not available and only partial data for this time period were available for gadolinium, holmium, and erbium oxides.

Several factors have contributed to price volatility for rare earth materials including political tensions, shifts in demand patterns, and supply disruptions. Concerns for the availability of rare earths rose to global attention in 2010. In July 2010, China's Ministry of Commerce announced that China would cut its exports of rare earth minerals by about 72% for 2010. Consequently, most of the 2010 quota had already been met. In September 2010, China's rare earth exports to Japan decreased dramatically following a territorial waters incident between the two countries (Grasso, 2013). By 2011, prices for rare earth materials had increased to record highs and there were widespread reports of illegal mining, processing, and exports of China's rare earths. Since 2010, prices of some REEs have experienced significant price volatility owing to shifting demand and limited availability for certain HREEs.

Growing demand for rare earth magnets have reshaped the market over the last decade. Since REEs are mined and extracted together, the supply of specific REEs related to magnet materials has outpaced demand for certain elements while being in deficit for other elements. Subsequently, prices of some materials have declined significantly over the 2010–2022 period (Argus Media Group, 2023).

1.4.1. Cerium and Lanthanum

Cerium and lanthanum are the most abundant REEs and are often used together in various applications. The applications that require the most cerium and lanthanum, by volume, include catalysts, glass additives, polishing, and metallurgical applications. These two elements also serve other diverse markets and are used in a variety of other applications such as magnets, lasers, and ceramics.

Prices for cerium and lanthanum oxides have followed very similar trends. Prior to 2007, average prices for both oxides were less than \$2/kg. China's restrictions in exports combined with other factors caused a spike in prices that peaked in 2011 at over \$100/kg for both oxides. In the following year, average prices decreased by about 77%. Excluding moderate increases in 2017 and 2018, prices have continued to decline post 2011. In 2022, the average annual prices for cerium and lanthanum oxides were about \$1.50/kg (Figure 1.6).

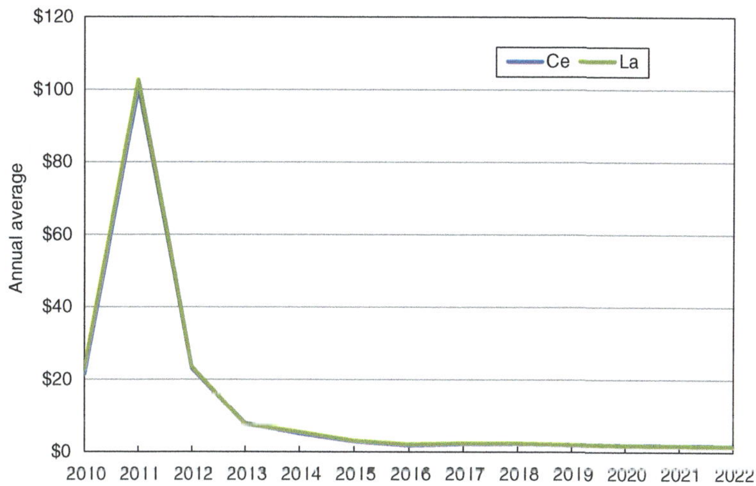

Figure 1.6 Cerium and lanthanum oxide prices in U.S. dollars per kilogram (FOB China). Source: Argus Media group (Argus Metals International), https://metals.argusmedia.com/.

1.4.2. Neodymium and Praseodymium

Neodymium–iron–boron (NdFeB) permanent magnets are the leading end use for both neodymium and praseodymium. Other end uses include catalysts, pigments, and other metallurgical uses typically as alloying elements. NdFeB magnets typically contain greater than 30% REE elements by weight with the majority of the REE content as neodymium and praseodymium.

Average prices for neodymium and praseodymium oxide peaked in 2011 at about \$233/kg and \$196/kg, respectively. Since 2011, fluctuations in prices have been mixed with numerous double-digit year-on-year increases and decreases. In 2022, the average prices for neodymium and praseodymium oxides significantly increased to \$134/kg and \$128/kg, respectively, from \$49/kg and \$47/kg in 2020 (Figure 1.7).

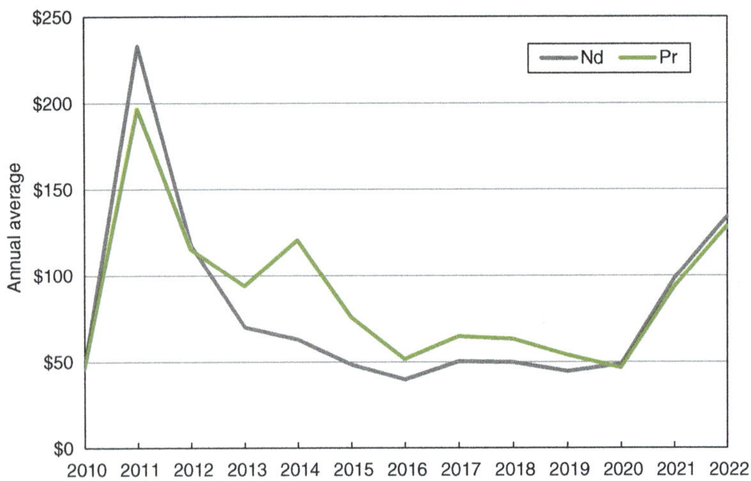

Figure 1.7 Neodymium and praseodymium oxide prices in U.S. dollars per kilogram (FOB China). Source: Argus Media group (Argus Metals International), https://metals.argusmedia.com/.

1.4.3. Samarium and Gadolinium

Samarium and gadolinium are used predominantly in magnetic applications. Samarium is used in samarium–cobalt (Sm–Co) permanent magnets, which have a lower energy product than neodymium magnets. Although they have a lower magnetic strength, the advantage of Sm–Co magnets over NdFeB magnets is that they are capable of being used at higher operating temperatures before they lose magnetism. In 2011, the annual average price for samarium oxide increased sixfold, compared with 2010. By 2013, the average price was less than that in 2010. Subsequently, oxide prices reached a low of less than \$2/kg in 2016 and have risen above \$3/kg in 2022.

For some grades of Sm–Co magnets, samarium is substituted with gadolinium for improved temperature stability. Gadolinium is also known for its use as a contrast agent in magnetic resonance imaging. Gadolinium oxide prices were not available prior to 2015. In 2015 the average price was about \$21/kg, and in 2022, the price had increased to \$79/kg (Figure 1.8).

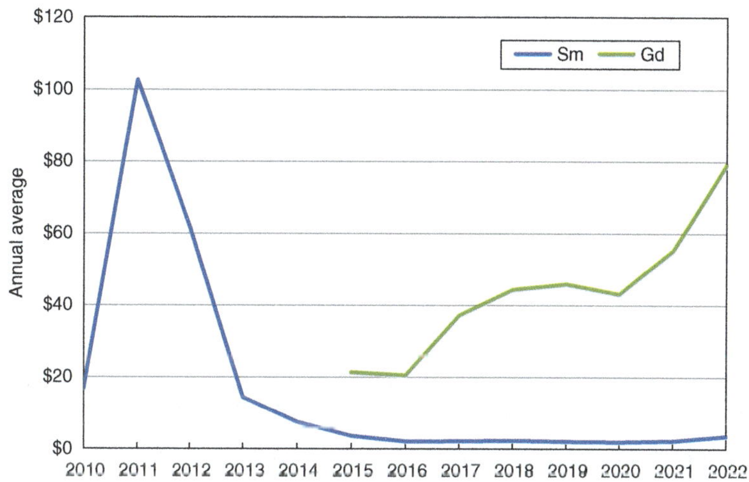

Figure 1.8 Samarium and gadolinium oxide prices in U.S. dollars per kilogram (FOB China). Source: Argus Media group (Argus Metals International), https://metals.argusmedia.com/.

1.4.4. Europium

Europium is known for its use in phosphor compounds used to produce red light. During the 1960s, much of production and demand for rare earths was related to REE europium and yttrium phosphor compounds used in color televisions. The transition from fluorescent lighting to solid state lighting has significantly reduced demand for these phosphors and depressed prices for europium relative to other REEs.

The annual average price for europium oxide peaked in 2011 at about $2,900/kg and in subsequent years decreased significantly. By 2013, the average price had decreased to $1,100/kg and had dropped to less than $100/kg by 2016. In 2022, the price was about $30/kg (Figure 1.9).

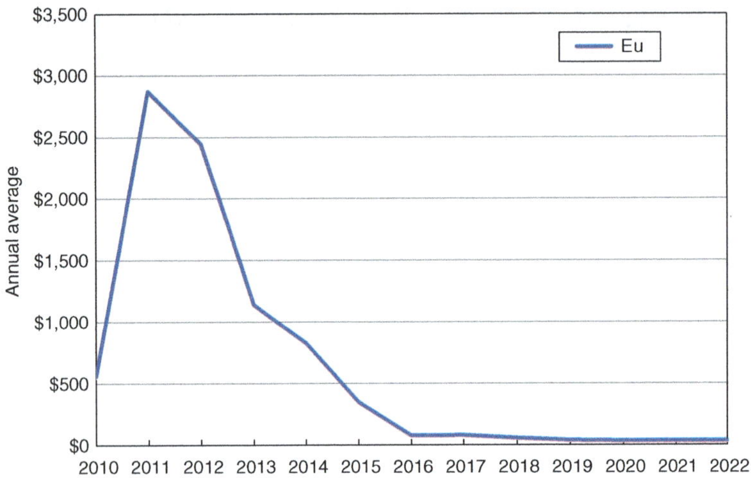

Figure 1.9 Europium oxide prices in U.S. dollars per kilogram (FOB China). Source: Argus Media group (Argus Metals International), https://metals.argusmedia.com/.

1.4.5. Dysprosium and Terbium

Dysprosium and terbium are HREEs that are used in NdFeB magnets to expand their thermal operating range. Terbium is also used in phosphors and in specialty alloys like Terfenol-D. Terfenol-D has the highest known magnetostriction of any alloy.

Prices for Dy and Tb peaked in 2011 at $1,450/kg and $2,350/kg, respectively. Prices for these materials decreased significantly following the spike in 2011. The terbium price increased significantly between 2020 and 2022 owing to limited supply of mineral concentrates high in HREEs and increased demand for magnets in transportation (Figure 1.10).

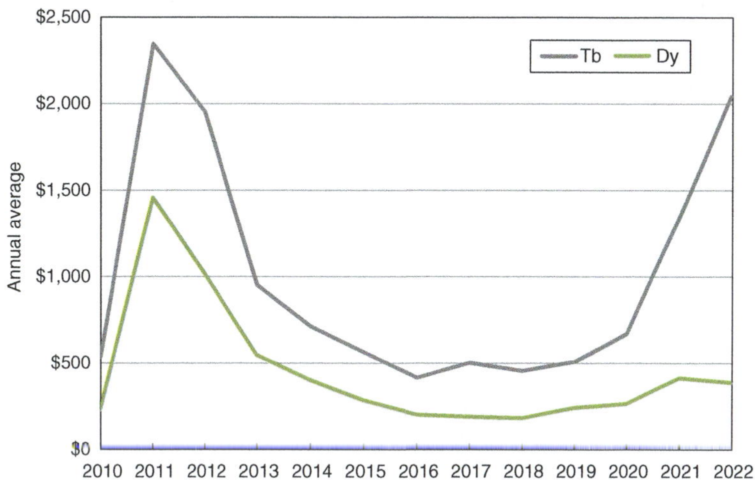

Figure 1.10 Dysprosium and terbium oxide prices in U.S. dollars per kilogram (FOB China). Source: Argus Media group (Argus Metals International), https://metals.argusmedia.com/.

1.4.6. Holmium and Erbium

Holmium is used in some NdFeB magnets to improve coercivity properties. Holmium is also used in phosphors and as a dopant in lasers. In the nuclear industry, holmium is used in the reactor control rods. Availability of price data for holmium began in 2020. The annual average price increased to $179/kg in 2022 from $66/kg in 2020.

Erbium is used for fiber optic signal amplification and as a dopant in lasers. Available price data for erbium began in 2015 at $37/kg. Prices reached a low of $23/kg in 2019 and a high of $58/kg in 2022 (Figure 1.11).

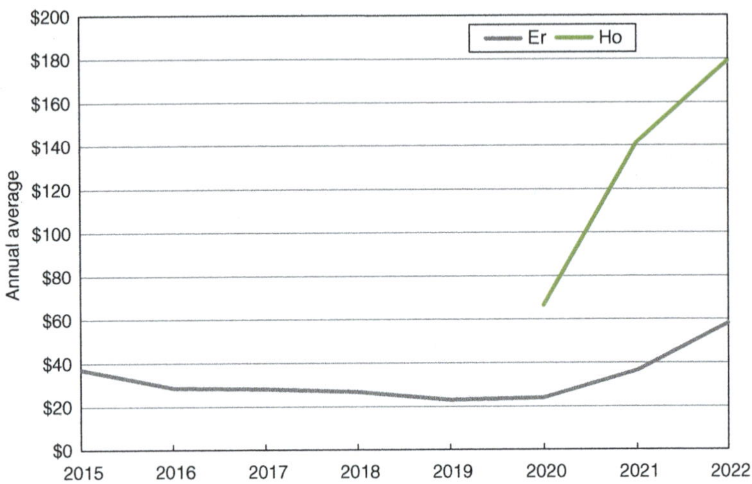

Figure 1.11 Holmium and erbium oxide prices in U.S. dollars per kilogram (FOB China). Source: Argus Media group (Argus Metals International), https://metals.argusmedia.com/.

1.4.7. Ytterbium

Ytterbium is similar to other HREEs in that it is used in niche technology applications such as optical fiber amplifiers, fiber optic technologies, thermal barrier coatings, and as a dopant in laser applications. Other uses include stainless steels and dental amalgams.

Prices for ytterbium oxide were not available prior to 2020. Annual average ytterbium oxide prices started at about \$15/kg in 2020 and decreased to about \$14/kg in 2022 (Figure 1.12).

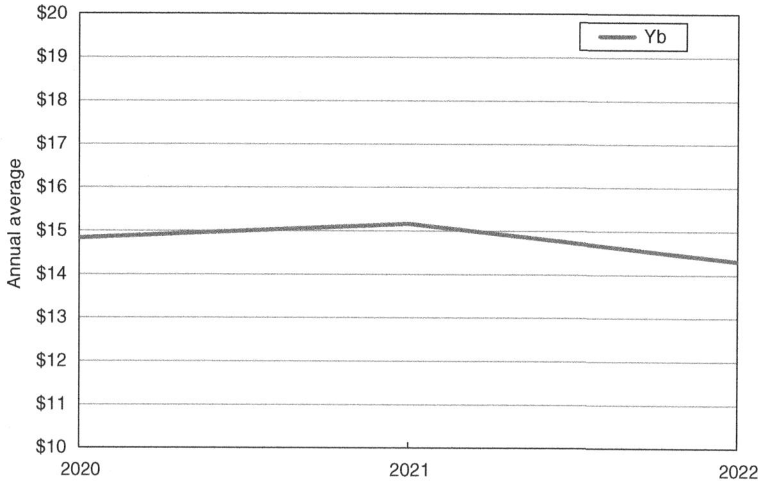

Figure 1.12 Ytterbium oxide prices in U.S. dollars per kilogram (FOB China). Source: Argus Media group (Argus Metals International), https://metals.argusmedia.com/.

1.4.8. Lutetium

Lutetium is the least abundant REE in the crust, with an average crustal abundance of less than 1 ppm. It is important for use in specialized applications including scintillators for medical imaging, certain electronics, and some cancer therapies.

Prices for lutetium oxide were not available prior to 2020. Annual average lutetium oxide prices started at about \$750/kg in 2020 and increased to about \$810/kg in 2021 and 2022 (Figure 1.13).

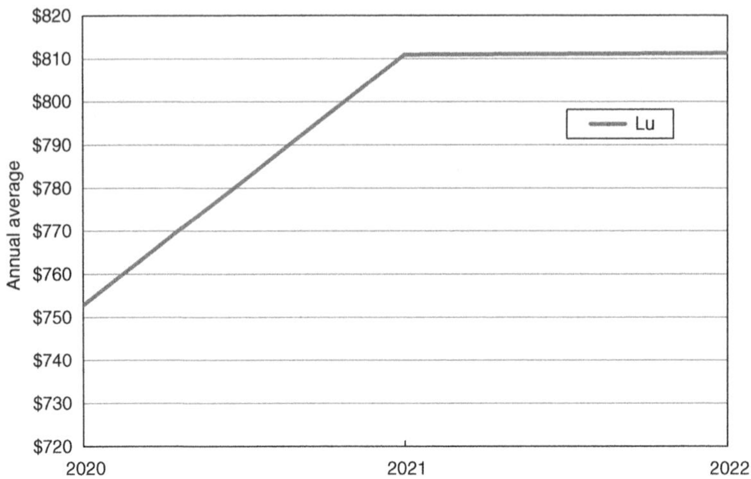

Figure 1.13 Lutetium oxide prices in U.S. dollars per kilogram (FOB China). Source: Argus Media group (Argus Metals International), https://metals.argusmedia.com/.

1.4.9. Yttrium

On a volume basis, yttrium is among the primary REEs used in ceramics and phosphors and has important uses in metallurgical applications such as superalloys. Yttrium oxide has been estimated to account for 87% of all the TREO in ceramics (Roskill Information Services Ltd., 2019). Similar to all REE prices, the average price of yttrium oxide peaked in 2011. In the years following 2011, the price declined precipitously reaching a low of about \$2.80/kg in 2019. In 2021, the price was about \$6.40/kg, more than double the price compared with the prior year (Figure 1.14). The rise in price in 2021 has been attributed in part to the scarcity of rare earth ores rich in yttrium (Argus Media Group, 2022b).

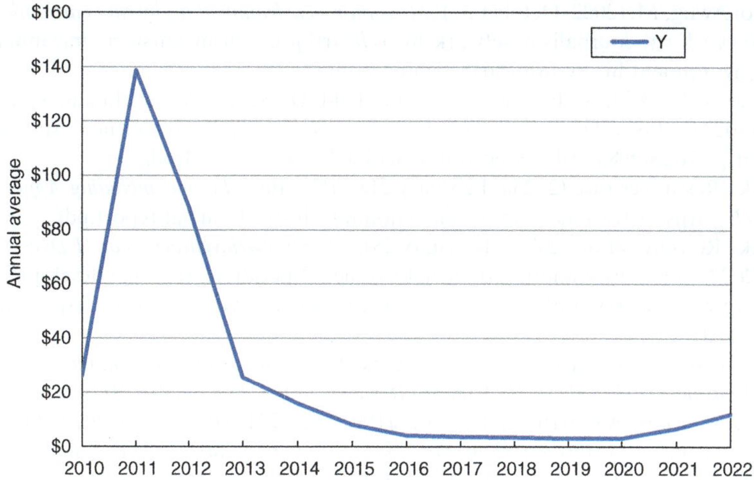

Figure 1.14 Yttrium oxide prices in U.S. dollars per kilogram (FOB China). Source: Argus Media group (Argus Metals International), https://metals.argusmedia.com/.

References

Andrade, R. H. P. (2019). Terras raras. *Sumário Mineral 2017*, *37*, 164–166. Agência Nacional de Mineração. https://www.gov.br/anm/pt-br/centrais-de-conteudo/publicacoes/serie-estatisticas-e-economia-mineral/sumario-mineral/sumario mineral_2017

Argus Media Group. (2020, January 21). *China's rare earth consolidation to cut supplies.* https://metals.argusmedia.com/newsandanalysis/article/2054597

Argus Media Group. (2022a, August 22). *China raises 2022 rare earth production quotas*. https://metals.argusmedia.com/newsandanalysis/article/2363144

Argus Media Group. (2022b, February 25). *China's heavy rare earth market to resume upward track.* https://metals.argusmedia.com/newsandanalysis/article/2305714

Argus Media Group. (2023). *Prices and data*. Via https://metals.argusmedia.com/.

Ferreira Neto, J. B. (2021, April). *Extração dos elementos de terras raras na CBMM* [Slide show]. Centro de Desenvolvimento da Tecnologia Nuclear. http://antigo.cdtn.br/images/eventos/material/slides_sbtr/12-15h30_CBMM-Extracao_dos_elementos_terras_raras_na_CBMM-Joao_Batista_CBMM.pdf

Frontier Rare Earths Ltd. (n.d.). *Zandkopsdrift project*. https://frontierrareearths.com/the-zandkopsdrift-project/

Galileo Resources Plc. (n.d.). *Glenover Rare Earth Project — Limpopo Province, South Africa*. https://galileoresources.com/glenover-rare-earth-project/

Grasso, V. (2013, December 23). *Rare earth elements in national defense: Background, oversight issues, and options for congress*. Congressional Research Service. https://sgp.fas.org/crs/natsec/R41744.pdf

Htoi Awng, M. (2022, October 31). *Myanmar's environment hit by rare earth mining boom.* Earth Journalism Network. https://earthjournalism.net/stories/myanmars-environment-hit-by-rare-earth-mining-boom

Hughes, A., Britt, A., Pheeney, J., Summerfield, D., Senior, A., Hitchman, A., et al. (2023). *Australia's identified mineral resources 2022.* Geoscience Australia. https://d28rz98at9flks.cloudfront.net/147673/147673_00_1.pdf

Iluka Resources Ltd. (2023a, February 21). *2022 Annual report including Appendix 4E.* https://www.iluka.com/media/h3jonm0z/ilu-fy22-annual-report.pdf

Iluka Resources Ltd. (2023b, February 28). *Ore reserve inventory as at 31 December 2022.* https://www.iluka.com/getattachment/operations-resource-development/ore-reserves-mineral-resources/resource-and-reserve-deposit.pdf.aspx?lang=en-AU

Indian Bureau of Mines. (2020). Rare earths (advance release). *Indian Minerals Yearbook 2019, Part-III Mineral Reviews,* (58), 24-1–24-10. Retrieved from https://ibm.gov.in/writereaddata/files/10012020172151RareEarth_2019_AR.pdf

ISR Capital Ltd. (2019). *Entry into memorandum of understanding with China Nonferrous Metal Industry's Foreign Engineering and Construction Co., Ltd.* https://links.sgx.com/FileOpen/ISRCL%20-%20Announcement%20on%20MOU%20with%20NFC%20-%2024%20June%202019.ashx?App=Announcement&FileID=564536.

Liedtke, S. (2019, December). Steenkampskraal now fully permitted for construction. *Mining Weekly*. https://www.miningweekly.com/article/steenkampskraal-receives-last-permit-now-fully-permitted-to-pursue-construction-and-production-2019-12-18/rep_id:3650

Lynas Rare Earths Ltd. (2021, January 22). *Lynas signs contract to build a light rare earths facility.* https://lynasrareearths.com/lynas-signs-contract-to-build-a-light-rare-earths-facility/

Lynas Rare Earths Ltd. (2022a, June 14). *Lynas awarded US$120M contract to build commercial heavy rare earths facility.* https://lynasrareearths.com/lynas-awarded-us120m-contract-to-build-commercial-heavy-rare-earths-facility/

Lynas Rare Earths Ltd. (2022b, February 24) Interim Unaudited consolidated financial report incorporating appendix 4D for the half year ended 31 December 2021.

Lynas Rare Earths Ltd. (2023). *2022 Annual report* (pp. 107–108). https://wcsecure.weblink.com.au/pdf/LYC/02581455.pdf

Lynas Rare Earths Ltd. (n.d.). *Mt Weld, Western Australia.* https://lynasrareearths.com/about-us/locations/mt-weld-western-australia/

Mason, B. H., & Moore, C. B. (1982). *Principles of Geochemistry.* New York, NY: John Wiley & Sons.

MP Materials Corp. (2023). *MP Materials reports fourth quarter and full year 2022 results.* https://investors.mpmaterials.com/investor-news/news-details/2023/MP-Materials-Reports-Fourth-Quarter-and-Full-Year-2022-Results/default.aspx

Natural Resources Canada. (2023, February 14). *Rare earth elements facts.* https://www.nrcan.gc.ca/our-natural-resources/minerals-mining/minerals-metals-facts/rare-earth-elements-facts/20522

Northern Minerals Ltd. (2022). *2022 Annual report* https://wcsecure.weblink.com.au/pdf/NTU/02585733.pdf

Peak Resources Ltd. (2022). *Annual report.* https://wcsecure.weblink.com.au/pdf/PEK/02584166.pdf

Rainbow Rare Earths Ltd. (2022). *Annual report & accounts* 2022. https://www.rainbowrareearths.com/wp-content/uploads/2022/10/RRE-2022-Annual-Report_Final.pdf

Rare Earth Industry Association. (n.d.). *About rare earth.* https://www.global-reia.org/about-rare-earth/

Rio Tinto plc. (2022). *Annual report 2021.*Retrieved from https://cdn-rio.dataweavers.io/-/media/content/documents/invest/reports/annual-reports/2021/rt-annual-report-2021.pdf?rev=d54c8b0596a44e159024ac7421a4d38d

Roskill Information Services Ltd. (2019). *Rare earths — Outlook to 2029* (19th ed.). London, United Kingdom: Roskill Information Services Ltd., 373 p.

Tantalus Rare Earths (2014). *Tantalus rare earths AG resource update published for the companies rare earths project in Madagascar with valueable results.* Retrieved from https://www.mining.com/web/tantalus-rare-earths-ag-ni-43-101-technical-report-and-updated-resource-estimate-by-sgs-published/#:~:text=The%20Tantalus%20project%20displays%20a%20pervasive%20and%20well-balanced,as%20defined%20by%20the%20U.S.%20Department%20of%20Energy

U.S. Geological Survey. (2023). *Mineral commodity summaries 2023: U.S. Geological Survey*, 210 p. https://doi.org/10.3133/mcs2023

Van Gosen, B. S., Verplanck, P. L., Seal, R. R., II, Long, K. R., & Gambogi, J. (2017). *Rare-earth elements* (Professional Paper 1802O). doi:10.3133/pp1802O

Vital Metals Ltd. (2022). *Annual report for the year ended 30 June 2022.* https://wcsecure.weblink.com.au/pdf/VML/02575884.pdf

Zen Innovations AG. (2023). *Global trade tracker.* [Data set]. https://www.globaltradetracker.com/

2

RARE EARTH ELEMENTS IN COAL FLY ASH AND THEIR POTENTIAL RECOVERY

James C. Hower[1], Allan Kolker[2], Heileen Hsu-Kim[3], and Desirée L. Plata[4]

[1] *Center for Applied Energy Research, University of Kentucky, Lexington, Kentucky, USA*
[2] *Geology, Energy & Minerals Science Center, U.S. Geological Survey, Reston, Virginia, USA*
[3] *Department of Civil & Environmental Engineering, Duke University, Durham, North Carolina, USA*
[4] *Department of Civil & Environmental Engineering, Massachusetts Institute of Technology, Cambridge, Massachusetts, USA*

Abstract

Coal fly ash is a potential resource of valuable elements, such as rare earth elements (REEs), which are retained and concentrated upon combustion. Understanding REE occurrence within fly ash is vital to developing recovery methods. Some of the highest REE contents occur in fly ash derived from U.S. Appalachian Basin coals, and coals influenced by input volcanic ash are especially enriched. Leaching studies of bulk fly ash show that, as a proportion of the total REEs present, samples from eastern U.S. coals are generally less extractable than fly ash derived from lower REEs western U.S. coals. Aluminosilicate glasses formed during combustion comprise

Rare Earth Elements: Sustainable Recovery, Processing, and Purification, Special Publications 79, First Edition. Edited by Athanasios K. Karamalidis and Roderick Eggert.

DOI: 10.1002/9781119515005.ch2

the largest mass fraction of coal fly ash. REE-enriched domains are present locally in fly ash at the nanometer scale. Annual fly ash production, combined with coal ash already in storage, makes up a large resource for potential recovery of rare earths and associated critical elements. Further extraction technology developments are needed to overcome difficulties in REE concentration and purification to produce coal-ash-derived REE materials of saleable purity.

2.1. Introduction

Rare earth elements (REEs) are critical in the production of a wide array of electronics, magnets, catalysts, metal alloys, optics, and other items needed in modern society (Basu, 2017; Dobransky, 2013; Greene, 2012; Hatch, 2012; Home, 2020; Van Gosen et al., 2014; Watson, 2018). Coal, coal waste, and coal combustion products make up one of the most actively investigated categories of secondary sources for potential REE recovery (U.S. Department of Energy, 2018). Coal use currently accounts for about 30% of the electric power generated in the United States (Energy Information Administration (EIA), 2018). Fly ash, produced during the burning of coal, is a fine-grained solid derived from non-combustible mineral constituents of coal. Coal combustion results in the retention and enrichment of REEs in fly ash (Clarke & Sloss, 1992; Ratafia-Brown, 1994). As a result, fly ash has long been considered a potential resource for recovery of REEs and other valuable elements (Goldschmidt, 1935; Seredin & Dai, 2012). The United States has the world's largest coal reserves (World Energy Council, 2016). World coal consumption for power generation is projected to increase from 2017 to 2040, with reductions in the United States, the European Union, and Japan offset by more or less steady trends in China and increases in coal-fired generation in India (International Energy Agency, 2018; Kaplan, 2020). Even as gas-fired power generation has increased significantly in the last decade, production of vast quantities of fly ash has continued. Advanced power generation systems will account for some of the growth in coal-fired power generation, resulting in greater generating efficiency, implying that about 1/3rd less coal will be required per megawatt hour (MWh) of generation (Sloss, 2019). About 55.5% of the 36.2 Mt of fly ash produced annually in the United States is beneficially reused (2018 data; American Coal Ash Association, 2019). The remainder is stored, mostly in landfills and impoundments. Thus, annual fly ash production, combined with fly ash already in storage, constitutes a huge potential resource (Hower, Groppo,

Henke, et al., 2017). For example, the annual production of the unused (and presumably discarded) portion of coal fly ash in the United States comprises a REE reserve that is similar in scale to the Mountain Pass mine (Taggart et al., 2016). Of coal-related materials investigated for potential REE recovery, fly ash has among the highest REE contents, and due to its fine grain size, ostensibly requiring minimal processing compared to mining and extraction of conventional REE ores. Viable and sustainable recovery of REEs from fly ash greatly depends on geochemical and physical properties of the ash matrix.

2.1.1. Nature of Fly Ash

During coal combustion for electric power generation, the inorganic constituents in the parent coal experience a variety of processes at boiler temperatures (1300–1600 °C) that largely govern the geochemical properties of fly ash collected from the flue gas. Clay minerals in coal melt to form a glass (i.e., an amorphous solid); quartz undergoes a phase transformation to the high-temperature β form, eventually becoming β-cristobalite; and pyrite is oxidized to iron oxides (Hower, 2012; Kolker, 2018; Kutchko & Kim, 2006; Ward & French, 2006). Melting temperatures of some of the most common REE-bearing minerals in coal, such as zircon and monazite, are higher than the boiler temperatures and, as a result, these trace phases should survive the combustion process. However, these phases are commonly dispersed throughout the fly ash matrix compared to their distribution in the feed coal (Section 2.2). In addition to the phase transformation of parent minerals, new (neoformed) minerals, such as mullite ($Al_6Si_2O_{13}$), form by high-temperature reactions in the boiler. Glass formed by quenching of the melt is the most abundant constituent of fly ash.

Classification of coal ash is based on its major element characteristics expressed as the sum of SiO_2 + Al_2O_3 + Fe_2O_3 compared to other major constituents. "Cementitious" ash is classified as Class C where SiO_2 + Al_2O_3 + Fe_2O_3 is greater than or equal to 50 weight percent. For fly ashes with the sum of these constituents greater than or equal to 70 weight percent, the "pozzolanic" ash is classified as Class F (ASTM International, 2017). Each classification has specific properties for use in concrete. Other beneficial reuses of coal fly ash include structural fill and other construction materials (American Coal Ash Association, 2019). Unused ash is stored as solid waste, often in landfills or surface water impoundments located adjacent to coal-fired power plants. In the United States, a ruling by the Environmental Protection Agency (EPA, 2014) upheld the "nonhazardous" classification for fly ash while also recommending improved monitoring of existing disposal impoundments. This policy allows for continued beneficial use of coal fly ash in construction materials and in other applications, while also encouraging exploration of new reuse opportunities (EPA, 2014).

2.1.2. REEs in Coal and Coal Ash

The International Union of Pure and Applied Chemistry (IUPAC; Connelly et al., 2005) defines the REEs as the lanthanides, plus scandium (Sc) and yttrium (Y). Apart from literature referring to other nomenclature, the REE designation for the lanthanides, Y, and Sc, as per Connelly et al. (2005), is followed here. The lanthanide series having its range in atomic numbers from 57 to 71 results from progressive filling of 4f valence electron orbitals; lanthanum (La, atomic number 57) does not have 4f electrons, whereas lutetium (Lu, atomic number 71) has fourteen and a full 4f level. As such, the 15 elements of the lanthanide series exhibit a continuous distribution of chemical and physical properties (e.g., atomic radii). Nonetheless, various classification approaches have been introduced to understand similarities and differences among the elements in the series. Traditionally, the lanthanides are divided into light, middle, and heavy rare earths (Hanson, 1980). For some studies involving coal, the series has been divided only into light rare earth elements (LREEs; La to Sm) and heavy rare earth elements (HREEs; Eu to Lu), with Y sometimes included among the HREEs (Dai, Liu, et al., 2016; Lin, Bank, et al., 2017; Hower et al., 1999; Hower, Granite, et al., 2016; Mardon & Hower, 2004; Seredin, 1996; among others). Others have used the LREE/HREE ratio in evaluations of REE distribution in coal and coal products (Seredin, 1996; Balashov, 1976). In other coal-related studies, REY (lanthanides + Y) have been divided into light, medium, and heavy fractions: LREY (La through Sm), MREY (Eu through Dy plus Y), and HREY (Ho through Lu). Following normalization of REE abundances to crustal averages (indicated here by the suffix "N"), further delineation can be made of L-type (light type; $La_N/Lu_N > 1$), M-type (medium type; $La_N/Sm_N < 1$, $Gd_N/Lu_N > 1$), and H-type (heavy type; $La_N/Lu_N < 1$) enrichment patterns (Seredin & Dai, 2012).

The concentration of REEs in fly ashes from three U.S. coal basins was shown by Taggart et al. (2016, their Figure 2). The ash-basis concentrations in the coal are similar to the ash-basis concentrations in the resulting fly ashes. For the samples discussed by Taggart et al., Central Appalachian Basin-derived fly ashes had the highest REE concentration followed by Illinois Basin-derived and Powder River Basin-derived fly ashes. Normalization of the REE series facilitates observations of enrichments or depletions relative to natural abundances, which display a sawtooth pattern where elements with even atomic numbers are more abundant than their nearest odd-numbered neighbors (Figure 2.1a). This pattern reflects the Oddo–Harkins effect, in which elements with even atomic numbers are more abundant in the universe than odd-numbered elements due to the enhanced stability of nuclei with spin-paired protons (i.e., an even number of protons: Harkins, 1917; Oddo, 1914). To compare the REE abundance distribution of different samples more easily and eliminate the sawtooth pattern resulting from the Oddo–Harkins effect, REEs are commonly "normalized" by dividing by a reference such as the REE content of the upper continental crust (UCC; Figure 2.1b), or by that of chondrites, a group of stony meteorites (Figure 2.1c). Chondrites were chosen by geochemists because they are among the most primitive materials in the solar system, and therefore, approximate the bulk Earth prior to its differentiation into a core, mantle, and crust. Relative to chondrites, the upper continental crust is enriched with light REEs, and most coals show a similar crustal REE distribution. This distribution pattern is retained during coal combustion, but with overall enrichment (Figure 2.1) or concentration. Besides normalization to UCC or to chondrites, in some REE studies, normalization to shale composites, such as the North American Shale Composite (NASC; Gromet et al., 1984), have been used. In presenting REE data, any of these standard normalization approaches are acceptable.

The inclusion of Y and Sc in many REE studies is motivated by their industrial utility rather than some geochemical motivation; Sc is seldom found in the same minerals as the other REEs. However, owing to the price commanded for Sc, several U.S.-based coal and coal-combustion fly-ash-based projects are relying on the co-extraction and concentration of Sc in order to underwrite the economics of their projects (Wraich, 2017). This is challenging because Sc exhibits a distinct behavior from the lanthanides, which behave similarly as a group, largely due to their common occurrence and trivalent form and similar ionic radii, with Y fitting in with the heavy REEs. Exceptions include Ce and Eu, which may also exist in tetravalent and divalent forms, respectively. As such, Ce and Eu are, in some cases, fractionated from the trivalent lanthanides under certain redox conditions (Hanson, 1980). This fractionation results in so-called anomalies, where the content of a single element diverges from the smooth pattern defined for all the remaining trivalent lanthanides on normalized plots. Europium commonly shows such anomalies on chondrite-normalized plots, but these may be absent in UCC-normalized plots because a negative europium anomaly (relative to chondrites) is inherent in the crustal distribution (Figure 2.1c). Surprisingly, in certain coals with UCC-normalization, similar anomalies have been shown for non-redox-sensitive trivalent-only lanthanides such as Gd (Dai, Graham, & Ward, 2016). If present in feed coals, such anomalous lanthanide distributions would be retained by the ash during coal combustion.

REE-bearing minerals in coals include major constituents such as clays, carbonates, and phosphate minerals, as well as REE trace phases

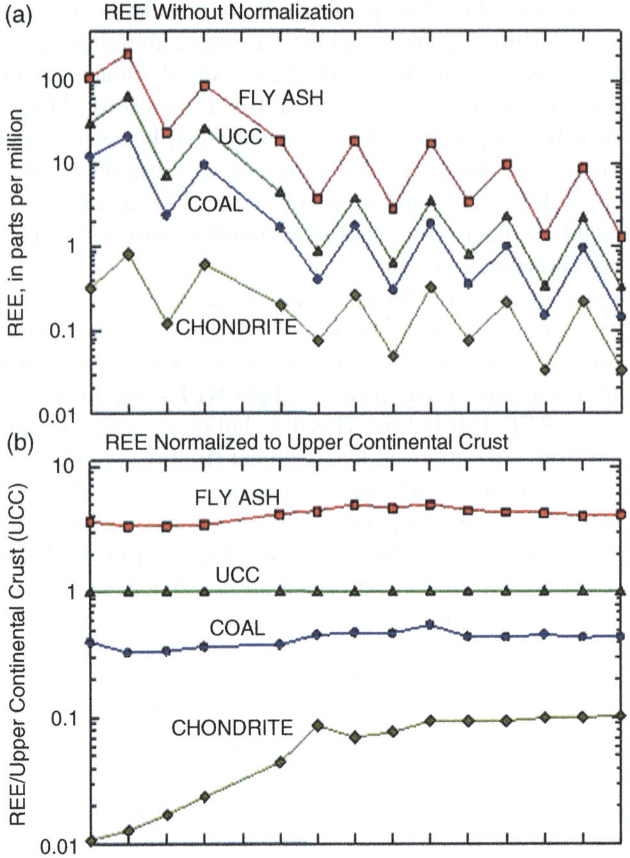

Figure 2.1 Plots showing how REE data are represented, including distributions for: (1) fly ash from a Kentucky power station burning Appalachian Basin Fire Clay coal (fly ash), data from Taggart et al. (2016); (2) the upper continental crust (UCC), data from Taylor and McLennan (1985); (3) U.S. coal in USGS COALQUAL database (coal), data from Finkelman (1993); and (4) chondrite normalized values (chondrite), as suggested by Korotev (2009). Plots show how normalization of REE concentration data results in smooth patterns for distribution of elements plotted in order of atomic number. (a) REE concentration data in parts per million showing a sawtooth pattern with even-atomic-numbered elements (Ce, Nd, Sm, Gd, Dy, Er, and Yb) more abundant than odd-numbered elements (Pr, Eu, Tb, Ho, Tm, and Lu) due to the Oddo–Harkins effect. Promethium (Pm, atomic number 61) is omitted because it is not stable in nature. (b) REEs normalized to UCC. Plot shows that REE distribution of fly ash and coal nearly parallel UCC, with greatest REE enrichment in fly ash. REEs in coal are less than in UCC. (c) REEs normalized to chondrite showing similar patterns for fly ash, UCC, and coal, with each showing enrichment in light rare earths relative to chondrite. The prominent dip in the patterns at europium (Eu, atomic number 63) is a europium anomaly (see text). Chondrite lacks this Eu anomaly. A chondrite-normalized plot is used later in this chapter.

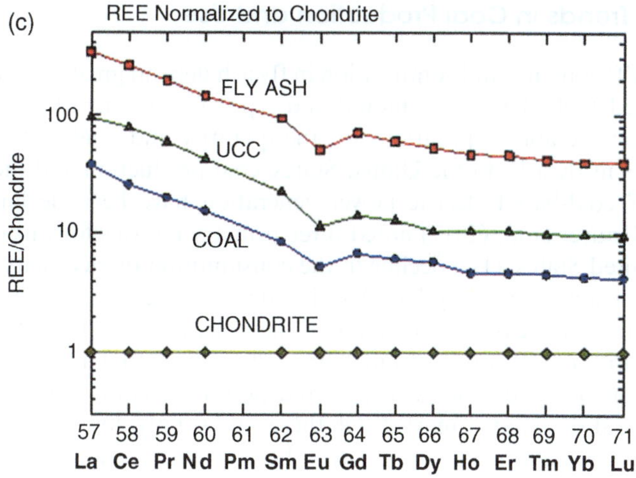

Figure 2.1 (Continued)

such as monazite ((Ce, La, Nd, Th) PO_4), allanite ((Ce, Ca, Y)$_2$ (Al, Fe^{3+})$_3$ (SiO_4)$_3$ (OH)), zircon ($ZrSiO_4$), and xenotime (YPO_4; Finkelman, 1981). Additional REE minerals reported in some coals include rhabdophane ((Ce, La, Y) $PO_4 \cdot H_2O$) (Dai, Luo, et al., 2014), florencite ($CeAl_3(PO_4)_2(OH)_6$) (Hower, Qian, et al., 2020), and Ce–Nd-bearing carbonates (Dai, Liu, et al., 2016; Dai et al., 2017; Zhao et al., 2017). Organic-hosted REEs are also indicated, especially for the heavy rare earths (Eskanazy, 1987a,b, 1999).

In some cases, REE minerals present in coal have been shown to survive combustion. For example, during combustion of Polish hard coal in the Upper Silesian region, Smolka-Danielowska (2010) found that monazite, zircon, and xenotime remained in fly ash following combustion at unspecified conditions. In a combined X-ray diffraction (XRD) and scanning electron microscopy (SEM) study, Brownfield et al. (1999, 2005) found that REE-bearing minerals (zircon, apatite, and monazite) were retained in the fly ash and bottom ash from the burning of a Powder River Basin (Wyoming) coal. An unidentified REE-bearing Ca–P-silicate was also found in the fly ash (Brownfield et al., 2005; their Figure 30). As noted in the earlier text, REE-bearing minerals are more dispersed in fly ash than in the respective feed coal (Section 2.2), despite the relative enrichment of REEs in fly ash. In SEM examination of a Kentucky coal and its corresponding ash, Zhang et al. (2015) found that REE-bearing trace phases were more apparent in the coal and the separation of trace phases from the ash did not result in increased REE contents in the trace mineral fraction, suggesting that these trace phases were not the primary mode of REE occurrence.

2.1.3. Trends in Coal Production and Use

REE contents and composition in fly ash depend greatly on the parent coal feedstock. Thus, the amount and type of coal burned for electricity have considerable implications for the quantity and quality of the REE reserves in fly ash. In the United States, coal production and the proportion of coal-based electric power generation have been declining since 2011. Houser et al. (2017) noted three primary reasons for this decline in the United States: (1) a decline in the consumption of electricity, with the 2016 consumption being less than in 2007 in spite of a larger economy; (2) the rapid increase in production of shale gas, resulting in a 70% decrease in the price of natural gas from 2008 to 2016 (in 2016 dollars); and (3) the incursion of wind- and solar-powered power generation into the previously coal-dominated market. Of these, the supply, low price, and price stability of natural gas dominated the transition away from coal-fired power generation. In addition to these factors, in some coalfields, particularly in the Appalachians, the impact of declining coal reserves and a general decrease in quality and thickness of the remaining reserves have led to a striking increase in the cost of mining and processing coal. As an example of a cost factor, the average amount of land surface required to extract each ton of coal mined in eastern Kentucky tripled from 2008 to 2015 (Wright, 2018).

The decline in coal-fired power generation versus the corresponding rise in gas-fired generation is reflected in a doubling of the U.S. gas capacity and a halving of the coal capacity from 2015 to 2017, with each accounting for about 85–90 million MWh/year of generation by 2017 (Hoff, 2017). The decline in coal-fired generation has also triggered other major operational changes, such as conversion of generation stations to natural gas, modifications to another coal source, or retirement/closure of power plants. For plants previously burning low-S Central Appalachian coal, such a switch has often been to high-S Illinois Basin coal because (1) the latter is substantially cheaper ($34.00/short ton vs. $56.90/short ton, for the week ending 12 June 2020, where a short ton is a U.S. ton, equivalent to 907.19 kg or 2,000 lbs; EIA, 2020) and (2) with the implementation of revised SO_2-emission and Hg-control standards (EPA, 2019, 2020) and the need to install some form of emission-control system, the choice of high-S coal became viable. Overall, 43 GW of coal-fired capacity was retired from 2012 to 2017, and this was somewhat offset by efficiencies of new coal-fired units; new units had a heat rate of 10,197 kJ/kWh compared to 10,912 kJ/kWh for the retired units, an added efficiency equivalent to 19.45 GW of capacity (EIA, 2017).

As an example of the cumulative impact of the factors noted in the earlier text, of the 58 power plants burning Central Appalachian coal in

10 eastern U.S. states in 2008 (information from Keystone Coal Industry Manual, 2009), roughly 22 were still using Central Appalachian coal in 2018 (based on personal knowledge and checks of publicly available information by the first author). Since Central Appalachian coals, particularly those from eastern Kentucky, have some of the best prospects for REE recovery among eastern U.S. coals (Hower et al., 1999; Hower, Eble, et al., 2016; Hower, Groppo, et al., 2020; Lin, Soong, & Granite, 2018; Mardon & Hower, 2004), the proportionately large decrease in Central-Appalachian-sourced coal means that active production of some of the inherently highest-REE-concentration (e.g., those derived from the Fire Clay coal) fly ashes has diminished. Discussions of the geology of REE incorporation and occurrence in coal are out of the scope of this chapter; global observations by Seredin and Dai (2012) and Appalachian-centered discussions by Hower, Eble, et al. (2016, 2020) cover aspects of the subject. Overall, the decrease in coal-fired generating capacity implies a decrease in the amount of fly ash being generated, from 78.454 Mt in 2008 to 36.2 Mt in 2017 (American Coal Ash Association, 2019). The long history of use of such coals, though, means that much of the ash generated over decades of coal utilization remains in pond or landfill storage throughout a large portion of the United States, but environmental and regulatory factors for recovering this material have not been addressed. Altogether, the coal power and coal ash industries continue to change at a rapid pace, and this evolution will alter the outlook for fly ash as an unconventional REE resource.

2.1.4. Critical Elements and U.S. Fly Ash Resources

Critical materials are defined as those that are essential to the economy but whose supply might be limited (Neuendorf et al., 2005). Apart from inclusion of all REEs among a group of 35 elements deemed critical to the United States (Fortier et al., 2018), division among REEs into critical (Nd, Eu, Tb, Dy, Y, and Er), uncritical (La, Pr, Sm, and Gd), and excessive (Ce, Ho, Tm, Yb, and Lu) groups (after Seredin, 2010; Seredin & Dai, 2012) is somewhat subjective and transient, contingent upon the current demands of the market and the supplies on hand (Hatch, 2012). For example, Nd and Pr are in demand (in 2018) for use in permanent magnets with Dy used to increase the performance of the magnets in high-T environments (Roskill, 2017, 2018). Potentially economic levels of REEs occur in many U.S. coals (data from Bragg et al., 1998; Palmer et al., 2015). However, because most coals and coal fly ashes are LREE-enriched, and the most abundant elements of the LREEs (La and Ce) are not in the critical category, the proportion of critical elements is limited to about 45% of the total REEs (using the classification of Seredin (2010) and Seredin and Dai (2012)). Nevertheless, a mix of elements tending toward the less-valuable uncritical and excessive categories can directly influence the economics of a mining or processing operation (Miller, 2015). Most of the REE mining and processing has been concentrated in China, and even newer, non-Chinese operations turn to the Chinese market for processing and downstream manufacturing (Merriman & Backeberg, 2018).

2.2. Geochemical Characterization of Fly Ash

Knowledge of the geochemical distribution of REEs in coal or fly ash can inform REE recovery strategies. Primary approaches to determine targeted REE mineral hosts in devising REE recovery strategies include extraction of bulk samples using a variety of techniques; elemental analysis methods, such as inductively coupled plasma mass spectrometry (ICP-MS), and grain-scale trace element microanalysis using either laser ablation or ion microprobe approaches. Speciation of relatively abundant REEs such as Ce can be accomplished using synchrotron-based X-ray absorption near-edge structure (XANES) on either the bulk scale or microscale. Transmission electron microscopy (TEM) is useful in characterizing nanometer-scale REE-bearing domains, which may differ considerably from macroscale and microscale REE hosts. We detail some of the major findings of these efforts in the following discussion.

2.2.1. Bulk Characterization of Rare Earth Elemental Contents via ICP-MS

The determination of trace element contents in coal fly ash generally requires a combination of robust methods to solubilize the target analytes from the ash followed by sensitive detection methods that can accommodate the matrix complexity of the digests.

For the analysis of REEs, the elements must first be liberated from the particles into an aqueous solution phase. Because these elements are mostly encapsulated within aluminosilicate glasses of fly ash particles and, further, can be difficult to dissolve, the extraction step generally requires reagents that can decompose this matrix. Common leaching methods entail heated wet-digestion methods developed for geological materials, including heated acid digestion with HF (hydrofluoric acid) alone or combined with other acids (e.g., HNO_3, H_2O_2, HCl, $HClO_4$) (Lichte et al., 1987; Taggart et al., 2016). The acid mixture is then dried and redissolved in a dilute acid (e.g., HNO_3/HCl) that is compatible with the sample introduction system of analytical instrumentation. Alternatives to acid digestion which are more effective in breaking down insoluble REE-bearing trace phases such as zircon include flux fusion methods with lithium metaborate

(Bank et al., 2016) and alkaline roasting with Na_2O_2 or NaOH (Taggart, Hower, & Hsu-Kim, 2018).

The analysis of REEs in the digestate solution requires highly sensitive analytical methods such as ICP-MS. Because REEs can be prone to polyatomic interferences, especially for complex matrices such as the acid leachates of coal fly ash, care must be taken to address such interferences. Collision or reaction cell technologies with a reaction gas (e.g., helium or hydrogen) for quadrupole ICP-MS can greatly reduce such polyatomic interferences and are now standard features in modern ICP-MS instrumentation. However, even with these improvements, other polyatomic interferences can greatly influence the accuracy of analysis for specific REEs. For example, coal fly ash often contains large amounts of barium (Ba), resulting in $^{135}Ba^{16}O^+$ interference of the mass isotope signal for $^{151}Eu^+$. The correction for this interference has been addressed through a variety of approaches that include steps to remove Ba from the sample (Yan et al., 2018) or the use of high-resolution mass spectrometry (Thompson et al., 2018). Other major constituents that might be relevant for digested coal fly ash include Ca, which interferes as $^{44}CaH^+$ with $^{45}Sc^+$, and oxide forms of light REEs that have the same mass as heavy REE isotopes (Thomas, 2013).

Inductively coupled plasma optical emission spectrometry (ICP-OES) has also been used to determine REEs. Hower, Groppo, et al. (2020) discussed the comparisons between ICP-OES and ICP-MS, showing that there was a good correlation between the techniques for the more abundant REEs, but that ICP-MS was more suitable for the HREEs, in particular, Tb and Ho through Lu.

2.2.2. Fly Ash and Bottom Ash Petrology

The various techniques for the examination of coal combustion ashes are all dependent on the scale of the investigation. Optical petrology is a valuable tool for the examination of plus-micron-size particles. The optical characterization of fly ash and bottom ash, typically done on polished, epoxy-bound particulate pellets (Hower, 2012; Suárez-Ruiz et al., 2017; Valentim, 2020), utilizes polarized reflected-light, oil-immersion optics in the analysis. While some studies, particularly for scanning electron microscopy, employ dispersed powder specimens (Valentim et al., 2009), those types of preparations cannot match the petrographic detail seen in the cross sections of particles. In general, petrographic classifications of the ash are its coal maceral-derived constituents (including other organic-derived material such as biomass, petroleum coke, and tires) and inorganic-derived portions of the ash. In general, submicron particles cannot be seen or identified by optical microscopy.

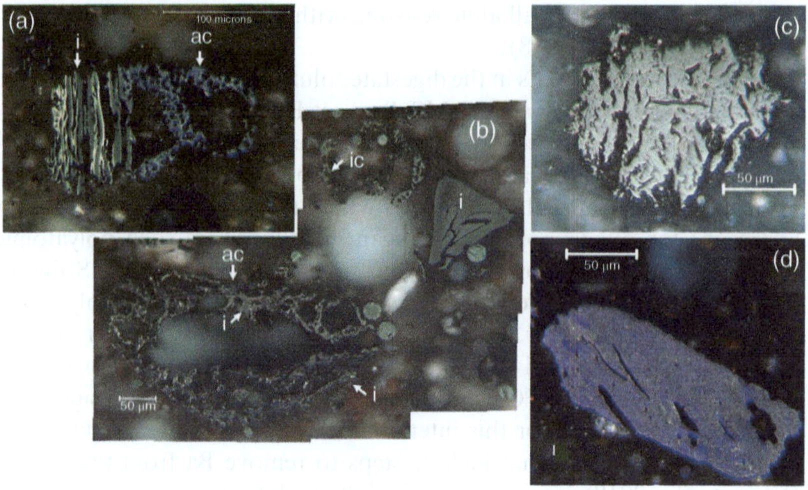

Figure 2.2 Coal-derived carbons from the combustion of eastern U.S. bituminous coals. (a) Inertinite (i) and anisotropic coke (ac). Image MD PR 01. (b) Inertinite (i), isotropic coke (ic), and anisotropic coke (ac). Image 93902 11. (c) U.S. San Juan Basin subbituminous coal-derived char. Image 92922 04 from Hower, Groppo, Graham, et al. et al. (2017). (d) Chinese Yunnan anthracite-derived carbon. Image Yunnan 2 02 from Silva et al. (2012).

Of the maceral-derived fly ash and bottom ash constituents, the products of the melting and repolymerization of bituminous vitrinite can be identified in both relatively unaltered inertinite and isotropic and anisotropic cokes (Figure 2.2a,b). Also illustrated are the chars derived from lignite and subbituminous coals and anthracitic-vitrinite-derived carbons (Figure 2.2c,d).

The inorganic fly ash and bottom ash constituents include glass and glass with heated, but not melted, rock and carbon (Figure 2.3). As will be discussed later in this section and in Section 2.6, fine-grained minerals, many of them below the resolution of optical petrology, can be enclosed within the glass. Spinel minerals (Figure 2.4), which include magnetite, are less abundant than glass, but can be common minerals in the combustion products of moderate-to-high-pyrite coals. Spinels are often found within a glassy matrix. In both the panels (b) and (c) of Figure 2.4, note that some grains have a few-micron-thick red (hematite) oxidation rim. While such a feature would be a dominant feature in a powder-dispersion specimen, the views on the images shown here demonstrate that exploring only the surface conceals the rich variety of the interior of the particles. Mullite (Figure 2.5) is one of the neoformed minerals found in both fly ash and,

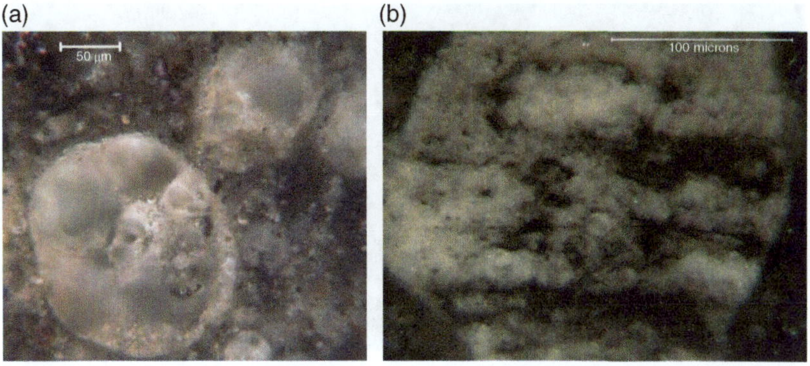

Figure 2.3 Glassy fly ash particles from the combustion of eastern U.S. bituminous coals. (a) Glass cenospheres. Image 93685 06. (b) Particle with glass and dark, partially vitrified shale or silt bands. Image MD PS 10.

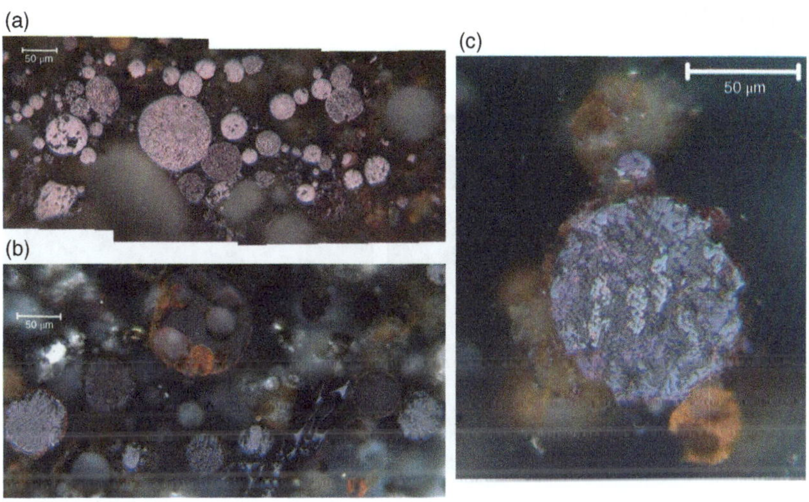

Figure 2.4 Spinel minerals from the combustion of eastern U.S. bituminous coals. (a) Spinel forms in a bottom ash. Image 93662 01. (b) Spinels in fly ash. Image 93614 04. (c) Spinel from a bottom ash. Image 93933 08.

more commonly, in bottom ash or stoker ash resulting from the combustion of high-Al–Si eastern U.S. bituminous coals. Mullite formation depends on the temperature and chemistry of the melt phase, with the mineral forming along with cristobalite during the cooling of an Al_2O_3-rich melt in the Al_2O_3–CaO–SiO_2 system (Alekseev & Vereshchagin, 1997; Ehlers, 1972;

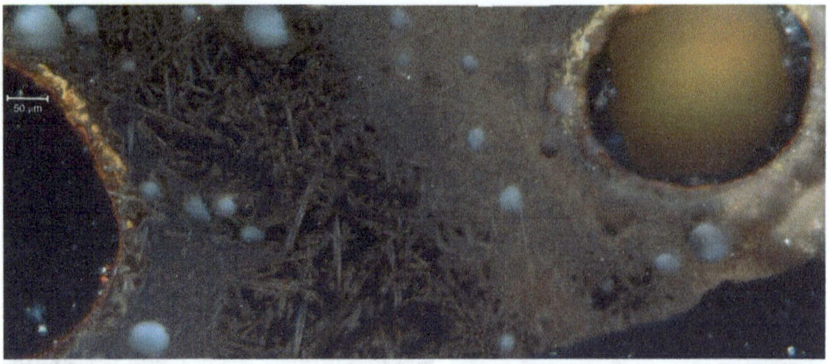

Figure 2.5 Stoker combustion ash from the burning of eastern U.S. bituminous coals. Much of the left portion of the image, the triangular lower-right areas, and the "bubble" in the upper right are void spaces. Much of the rest of the particle is packed with varying densities of mullite crystals in a glassy matrix. The left, rounded edge of the particle has an oxidation rim. Image 93962 06 from Hower, Qian, et al. (2018).

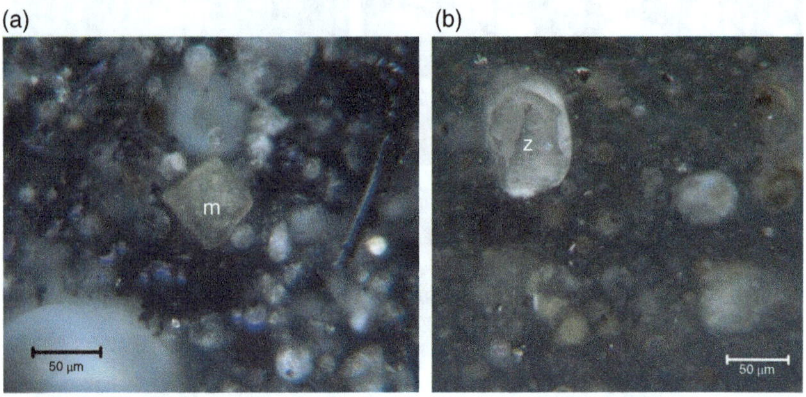

Figure 2.6 (a) Monazite (m) with glassy particles in an eastern U.S. bituminous coal-derived fly ash. Image 93954 04 from Hood et al. (2017). (b) Zircon (z) in fly ash from Powder River Basin subbituminous coal-derived fly ash. Image 93971 02.

Osborn & Muan, 1960). Increased concentrations of CaO in the melt, as would be common in the combustion of western U.S. low-rank coals, would result in the precipitation of neoformed anorthite from the melt (Hower, Henke, et al., 2017).

Although not common, optical microscopy reveals REE-rich minerals such as monazite in fly ash (Figure 2.6a,b). So, if monazite and other REE-rich minerals are seen by optical microscopy, where are the REEs in the fly ashes with over 1,000 ppm REE? As the melting temperatures of pure phases of monazite (±2000 °C) and zircon (1690 °C) are well above normal combustion temperatures (1400–1500 °C), it is not likely that there was much, if any, melting of these minerals (Finch & Hanchar, 2003; Hikichi & Nomura, 1987; Quercia et al., 2007; Seydoux-Guillaume et al., 2002). Interestingly, Hood et al. (2017) demonstrated that monazite will shatter by 1400 °C, well below its melting temperature, a mechanism attributed to the thermal expansion of helium, a product of the radioactive decay of the Th in the monazite (Cherniak et al., 2009; Cherniak & Watson, 2013; Farley, 2007; Guenthner et al., 2013).

In a study of sized fly ashes from the economizer, mechanical, and electrostatic precipitator (ESP) rows at a power plant burning eastern U.S. bituminous coal, Liu et al. (2017) demonstrated that the size-dependent variations in fly ash petrology influenced the distribution of the REEs. Fly ash petrology and chemistry vary between the rows of the pollution control system, usually an array of a combination of economizer, cyclone (mechanical), ESP, and/or baghouse (fabric filter) units. Each successive row away from the boiler experiences a concomitant reduction in temperature and, usually, a decrease in the particle size of the ash (Hower, Henke, et al., 2017). Unlike volatile elements such as Zn or As, the concentration of REEs generally shows little variation between the progressive treatment rows, although the LREE/HREE ratio can vary (Hower, Dai, et al., 2013).

2.2.3. Trace Element Microanalysis

Laser-Ablation ICP-MS and Secondary Ion Mass Spectrometry

Instrumentation for REE analysis on the scale of individual fly ash particles includes laser ablation ICP-MS (LA-ICP-MS) and secondary ion mass spectrometry (SIMS; also called ion microprobe analysis). Each approach is discussed in detail elsewhere: for LA-ICP-MS, these include Sylvester (2001, 2008) and Longerich (2008); for SIMS, these include Stern (2009) and Ireland (2014). This section will therefore emphasize cases with application of trace element microanalysis to coal ash. A fundamental difference in these microanalysis techniques is the rate and depth of penetration within the sample. In LA-ICP-MS, geologic samples are ablated to tens of micrometers in a matter of seconds. As a result, LA-ICP-MS is useful in studies of compositional variation with sample depth. In SIMS, a primary beam of incident ions sputters secondary ions from the sample at a much slower rate, typically leading to penetration depths of only about 1–5 μm, despite much longer analysis times. Compared to smaller ion microprobes, the sensitive high-resolution ion microprobe with reversed geometry (SHRIMP-RG) and other SHRIMP configurations (Section "SHRIMP-RG Ion Microprobe") employ a large-format magnetic sector mass spectrometer with high mass resolution, thereby minimizing the effect of isobaric interferences in SIMS. In both LA-ICP-MS and SIMS, it is helpful to obtain standards that are homogenous on a grain scale and a good matrix match to the material being analyzed (Stern, 2009; Sylvester, 2001, 2008).

Earlier applications of LA-ICP-MS to coal and coal ash focused on quantifying the distribution of elements of environmental interest, such as As and transition metals (Chenery et al., 1995; Querol & Chenery, 1995). In the fly ash study by Spears (2004), these also included Pb, Se, Ga, and Ge. More recently, LA-ICP-MS has been used to map major and minor element distribution in fly ash particles (Bauer et al., 2017), and to compare results obtained by LA-ICP-MS to results by scanning electron micrograph with energy dispersive x-ray spectroscopy (SEM-EDS) (Piispanen et al., 2009) for major fly ash constituents such as Si, Al, Ca, and Fe. Kostova et al. (2016) used LA-ICP-MS to determine bulk fly ash concentrations for a large range of elements by analyzing prepared fused glass discs of whole sample material. In a study of fly ash samples derived from the Fire Clay coal (Kentucky), Hood et al. (2017) used reconnaissance LA-ICP-MS to obtain semi-quantitative results for the REEs together with bulk major and trace element analysis, grain size distributions, and TEM investigation of samples collected from mechanical and ESP hoppers. However, the available LA-ICP-MS spot size (100 μm) was too large to analyze REE-bearing domains found within these samples by SEM and TEM, and the matrix of NIST (National Institute of Standards and Technology, Gaithersburg, Maryland) standard glasses used for standardization differed from that of measured particles. Thompson et al. (2016, 2018) emphasized ablation of fine-grained (less than 2 μm) REE-bearing trace phases such as zircon, apatite, and monazite found to be embedded in aluminosilicate glasses in two U.S. fly ash samples; the presence of REE trace phases was revealed both in LA-ICP-MS depth profiles and corresponding SEM-EDS studies. LA-ICP-MS quantification for 36 elements, including REE, in randomly selected spots was achieved by internal summation of results to 100%, rather than by use of external standards, with excellent results for all 36 elements determined in NIST SRM 610 run as an unknown (Thompson et al., 2018).

SHRIMP-RG Ion Microprobe

Compared to smaller SIMS instruments, and especially LA-ICP-MS, deployment of SHRIMP facilities is much more limited. These large-format instruments were developed in the late 1970s at Australian National University in order to conduct grain-scale isotopic studies of geologic materials (Ireland et al., 2008). The primary application of this instrumentation is for radiometric dating using U–Th–Pb isotopes. A later generation of the SHRIMP-RG allowed ultrahigh mass resolution, especially useful in isotopic studies, and providing superior resolution of isobaric interferences in trace element determinations. Using electron beam methods and the Stanford/United States Geological Survey (USGS) SHRIMP-RG ion microprobe (Bacon et al., 2012), Kolker et al. (2017) studied a suite of fly ash samples from eastern, Illinois Basin, and western U.S. coals and determined REE contents of ash constituents on a grain scale. These authors were especially interested in determining the distribution of lanthanide REEs in fly ash glasses, following from Hower, Groppo, et al. (2013), who obtained counts for Ce in such glasses, but were limited by detection limits of the electron beam instrumentation. Results from the SHRIMP-RG show that glasses consisting almost entirely of Al and Si (aluminosilicates) had REE distributions comparable to or slightly below the bulk fly ash, whereas aluminosilicate glasses that also contained Fe or Ca trended toward higher REE contents (Figure 2.7). SHRIMP-RG

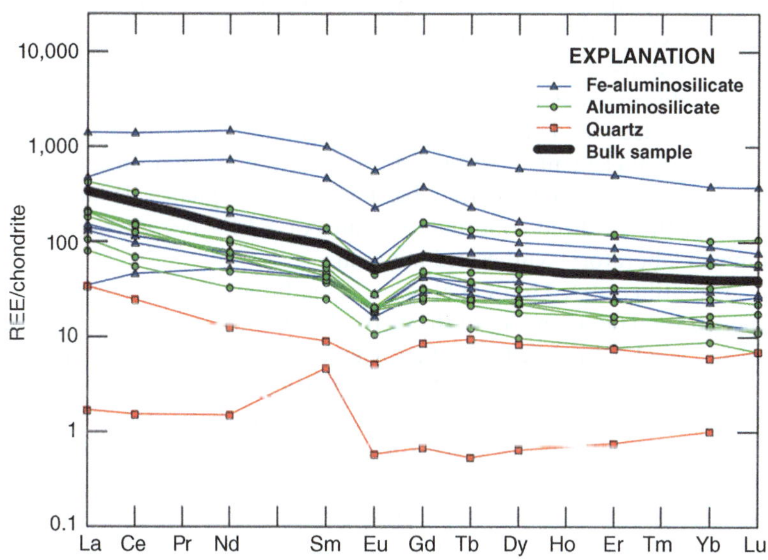

Figure 2.7 Chondrite-normalized plot of SHRIMP-RG results for constituents of a fly ash sample derived from Appalachian Basin Fire Clay coal. Figure is modified after Kolker et al. (2017) and Scott and Kolker (2019). Bulk REE distribution (heavy solid line) is from Taggart and others (2016). Plot demonstrates the relative enrichment of REEs in Fe-bearing aluminosilicate glass relative to Fe-poor glasses. Quartz shows relative REE depletion or is below detection. Promethium (Pm, atomic number 61) is omitted because it is not stable in nature.

analysis of co-occurring Fe-oxides showed that these also contain REEs, confirming, on a grain scale, the findings of Yang et al. (2014) for Fe-oxides separated from fly ash. As might be predicted from its crystal chemistry, quartz present in fly ash contains little or no REEs and is essentially a diluent in efforts to recover REEs from fly ash.

2.2.4. Synchrotron-Based Techniques

Another strategy to discern the chemical form of individual REEs in coal fly ash is to employ synchrotron X-ray spectroscopy-based techniques. For those REEs of relatively high abundance in fly ash samples (such as La, Ce, Nd, and Y), X-ray absorption spectroscopy of a fly ash sample allows the assessment of bulk-level element speciation. For example, bulk XANES has been used to identify major phases of Ce, Nd, and Y, primarily through linear combination fitting models of the sample spectrum with spectra of reference compounds (Liu et al., 2019; Stuckman et al., 2018; Taggart, Rivera, et al., 2018). The analyses generate relative percentages of the element as different reference species. This approach generally requires careful selection of known reference materials that could be reasonable approximations of the element chemical species in fly ash. Thus, the selection of reference materials must consider the conditions of fly ash formation and collection. This selection process can be especially challenging if the individual elements distribute diffusely within aluminosilicate glasses. In this case, commercially available REE mineral phases do not readily represent the localized coordination states of REEs, nor are they easy to mimic in laboratory-synthesized materials.

The utility of XANES analyses is often limited by the relatively high concentration requirements of the method, particularly as coal-combustion fly ash is also enriched in other elements that might be interfering (Table 2.1). Quantitative identification of species also requires sufficient differences in the spectral features of candidate reference compounds, which can be difficult to achieve *a priori*. In a previous application of yttrium K-edge XANES for coal fly ash (Taggart, Rivera, et al., 2018), the spectral features of a yttrium carbonate mineral ($Y_2(CO_3)_3 \cdot 3H_2O$) were very similar to Y-doped glass and monazite reference materials. Thus, in the fitting of XANES spectra for coal fly ash samples, the replacement of one of these references with the other did not substantially alter the quality of the model fits. These subtleties emphasize the need for careful and cautious interpretation of XANES results. In other words, often the results are limited to qualitative descriptions (e.g., samples A and B are similar or different in Y speciation), rather than quantitative interpretations (e.g., the 20% $Y_2(CO_3)_3$ in sample A is significantly different from 25% $Y_2(CO_3)_3$ in sample B).

Table 2.1 Rare earth elements in coal fly ash that could be analyzed for bulk speciation by X-ray absorption spectroscopy and potential interfering elements to consider in coal fly ash. The concentration values (from Taggart et al., 2016) correspond to fly ashes generated from coals of the Central Appalachian Basin in the United States (generally the more enriched of U.S. coal fly ashes). Fly ash samples are also enriched in other elements with absorption or emission energies that could interfere with the target element energies (Kortright & Thompson, 2009; Williams, 2009).

Element	Concentration in fly ash	Absorption energy	Emission energy	Potential interfering elements
Cerium (Ce)	100–200 mg kg^{-1}	5,732 eV (L_3-edge)	4,840.2 eV ($L\alpha_1$)	V, Ba
Lanthanum (La)	70–110 mg kg^{-1}	5,483 eV (L_3-edge)	4,651.0 eV ($L\alpha_1$)	Ti
Neodymium (Nd)	60–90 mg kg^{-1}	6,208 eV (L_3-edge)	5,230.4 eV ($L\alpha_1$)	Possibly Ce
Yttrium (Y)	80–100 mg kg^{-1}	17,038 eV (K-edge)	14,958 eV ($L\alpha_1$)	None likely

For elements that can exist in multiple oxidation states (e.g., Ce), bulk XANES can be especially useful for quantitative analysis of chemical form. For example, a recent study by Stuckman et al. (2018) demonstrated that greater than 85% of the Ce is in the Ce(III) form, suggesting that Ce is mostly distributed within the glassy matrix of fly ash and less frequently as distinct Ce minerals (e.g., CeO_2). The predominance of Ce(III) over Ce(IV) forms also supports the use of Ce as a proxy for other REEs that share the trivalent ionic form.

The analysis of extended X-ray absorption fine structure (EXAFS) data can be used to determine the local bonding environment of the absorber atom (generally up to 5 Å). Insights provided by EXAFS techniques such as coordination number and bond distance can further help differentiate compounds (i.e., between Y–O and Y–P structures) and complement results of the XANES analysis. However, in the recent studies that employed bulk X-ray absorption spectroscopy for Ce and Y (Liu et al., 2019; Stuckman et al., 2018; Taggart, Rivera, et al., 2018), the samples were not sufficiently concentrated to allow for EXAFS analysis. Thus, more work with EXAFS or other similar bulk elemental speciation methods is needed to improve the quantitative determination of REE chemical forms in fly ash.

X-ray microprobe configurations at synchrotron beamlines enable similar analyses at 1–10 μm spatial resolution, and more recent advances have improved this resolution to the nanometer scale (Ice et al., 2011). Spatially resolved approaches allow for “mapping” of element composition (via microprobe X-ray fluorescence spectroscopy (μXRF)). Element maps via μXRF require careful interpretation due to overlapping signals of multiple elements (i.e., nonunique assignments are possible). For example, V and Ba emission signals were previously observed to overlap with Ce emission energies (Stuckman et al., 2018). Regardless, for spots of sufficient concentration, micro-XANES and micro-EXAFS can discern the chemical form of the target element. For REEs in coal fly ash, this approach can help identify "hotspots," that is, discrete particles of REE-bearing minerals associated with or encapsulated in the ash matrix. While developments at microprobe beamlines represent a forefront of synchrotron-based techniques, their application for understanding REE recovery potential from fly ash is somewhat limiting because these hotspots represent a small proportion of the total element in fly ash particles.

2.2.5. Electron Microscopy

Electron microscopic examination of fly ash and bottom ash generally begins with SEM and energy-dispersive X-ray spectroscopy (EDX) examination of either powder samples or epoxy-mounted and polished pellets (as were used for optical petrography) (for example, Hood et al., 2017; Hower, Groppo, Henke, et al. 2017, Hower, Qian, et al., 2018; Hower et al., 2019). Individual minerals or mineral assemblages of interest are isolated and extracted by a dual-beam focused ion beam (FIB)/SEM in order to monitor the chemistry of the minerals during the extraction process. Following FIB extraction, samples are thinned to less than 100 μm for examination by TEM and related techniques. Selected area electron diffraction (SAED) is employed for mineralogical identification of grains. Either in conjunction with SAED or for grains too small for SAED, mineral identification can be accomplished with high-resolution TEM (HRTEM). Fast Fourier transform (FFT), computed by image analysis software, is used to determine the lattice spacing in HRTEM micrographs. Chemical identification in TEM is available in some units via electron energy-loss spectroscopy (EELS).

REE-bearing minerals observed by TEM in select eastern Kentucky coals include a mixed monazite/kaolinite grain from the Fire Clay coal (Figure 2.8) (Hower, Berti, Hochella, & Mardon, 2018; Hower, Berti, Hochella, Rimmer, & Taulbee, 2018). The Fire Clay coal is of particular interest as a REE resource as the coal, as the coal preparation products (Honaker et al., 2018; Huang et al., 2018; Zhang & Honaker, 2018; Zhang et al., 2018), and as the fly ash resulting from the coal utilization (Hower et al., 2015; Hood et al., 2017; Hower, Groppo, Henke, et al., 2017; Hower, Groppo, et al., 2020; Zhang et al., 2015). In the case illustrated here, the original layered monazite and volcanic glass from the volcanic ash fall (Hower, Eble, et al., 2016) were altered to a monazite/kaolinite mixed

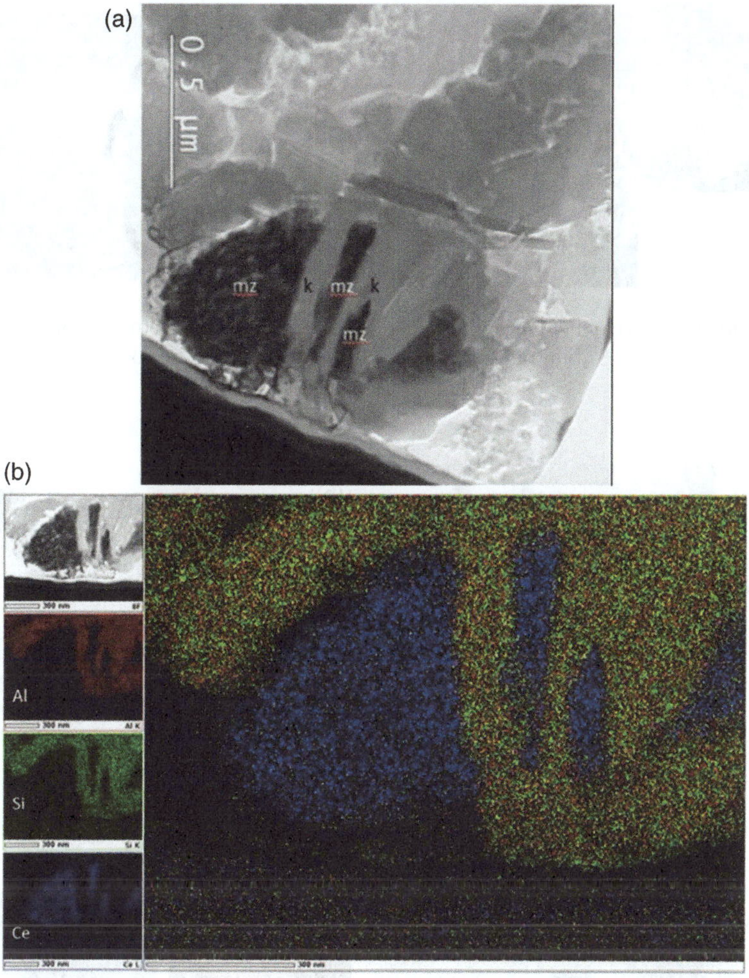

Figure 2.8 (a) Mixed monazite (mz) and kaolinite (k) grain in Fire Clay coal. (b) The EDX scan shows the segregation of the kaolinite (Al and Si) from the monazite (using Ce as a proxy for the REEs in monazite). After Hower, Berti, Hochella, & Mardon (2018).

grain. The Manchester coal, Clay County, eastern Kentucky, contains florencite ($(REE)Al_3(PO_4)_2(OH)_6$) nodules (Hower, Qian, et al., 2020).

As shown by optical petrology (Section 2.2), monazite and other REE-bearing minerals occur in combustion ashes. Figure 2.9 illustrates a relatively large La–Ce–Nd monazite grain with a Y-rich xenotime inclusion. Small (tens of nanometers) REE-phosphate grains on a mullite crystal are shown in Figure 2.10. Mullite is a more important constituent of this ash as

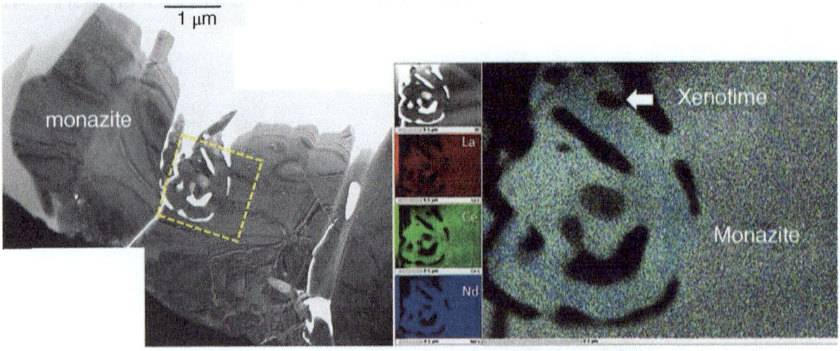

Figure 2.9 Monazite with included xenotime in the fly ash from the combustion of an eastern U.S. bituminous coal blend. After Hower, Groppo, Henke, et al. (2017).

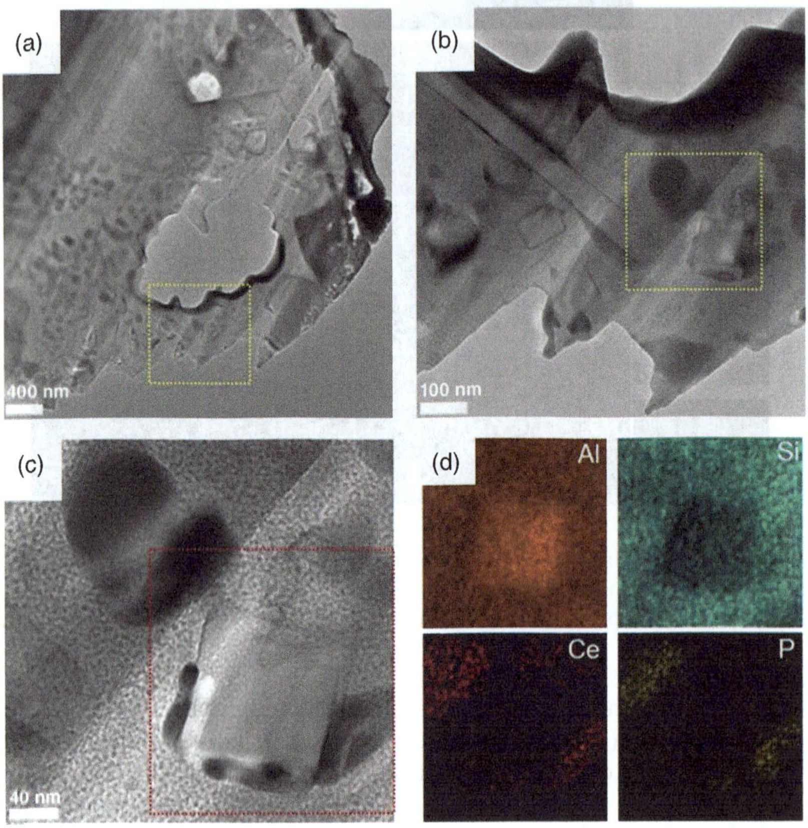

Figure 2.10 (a)–(d) Ce-phosphate on mullite in stoker ash from the combustion of an eastern U.S. bituminous coal blend. After Hower, Qian, et al. (2018).

compared to most fly ashes due to the larger particle size and the consequent longer cooling time of the stoker ash; the mineral has the time required to crystallize versus the formation of an Al-Si glass (also see Figure 2.5 for an optical view of the same ash). In contrast to the visible minerals in the latter ashes, other REE-rich areas have no discernable minerals (Figure 2.11). In this case, while there is an indication (via SAED) of crystallinity, the minerals are too small, overlap, and/or have disparate orientations, making mineral identification impossible with current techniques. In addition to monazite and Y-bearing zircon, davidite-Ce ((Ce, La)(Y, U)(Ti, Fe)$_{20}$O$_{38}$) or ((Ce, La)(Y, U, Fe)(Ti, Fe)$_{20}$(OH, O)$_{38}$) was found in a stoker ash from the combustion of the Fire Clay coal (Hower et al., 2019).

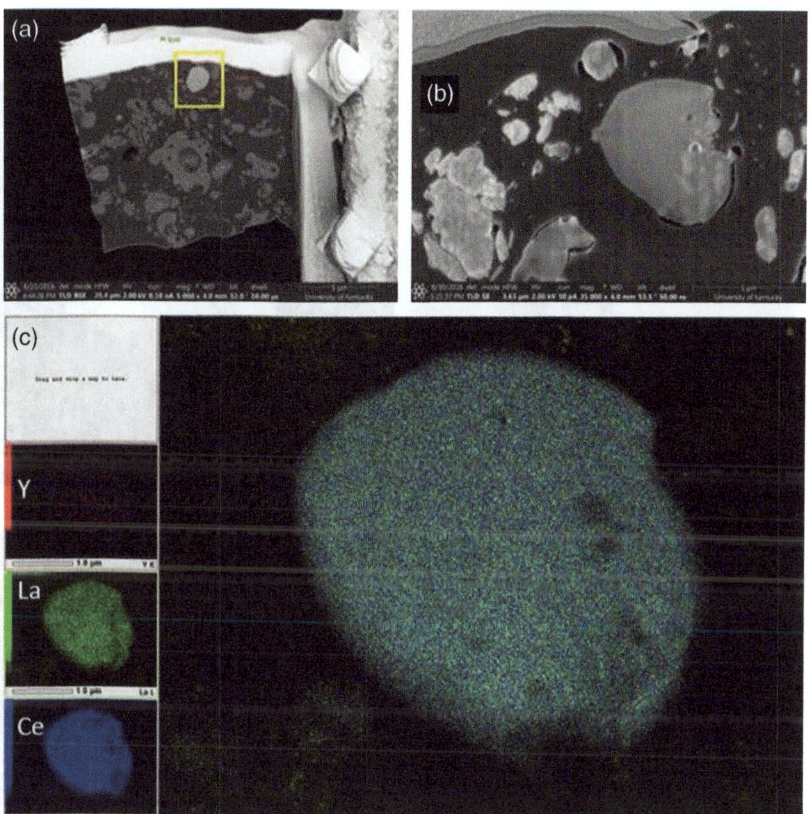

Figure 2.11 (a) and (b) Opposing sides and different sizes of the FIB lift-out slice. (c) EDX image for Y, La, and Ce showing the diffuse nature of the REE distribution. From Hood et al. (2017).

Studies of fly ash carbons (Hower et al., 2008; Hower, Groppo, Graham, et al., 2017; Hower, Groppo, Henke, et al., 2017; Hower & Groppo, 2020; Hood et al., 2017; Wilcox et al., 2015) indicate that graphitic and fullerene-like carbons, often with included/entrained few-nanometer minerals and/or metals, occur in the fly ash resulting from the combustion of bituminous coals. Spinel minerals (Figures 2.12 and 2.13) occur in association with REEs; for example, Ce found within the carbon surrounding the Fe-spinel (Figure 2.12c). Similarly, Nd, Ce, Sm, and Y occur in the carbon surrounding a spinel from the same fly ash (Figure 2.13). Figure 2.14 illustrates an Al-Si glass with an Fe–O core, possibly spinels, surrounded by Ce–Nd–Sm–Y–graphitic carbon. Another grain showing the dispersion of the Nd and Ce in the carbon surrounding the Al-Si glass

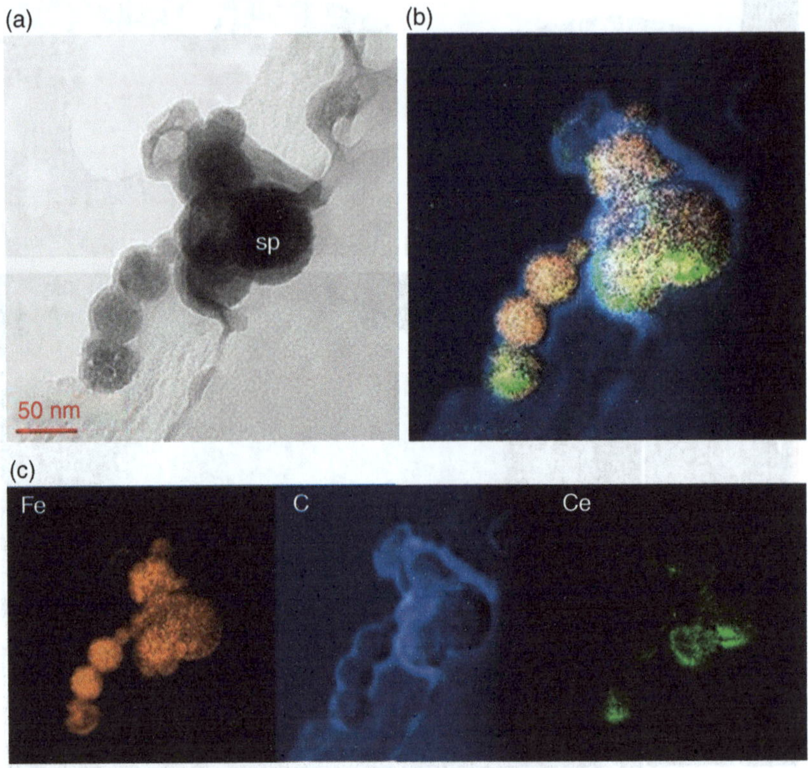

Figure 2.12 (a) TEM image of Fe-spinel minerals (sp) in fly ash from the combustion of an eastern U.S. bituminous coal blend. (b) Composite EDX image based on the elements mapped individually in (c). After Hower, Groppo, Henke, et al. (2017).

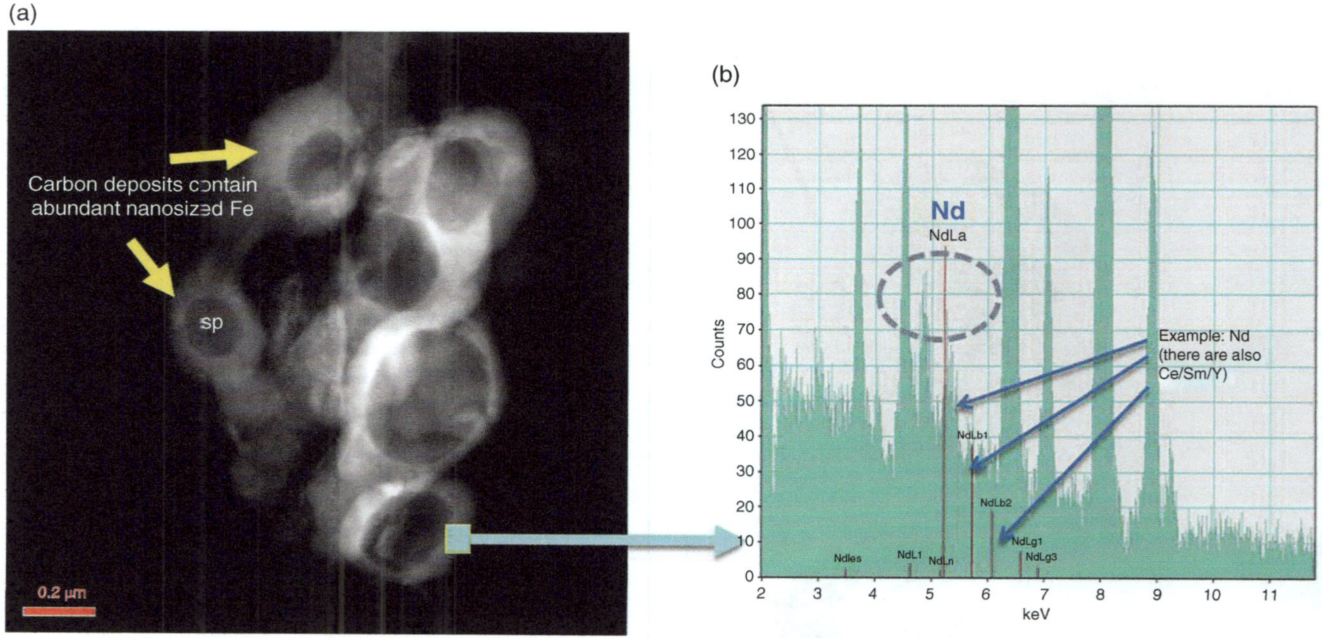

Figure 2.13 (a) TEM image of spinel (sp) surrounded by carbon. The area scanned by EDX is shown by the rectangle on the lower right of (a). The fly ash is from the combustion of an eastern U.S. bituminous coal blend. (b) EDX scan of area shown in (a), indicating that the carbon deposit on the surface of the spinels includes/entrains Nd, Ce, Sm, and Y. Adapted from Hower, Groppo, Henke, et al. (2017).

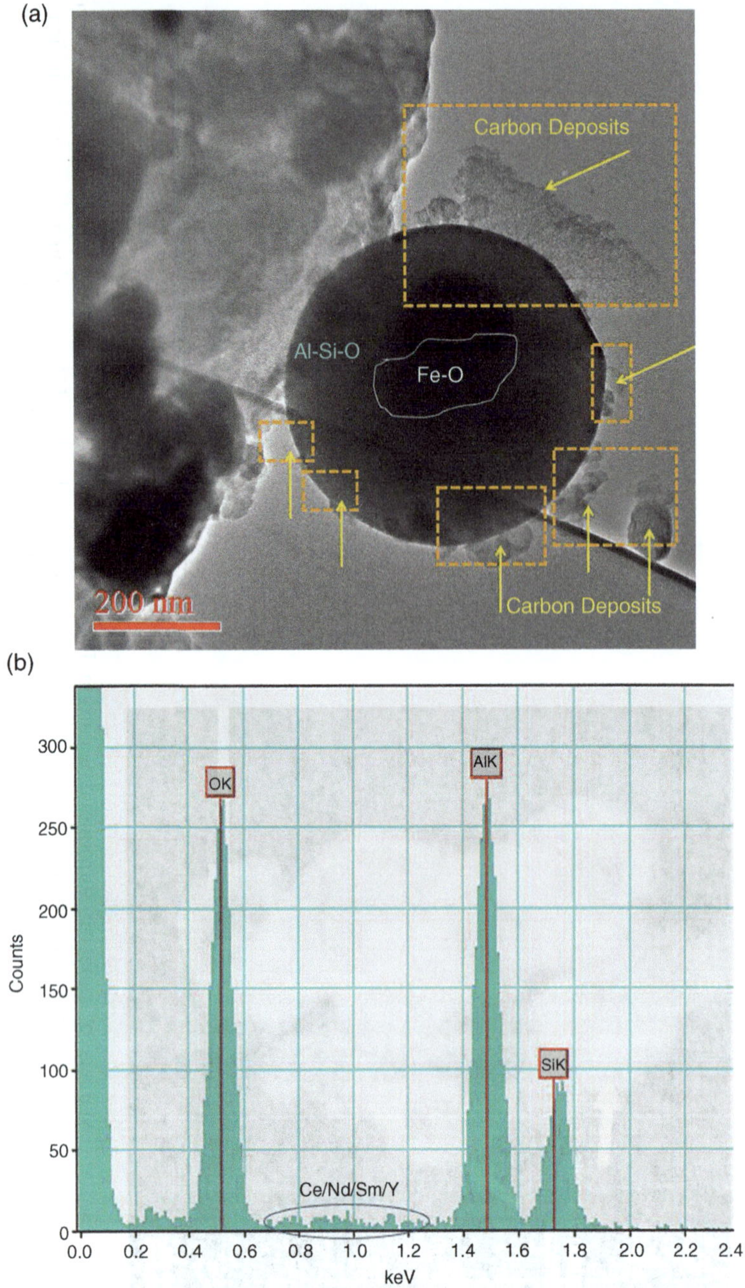

Figure 2.14 High-resolution TEM image (a) of Al-Si glass sphere surrounded by graphitic carbon deposits. The fly ash is from the combustion of an eastern U.S. bituminous coal blend. The composite of 50 EDX scans (b) indicates that the carbon contains Ce, Nd, Sm, and Y. Hower et al. (2017b) / The University of Kentucky Center for Applied Energy Research and the American Coal Ash Association.

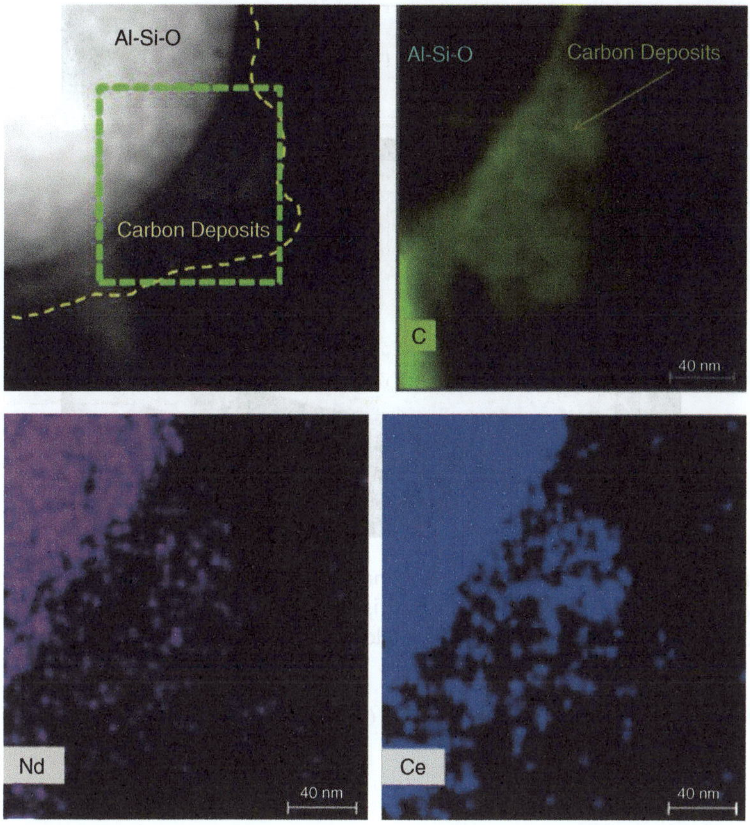

Figure 2.15 High-resolution TEM image (upper left) of Al-Si glass sphere surrounded by graphitic carbon deposits. EDX images of C, Nd, and Ce are shown in the upper right, lower left, and lower right images, respectively. The EDX C overlapping the Al-Si glass is blacked out on the upper right image, concealing the basis for the dense Nd and Ce concentrations coincident with the glass. The fly ash is from the combustion of an eastern U.S. bituminous coal blend. After Hower, Groppo, Henke, et al. (2017).

is shown on Figure 2.15. The EDX carbon overlapping the Al-Si glass is blacked out on the upper right image, concealing the dense Nd and Ce concentrations coincident with the glass. The TEM image and the derivative EDX scans are really capturing some semblance of three dimensions; therefore, the REE elements overlying the glass appear to be associated with the carbon, just as they are in the carbon corona adjacent to the glass. EELS imaging of a fly ash particle shows Ce to be associated with

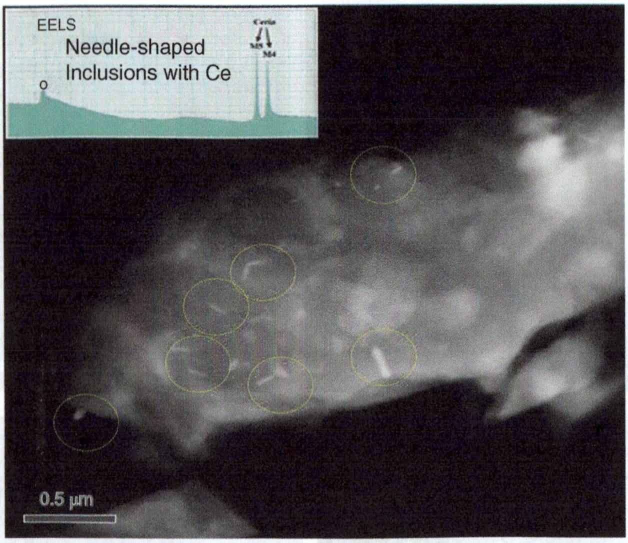

Figure 2.16 EELS imaging indicating the presence of Ce associated with needle-shaped minerals (in circles). The fly ash is from the combustion of an eastern U.S. bituminous coal blend. Adapted from Hower et al. (2017b).

needle-shaped minerals (Figure 2.16), an occurrence similar to that in Figure 2.10.

2.3. Extractions and Separations for REE Recovery from Fly Ash

The extraction of REEs and other critical metals from fly ash and bottom ash requires a combination of acid leaching (or roasting followed by acid leaching), physical separations, and ion-separation processes (Rybak & Rybak, 2021; Zhang et al., 2020). Economically feasible methods to produce saleable REE products from coal fly ash have yet to be fully proven and research on methods remains in early stages of development. In the development of REE extraction methods, lessons can be learned from past studies on recovery of other elements (e.g., aluminum and germanium) from fly ash and from the processes used to recover REEs at conventional mining operations (Arroyo & Fernández-Pereira, 2008; Bai et al., 2011; Dai, Seredin, et al., 2014; Dai, Zhao, et al., 2014; Fang & Gesser, 1996; Gilliam et al., 1982; Kelmers et al., 1982; Meawad et al., 2010; Molycorp Minerals LLC, 2010; Rayzman et al., 1997).

At traditional mining sites, ores are first excavated and then processed to finer particles using milling, crushing, and grinding units (Molycorp Minerals LLC, 2010; Peak Resources Limited, 2017) (Figure 2.17). As this step would not be necessary for coal ash, the first step involves physical separations such as sieving, magnetic separation, or froth flotation to remove materials with low-REE content and/or materials which would interfere with the downstream processing (e.g., high Fe-content and/or high-C-content particles). For example, slight enrichment of REE contents has been observed in the nonmagnetic fraction and small-size fraction (<200 mesh) of coal fly ashes (Blissett et al., 2014; Lin, Howard, et al., 2017). Thus, preprocessing steps to remove magnetic and coarse material aid in the concentration of REEs as well as remove interferences (such as iron) for downstream ion separations.

Physical separations are followed by a leaching step that typically requires strong acids (e.g., sulfuric, nitric, or hydrochloric acids) to solubilize the metals from the material. For a subset of coal fly ashes, such as Ca-rich fly ashes that are characteristic of western U.S. coals, acid leaching is usually sufficient for full extraction of REEs (Cao et al., 2018; King et al., 2018). For other coal fly ashes, an alternative process could entail alkaline treatment that decomposes aluminosilicate glasses (Bai et al., 2010; McDowell & Seeley, 1981; Padilla & Sohn, 1985; Querol et al., 2002). This approach might be implemented as a wet-alkaline digestion or a dry roasting process as a way to convert the REEs in coal fly ash to an acid-soluble phase (Honaker et al., 2019; King et al., 2018; Lin, Stuckman, et al., 2018; Taggart, Hower, & Hsu-Kim, 2018; Wang et al., 2019; Yakaboylu et al., 2019).

After the extraction of REEs from the fly ash, the REEs need to be further purified from the aqueous acidic mixture. In conventional REE hydrometallurgical processes, this step often entails liquid–liquid separations, selective precipitation, or solid phase extraction (Hu et al., 2018). For leachates of coal-derived fly ash and other geologically derived brine solutions, researchers have proposed a wide variety of approaches to purify REEs. In one patented method (Joshi et al., 2013), researchers proposed a combination of an organophosphate chelating compound (e.g., tributylphosphate) saturated in an organic phase (e.g., kerosene) to selectively separate the trivalent rare earth cations over other ions, followed by back extraction into the aqueous phase and recycling of the organic solvent. This liquid–liquid (solvent) extraction approach closely mimics methods utilized for traditional recovery of uranium and REEs from ores (Brown & Sherrington, 1979; Flett, 2005). Other methods for purification could also include selective precipitation of Fe and Al hydroxides through the addition of sodium hydroxide or lime (Middleton et al., 2020; Mutlu et al., 2018),

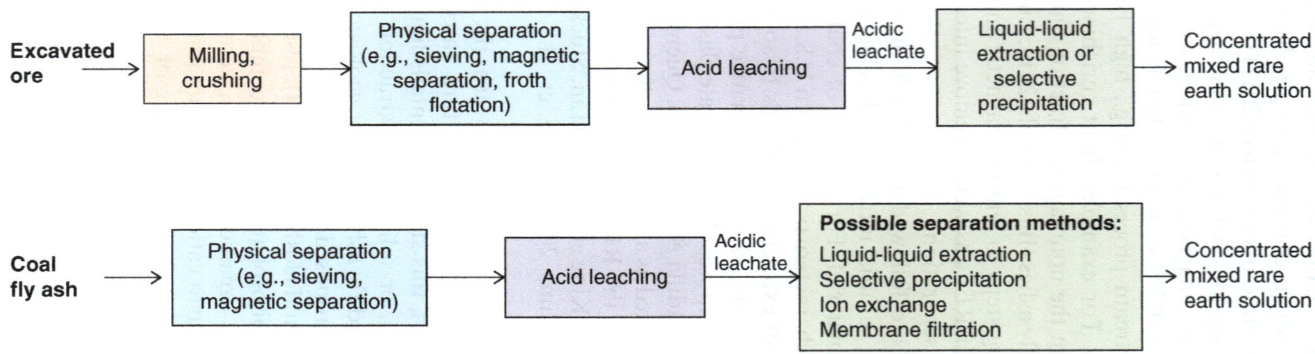

Figure 2.17 Example sequence of unit processes for the recovery and separation of REEs at conventional mining operations (top schematic) and potential modifications to the recovery process for REE-enriched coal fly ash (bottom schematic).

and while these approaches are favorable from a scalability perspective, they tend to have relatively high operating expenses due to substantial requirements for reagents to adjust the pH.

Several other ion-separation methods have been proposed for the separation of coal fly ash and other geological waste materials, including ion exchange, sorption, and membrane filtration (Das & Das, 2013; Lo et al., 2014; Callura et al., 2018; Kose-Mutlu et al., 2020; Mutlu et al., 2018; Park et al., 2017, 2020; Smith et al., 2019; Liu et al., 2023). Other REE purification technologies that have been directly tested on fly ash leachates include biosorption with bacterial cells that have been genetically engineered to express lanthanide-binding tags on their cell envelope (Park et al., 2017, 2020). Others have attempted supported liquid membranes (Smith et al., 2019), a liquid–liquid extraction approach that entails the use of a hydrophobic membrane embedded with an organic solvent/extractant mixture. Likewise, nanomaterial-enabled electrochemical methods for REE separation and recovery have the potential to selectively precipitate REEs due to local generation of hydroxide (from water and oxygen) that serves to precipitate mixed rare earth oxide and hydroxide cakes (O'Connor et al., 2018). Other novel, nano-enabled sorbents or chelating agents that have the potential to separate REEs selectively are being explored but have not yet been assayed against the complex mixture of coal ash (Bie et al., 2020; Chen et al., 2020; Hamza et al., 2021; Rahman et al., 2020). While filtration-based methods could provide advantages relative to conventional solvent-extraction methods by reducing liquid chemical consumption, decreasing the footprint of the process and enhancing modularity, the primary barriers toward applying these technologies for coal ash are the relatively low-purity REE content and the high abundance of interfering ions (e.g., Si, Al, and Fe) that co-leach with the REEs during processing (see Figure 2.18). Nevertheless, selective pre-treatments or future innovations may help surmount those challenges.

2.4. Outlook and Research Needs

The realization that coal fly ash can be a large and untapped reserve for REEs has sparked great interest for recovery technologies, particularly with motivations to increase the diversity of the global REE supply market. Furthermore, additional reuse markets for coal fly ash could decrease disposal needs for coal ash, the largest industrial-waste streams for nations heavily reliant on coal energy. As noted previously, existing hydrometallurgical methods for REE separations are now in early stages of development. However, scale-up of these processes for profitable recovery of REE products remains elusive. Ongoing challenges in the recovery and

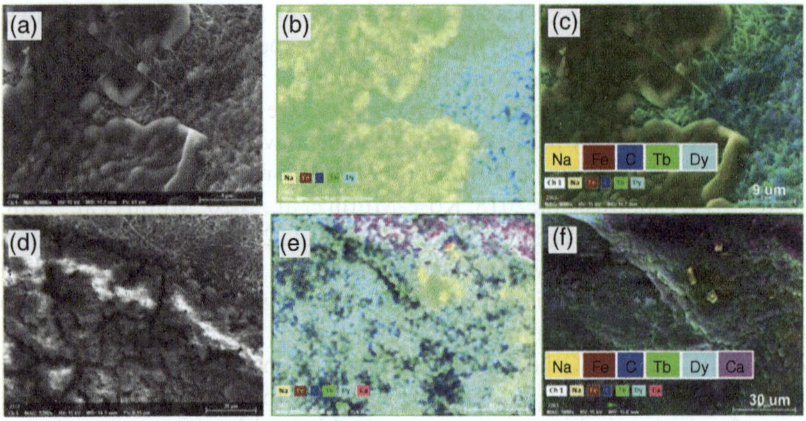

Figure 2.18 (a)–(f) Scanning electron micrograph with energy dispersive x-ray spectroscopy (SEM-EDS) of REEs separated from coal ash leachate using electrochemical deposition on carbon nanotube filters. Pre-treatments included an alkaline sintering of the coal ash, followed by a nitric acid extract and a pH titration using sodium hydroxide. Residual sodium is visible on top of the rare earth element oxide "cake," where terbium and dysprosium were the main REEs collected. The enrichment of the final cake was approximately 1.8% total REE.

concentration of REEs from fly ash could offer another benefit—innovations in separation technologies (such as those described in later chapters of this volume)—that could greatly advance the overall field of metals separation and waste reuse.

Major challenges currently hinder successful and sustainable recovery of REEs from coal fly ash. First, the enrichment of REEs in coal fly ash is "low" relative to REE contents of ores found at rare earth mines or REE contents in other secondary feedstocks (e.g., electronic wastes, where there has been much development in extraction efforts) (Bahaloo-Horeh et al., 2020; Makarova et al., 2020). As such, recovery operations must overcome high hurdles in extraction and separation efficiency. Potential processes must be especially selective for rare earth metal ions because other major elements such as Na, K, Ca, Mg, Fe, Al, and Si are 10^4 to 10^6 times more concentrated than individual rare earths in fly ash leachates (King et al., 2018; Taggart, Hower, & Hsu-Kim, 2018). Metal ions of Fe, Ca, and Al could be especially problematic given their high abundance as well as moderate-to-high affinity for ion-exchange resins and chelating agents used to purify REEs (King et al., 2018; Taggart, Hower, & Hsu-Kim, 2018). Elements of health concern, such as uranium (U) and thorium (Th), also require strategic separation and present a technological challenge that has long been identified but remains in nascent stages of investigation (Bie et al., 2020). Thus, preprocessing steps may be essential to reduce these interfering ions (and pH adjustments with NaOH, which adds interfering ions, may be counterproductive). Moreover, the challenge of moderate levels of enrichment requires novel separations that improve REE selectivity as a means to recover REE materials of saleable purity.

Second, development of an environmentally sustainable REE recovery process is a major challenge in REE separations from fly ash. With previous work demonstrating that complete dissolution of fly ash particles may be required to fully liberate REEs into solution (Hower, Groppo, Henke, et al., 2017), large volumes of chemicals including strong acids and organic solvents may be required for REE recovery operations. New innovations in process engineering are needed to facilitate recycling these chemicals during the recovery process, both to keep operating expenses low and to avoid the generation of another large-volume waste stream. Moreover, novel REE separation approaches could aim to reduce the usage of chemicals, especially organic solvents. Advanced sorbents, ion-exchange resins, electrochemical or biosorption methods, or new configurations of liquid–liquid separations might offer benefits in this respect (Callura et al., 2018; Couto et al., 2020; Kim et al., 2015; O'Connor et al., 2018; Park et al., 2020; Smith et al. 2019), as long as they can overcome needs for REE selectivity.

Finally, with REE recovery technologies from fly ash still in their infancy, current techno-economic analyses are fraught with high degrees of uncertainty. Chemical consumption and waste handling are likely to be major contributors to costs (Das et al., 2018; Hendren et al., 2017), and therefore, these need careful consideration as innovations progress. As noted previously, revenue estimates in these assessments are highly reliant upon the value of Sc in fly ash. Of the REE elements, Sc is currently one of the highest valued on a price per kilogram basis, but it is also produced at low volumes. For example, on 28 February 2020, Sc was priced at $3,487/kg (Berlet & Samnani, 2020) and global production was 15–20 t (Gambogi, 2020). In contrast, the market prices on the element basis of Ce and Nd, with 24,000 and 33,000 t of global production, respectively, are $4.59/kg and $52.66/kg, respectively (28 February 2020 from Berlet & Samnani, 2020; based on 22 June 2020 Chinese yuan to U.S. dollar conversion). The recovery of Sc from coal fly ash may inundate the Sc supply market, resulting in a decrease in the market price for this commodity and much lower revenue potential than values in current estimates. These uncertainties suggest that the design of REE recovery technologies would greatly benefit by other saleable products generated in the process. Thus, engineering for co-production is likely to be essential for overcoming these challenges and the attainment of REE production from coal combustion residuals.

Acknowledgments

Portions of this study were completed as part of U.S. Department of Energy contracts DE-FE0027167 and DE-FE0026952 and were further supported by National Science Foundation grants CBET-1510965 and CBET-1510861 to Duke University and the University of Kentucky, respectively. Hower's work was further supported by U.S. Department of Energy contract DE-FE0029007 to the University of North Dakota Energy & Environmental Research Center with a subcontract to the University of Kentucky.

This work used shared facilities at the Virginia Tech National Center for Earth and Environmental Nanotechnology Infrastructure (NanoEarth), a member of the National Nanotechnology Coordinated Infrastructure (NNCI), supported by NSF (ECCS 1542100). Michael Hochella, Jr., Christopher Winkler, Debora Berti, Elizabeth Cantando, and Mitsu Murayama assisted us at Virginia Tech.

Access to characterization instruments and staff assistance at the University of Kentucky was provided by the Electron Microscopy Center, supported in part by the National Science Foundation/EPSCoR Award No. 1355438 and by the Commonwealth of Kentucky. Uschi Graham, Madison Hood, Dali Qian, Nicolas Briot, Jason Unrine, Jack Groppo, Tristana Duvallet, and Bob Jewell assisted us at the University of Kentucky.

Clint Scott and Jorge Vazquez (USGS), and Matt Coble (Stanford University) are acknowledged for their help in acquiring SHRIMP-RG data. Use of the SHRIMP-RG was supported by a Technical Assistance Agreement between the USGS and the University of Kentucky, with funding from U.S. Department of Energy contract DE-FE0027167, and from the USGS Energy Resources Program. Any use of trade, firm, or product names is for descriptive purposes and does not imply endorsement by the U.S. Government.

References

Alekseev, Yu. I., & Vereshchagin, V. I. (1997). Formation of crystalline phases in electroceramics of the system $CaO–MgO–Al_2O_3–SiO_2$ (A review). *Glass and Ceramics, 54*, 340–343.

American Coal Ash Association (2019). *Coal ash recycling rate declines amid shifting production and use patterns.* https://www.prnewswire.com/news-releases/coal-ash-recycling-rate-declines-amid-shifting-production-and-use-patterns-300961238.html. (accessed 7 March 2024).

Arroyo, F., & Fernández-Pereira, C. (2008). Hydrometallurgical recovery of germanium from coal gasification fly ash. Solvent extraction method. *Industrial & Engineering Chemistry Research, 47*, 3186–3191.

ASTM International (2017). *ASTM C618-17, Standard specification for coal fly ash and raw or calcined natural pozzolan for use in concrete.* ASTM International website, https://www.astm.org/Standards/C618.htm. (accessed 7 March 2024)

Bacon, C. R., Grove, M., Vazquez, J. A., & Coble, M. A. (2012). The Stanford-U.S. Geological Survey SHRIMP ion microprobe—A tool for micro-scale chemical and isotopic analysis. *U.S. Geological Survey Fact Sheet* 2012-3067, 4 p., https://pubs.usgs.gov/fs/2012/3067/fs2012-3067.pdf (accessed 7 March 2024)

Bahaloo-Horeh, N., & Mousavi, S. M. (2020). Comprehensive characterization and environmental risk assessment of end-of-life automotive catalytic converters to arrange a sustainable roadmap for future recycling practices. *Journal of Hazardous Materials*, *400*, 123186.

Bai, G.-H., Teng, W., Wang, X.-G., Qin, J.-G., Xu, P., & Li, P.-C. (2010). Alkali desilicated coal fly ash as substitute of bauxite in lime-soda sintering process for aluminum production. *Transactions of Nonferrous Metals Society of China*, *20*, 169–175.

Bai, G., Qiao, Y., Shen, B., & Chen, S. (2011). Thermal decomposition of coal fly ash by concentrated sulfuric acid and alumina extraction process based on it. *Fuel Processing Technology*, *92*, 1213–1219.

Balashov, Y. A. (1976). *Geochemistry of rare earth elements*. Nauka, Moscow, 267 p. (in Russian).

Bank, T. L., Roth, E. A., Tinker, P., & Granite, E. (2016). Analysis of Rare Earth Elements in Geologic Samples using Inductively Coupled Plasma Mass Spectrometry. U.S. DOE Topical Report—DOE/NETL-2016/1794.

Basu, Z. (2017). In the heart of coal country, state officials bet on renewable energy. *CNBC*, 18 July 2017, http://www.cnbc.com/2017/07/18/post-coal-country-kentucky-bets-on-renewable-energy-metals-mining.html (accessed 7 March 2024).

Bauer, G., Achleitner, B., Bonta, M., Friedbacher, G., & Limbeck, A. (2017). Analysis of single fly ash particles using laser ablation ICP-MS-an approach achieving lateral elemental distribution information via imaging. *RSC Advances*, *7*(33), 20510–20519. doi:10.1039/C7RA00975E

Berlet, C. J., & Samnani, K. (2020). Mineral Prices.com. https://mineralprices.com/rare-earth-metals/ (accessed 22 June 2020; no longer available).

Bie, C., Gao, Y., Su, J., Dong, Y. M., Guo, X. G., & Sun, X. Q. (2020). The efficient separation of thorium from rare earth using oxamic acid in hydrochloric acid medium. *Separation and Purification Technology*, *251*, 117358.

Blissett, R. S., Smalley, N., & Rowson, N. A. (2014). An investigation into six coal fly ashes from the United Kingdom and Poland to evaluate rare earth element content. *Fuel*, *119*, 236–239.

Bragg, L. J., Oman, J. K., Tewalt, S. J., Oman, C. L., Rega, N. H., Washington, P. M., & Finkelman, R. B. (1998). *U.S. Geological Survey Coal Quality (COALQUAL) database: Version 2.0*. U.S. Geological Survey Open-file Report 97-134, CD-ROM.

Brown, C. G., & Sherrington, L. G. (1979). Solvent extraction used in industrial separation of rare earths. *Journal of Chemical Technology and Biotechnology*, *29*, 193–209.

Brownfield, M. E., Affolter, R. H., Brownfield, I. K., Hower, J. C., & Stricker, G. D. (1999). *Dispersed volcanic ash in feed coal and its influence on coal combustion products*. International Ash Utilization Symposium, Lexington, KY, 18–20 October 1999, paper #61. http://www.flyash.info (accessed 7 March 2024).

Brownfield, M. E., Cathcart, J. D., Affolter, R. H., Brownfield, I. K., Rice, C. A., O'Connor, J. T., et al. (2005). *Characterization and modes of occurrence of elements in feed coal and coal combustion products from a power plant utilizing low-sulfur coal from the Powder River Basin, Wyoming*. U.S. Geological Survey Scientific Investigations Report 2004–5271, 36 p. https://pubs.usgs.gov/sir/2004/5271/. (accessed 7 March 2024)

Callura, J. C., Perkins, K. M., Noack, C. W., Washburn, N. R., Dzombak, D. A., & Karamalidis, A. K. (2018). Selective adsorption of rare earth elements onto functionalized silica particles. *Green Chemistry*, *20*, 1515–1526.

Cao, S., Zhou, C., Pan, J., Liu, C., Tang, M., Ji, W., et al. (2018). Study on influence factors of leaching of rare earth elements from coal fly ash. *Energy & Fuels*, *32*, 8000–8005.

Chen, Z. L., Yang, X. Y., Song, L. J., Wang, X. Y., Xiao, Q., Xu, H. W., et al. (2020). Extraction and complexation of trivalent rare earth elements with tetraalkyl diglycolamides. *Inorganica Chimica Acta*, *513*, 119928.

Chenery, S., Querol, X., & Fernández-Turiel, J. L. (1995). Quantitative determination of trace element affinities in coal and combustion wastes by Laser Ablation Microprobe- Inductively Coupled Plasma—Mass Spectrometry. In J. A. Pajares & J.M.D. Tascón (Eds.), *Coal Science: Proceedings of the Eighth International Conference on Coal Science* (pp. 327–330). Amsterdam: Elsevier.

Cherniak, D. J., & Watson, E. B. (2013). Diffusion of helium in natural monazite, and preliminary results on He diffusion in synthetic light rare earth phosphates. *American Mineralogist*, *98*, 1407–1420.

Cherniak, D. J., Watson, E. B., & Thomas, J. B. (2009). Diffusion of helium in zircon and apatite. *Chemical Geology*, *268*, 155–166.

Clarke, L. B., & Sloss, L. L. (1992). Trace elements—Emissions from coal combustion and gasification. *IEA Coal Research, IEACR/49*, 111 p.

Connelly, N. G., Hartshorn, R. M., Damhus, T., & Hutton, A. T. (Eds.). (2005). *Nomenclature of Inorganic Chemistry: IUPAC recommendations 2005*. Royal Society of Chemistry.

Couto, N., Ferreira, A. R., Lopes, V., Peters, S. C., Mateus, E. P., Ribeiro, A. B., & Pamukcu, S. (2020). Electrodialytic recovery of rare earth elements from coal ashes. *Electrochimica Acta*, *359*, 136934.

Dai, S., Luo, Y., Seredin, V. V., Ward, C. R., Hower, J. C., Zhao, L., et al. (2014). Revisiting the late Permian coal from the Huayingshan, Sichuan, southwestern China: Enrichment and occurrence modes of minerals and trace elements. *International Journal of Coal Geology*, *122*, 110–128.

Dai, S., Seredin, V. V., Ward, C. R., Jian, J., Hower, J. C., Song, X., et al. (2014). Composition and modes of occurrence of minerals and elements in coal combustion products derived from high-Ge coals. *International Journal of Coal Geology*, *121*, 79–97.

Dai, S., Zhao, L., Hower, J. C., Johnston, M. N., Song, W., Wang, P., & Zhang, S. (2014). Petrology, mineralogy, and chemistry of size-fractioned fly ash from the Jungar power plant, Inner Mongolia, China, with emphasis on the distribution of rare earth elements. *Energy & Fuels*, *28*, 1502–1514.

Dai, S., Graham, I. T., & Ward, C. R. (2016). A review of anomalous rare earth elements and yttrium in coal. *International Journal of Coal Geology*, *159*, 82–95.

Dai, S., Liu, J., Ward, C. R., Hower, J. C., French, D., Jia, S., et al. (2016). Mineralogical and geochemical compositions of Late Permian coals and host rocks from the Guxu Coalfield, Sichuan Province, China, with emphasis on enrichment of rare metals. *International Journal of Coal Geology*, *166*, 71–95.

Dai, S., Xie, P., Jia, S., Ward, C. R., Hower, J. C., Yan, X., & French, D. (2017). Enrichment of U-Re-V-Cr-Se and rare earth elements in the Late Permian coals of the Moxinpo Coalfield, Chongqing, China: Genetic implications from geochemical and mineralogical data. *Ore Geology Reviews*, *80*, 1–17.

Das, N., & Das, D. (2013). Recovery of rare earth metals through biosorption: An overview. *Journal of Rare Earths*, *31*, 933–943.

Das, S., Gaustad, G., Sekar, A., & Williams, E. (2018). Techno-economic analysis of supercritical extraction of rare earth elements from coal ash. *Journal of Cleaner Production*, *189*, 539–551.

Dobransky, S. (2013). Rare earth elements and U.S. Foreign Policy: The critical ascension of REEs in Global Politics and U.S. *National Security. American Diplomacy*, https://americandiplomacy.web.unc.edu/2013/10/rare-earth-elements-and-u-s-foreign-policy/ (accessed 7 March 2024)

Ehlers, E. G. (1972). *The interpretation of geological phase diagrams*. San Francisco: W.H. Freeman & Co., 280 p.

Energy Information Administration (EIA). (2017). *Electricity*. https://www.eia.gov/electricity/monthly/update/archive/july2017/ (accessed 7 March 2024).

Energy Information Administration (EIA). (2018). *U.S. electricity data*. https://www.eia.gov/electricity/data.php (accessed 7 March 2024).

Energy Information Administration (EIA). (2020). *Coal*. https://www.eia.gov/coal/ (accessed 17 June 2020; current version accessed 7 May 2024).

Environmental Protection Agency (EPA). (2014). *Final Rule—Disposal of coal combustion residuals from electric utilities*. https://www.epa.gov/coalash/coal-ash-rule. (accessed 7 March 2024).

Environmental Protection Agency (EPA). (2019). *Final Cross-State Air Pollution Rule Update*. https://www.epa.gov/Cross-State-Air-Pollution/final-cross-state-air-pollution-rule-update (accessed 7 March 2024).

Environmental Protection Agency (EPA). (2020). *Regulatory Actions—Final Mercury and Air Toxics Standards (MATS) for Power Plants*. https://www.epa.gov/mats/regulatory-actions-final-mercury-and-air-toxics-standards-mats-power-plants (accessed 7 March 2024)

Eskanazy, G. M. (1987a). Rare earth elements in a sampled coal from the Pirin deposit, Bulgaria. *International Journal of Coal Geology*, *7*, 301–314.

Eskanazy, G. M. (1987b). Rare earth elements and yttrium in lithotypes of Bulgarian coal. *Organic Geochemistry*, *11*, 83–89.

Eskanazy, G. M. (1999). Aspects of the geochemistry of rare earth elements in coal: An experimental approach. *International Journal of Coal Geology*, *38*, 285–295.

Fang, Z., & Gesser, H. (1996). Recovery of gallium from coal fly ash. *Hydrometallurgy*, *41*, 187–200.

Farley, K. A. (2007). He diffusion systematics in minerals: Evidence from synthetic monazite and zircon structure phosphates. *Geochimica et Cosmochimica Acta*, *71*, 4015–4024.

Finch, R. J., & Hanchar, J. M. (2003). Structure and chemistry of zircon and zircon group minerals. In J. M. Hanchar & P. W. O. Hoskin (Eds.), *Zircon. Mineralogical Society of America, Reviews in Mineralogy and Geochemistry*, *53*, 1–25.

Finkelman, R. B. (1981). Modes of occurrence of trace elements in coal. *U.S. Geological Survey Open File Report 81-99*, 301 p. https://pubs.usgs.gov/of/1981/0099/report.pdf (accessed 7 May 2024)

Finkelman, R. B. (1993). Trace and minor elements in coal. In M. H. Engel & Macko S. A., (Eds.), *Organic Geochemistry: Principles and applications* (pp. 593–607). New York: Plenum Press. https://doi.org/10.1007/978-1-4615-2890-6_28

Flett, D. S. (2005). Solvent extraction in hydrometallurgy: The role of organophosphorus extractants. *Journal of Organometallic Chemistry*, *690*, 2426–2438.

Fortier, S. M., Nassar, N. T., Lederer, G. W., Brainard, J., Gambogi, J., & McCullough, E. A., (2018). Draft critical mineral list—Summary of methodology and background information—U.S. Geological Survey technical input document in response to Secretarial Order No. 3359: U.S. Geological Survey Open-File Report 2018–1021, 15 p. https://doi.org/10.3133/ofr20181021 (accessed 7 March 2024)

Gambogi, J. (2020). *Mineral commodity summaries, January 2020*. U.S. Geological Survey. https://minerals.usgs.gov/minerals/pubs/commodity/scandium/mcs-2020-scand.pdf (accessed 7 March 2024).

Gilliam, T., Canon, R., Egan, B., Kelmers, A., Seeley, F., & Watson, J. (1982). Economic metal recovery from fly ash. *Resources and Conservation*, *9*, 155–168.

Goldschmidt, V. M. (1935). Rare elements in coal ashes. *Industrial and Engineering Chemistry*, *27*, 1100–1102. https://doi.org/10.1021/ie50309a032

Greene, J. (2012). *Digging for rare earths: The mine where iPhones are born*. http://www.cnet.com/news/digging-for-rare-earths-the-mines-where-iphones-are-born/ (accessed 7 March 2024).

Gromet, L. P., Dymek, R. F., Haskin, L. A., & Korotev, R. L. (1984). The "North American shale composite": Its compilation, major and trace element characteristics. *Geochimica et Cosmochimica Acta*, *48*, 2469–2482.

Guenthner, W. R., Reiners, P. W., Ketcham, R. A., Nasdala, L., & Giester, G. (2013). Helium diffusion in natural zircon: Radiation damage, anisotropy, and the interpretation of zircon (U-Th)/He thermochronology. *American Journal of Science*, *313*, 145–198.

Hamza, M. F., Salih, K. A. M, Abdel-Rahman, A. A. H., Zayed, Y. E., Wei, Y. Z., Liang, J., & Guibal, E. (2021). Sulfonic-functionalized algal/PEI beads for scandium, cerium and holmium sorption from aqueous solutions (synthetic and industrial samples). *Chemical Engineering Journal*, *403*, 126399.

Hanson, G. N. (1980). Rare earth elements in petrogenetic studies of igneous systems. *Annual Reviews of Earth and Planetary Sciences*, *8*, 371–406.

Harkins, W. D. (1917). The evolution of the elements and the stability of complex atoms. *Journal of the American Chemical Society*, *39*, 856. doi:10.1021/ja02250a002

Hatch, G. P. (2012). Dynamics in the global market for rare earths. *Elements*, *8*, 341–346.

Hendren, A., Choi, Y. C., Hsu-Kim, H., Hower, J. C., Plata, D., & Wiesner, M. (2017). Preliminary techno-economic evaluation of a novel membrane based separation and recovery process for rare earth elements from coal combustion residues. *World of Coal Ash Proceedings*, 9–11 May 2017, Lexington, KY, paper 149, http://www.flyash.info/AshSymposium/AshLibraryAuthors.asp#H (accessed 7 March 2024).

Hikichi, Y., & Nomura, T. (1987). Melting temperatures of monazite and xenotime. *Journal of the American Ceramic Society*, *70*, C252–C253.

Hoff, S. (2017). *Competition between coal and natural gas affects power markets*. U.S. Energy Information Administration, 16 June 2017, https://www.eia.gov/todayinenergy/detail.php?id=31672 (accessed 7 March 2024).

Home, A. (2020). U.S. finds its Chinese rare earth dependency hard to break. *Reuters*. https://www.reuters.com/article/idUSKCN24T22A/ (accessed 7 March 2024).

Honaker, R. Q., Zhang, W., Yang, X., & Rezaee, M. (2018). Conception of an integrated flowsheet for rare earth elements recovery from coal coarse refuse. *Minerals Engineering*, *122*, 233–240. doi: 10.1016/j.mineng.2018.04.005

Honaker, R. Q., Zhang, W., & Werner, J. (2019). Acid leaching of rare earth elements from coal and coal ash: Implications for using fluidized bed combustion to assist in the recovery of critical materials. *Energy & Fuels*, *33*(7), 5971–5980.

Hood, M. M., Taggart, R. K., Smith, R. C., Hsu-Kim, H., Henke, K. R., Graham, U. M., et al. (2017). Rare earth element distribution in fly ash derived from the Fire Clay coal, Kentucky. *Coal Combustion & Gasification Products*, *9*, 22–33. doi:10.4177/CCGP-D-17-00002.1

Houser, T., Bordoff, J., & Marsters, P. (2017). *Can coal make a comeback?* Columbia University School of International and Public Affairs, Center on Global Energy Policy, https://www.energypolicy.columbia.edu/publications/can-coal-make-comeback (accessed 7 March 2024).

Hower, J. C. (2012). Petrographic examination of coal-combustion fly ash. *International Journal of Coal Geology*, *92*, 90–97.

Hower, J. C., Berti, D., Hochella, M. F., Jr., & Mardon, S. M. (2018). Rare earth minerals in a "no tonstein" section of the Dean (Fire Clay) coal, Knox County, Kentucky. *International Journal of Coal Geology*, *193*, 73–86. https://doi.org/10.1016/j.coal.2018.05.001

Hower, J. C., Berti, D., Hochella, M. F., Jr., Rimmer, S. M., & Taulbee, D. N. (2018). Submicron-scale mineralogy of lithotypes and the implications for trace element associations: Blue Gem coal, Knox County, Kentucky. *International Journal of Coal Geology*, *192*, 73–82. https://doi.org/10.1016/j.coal.2018.04.006

Hower, J. C., Cantando, E., Eble, C. F., & Copley, G. C. (2019). Characterization of stoker ash from the combustion of high-lanthanide coal at a Kentucky bourbon distillery. *International Journal of Coal Geology*, *213*, 103260. https://doi.org/10.1016/j.coal.2019.103260

Hower, J. C., Dai, S., Seredin, V. V., Zhao, L., Kostova, I. J., Silva, L. F. O., et al. (2013). A note on the occurrence of Yttrium and Rare Earth Elements in coal combustion products. *Coal Combustion and Gasification Products*, *5*, 39–47.

Hower, J. C., Eble, C. F., Backus, J. S., Xie, P., Liu, J. Fu, B., & Hood, M. M. (2020). Aspects of rare earth element enrichment in Central Appalachian coals. *Applied Geochemistry 120*, 104676. https://doi.org/10.1016/j.apgeochem.2020.104676

Hower, J. C., Eble, C. F., Dai, S., & Belkin, H. E. (2016). Distribution of rare earth elements in eastern Kentucky coals: Indicators of multiple modes of enrichment? *International Journal of Coal Geology*, *160–161*, 73–81. 10.1016/j.coal.2016.04.009

Hower, J. C., Graham, U. M., Dozier, A., Tseng, M. T., & Khatri, R. A. (2008). Association of sites of heavy metals with nanoscale carbon in a Kentucky electrostatic precipitator fly ash. *Environmental Science & Technology*, *42*, 8471–8477.

Hower, J. C., Granite, E. J., Mayfield, D., Lewis, A., & Finkelman, R. B. (2016). Notes on contributions to the science of rare earth element enrichment. *Minerals*, *6*. doi:10.3390/min6020032

Hower, J. C., & Groppo, J. G. (2020). Rare Earth-bearing particles in fly ash carbons: Examples from the combustion of eastern Kentucky coals. *Energy Geoscience*. https://doi.org/10.1016/j.engeos.2020.09.003

Hower, J. C., Groppo, J. G., Graham, U. M., Ward, C. R., Kostova, I. J., Maroto-Valer, M. M., & Dai, S. (2017). Coal-derived unburned carbons in fly ash: A review. *International Journal of Coal Geology*, *179*, 11–27. http://dx.doi.org/10.1016/j.coal.2017.05.007

Hower, J. C., Groppo, J. G., Henke, K. R., Graham, U. M., Hood, M. M., Joshi, P., & Preda, D. V. (2017). Ponded and landfilled fly ash as a source of rare earth elements from a Kentucky power plant. *Coal Combustion & Gasification Products*, *9*, 1–21. doi:10.4177/CCGP-D-17-00003.1

Hower, J. C., Groppo, J. G., Henke, K. R., Hood, M. M., Eble, C. F., Honaker, R. Q., et al. (2015). Notes on the potential for the concentration of rare earth elements and Yttrium in coal combustion fly ash. *Minerals*, *5*, 356–366. doi:10.3390/min50x000x

Hower, J. C., Groppo, J. G., Joshi, P., Dai, S., Moecher, D. P., & Johnston, M. N. (2013). Location of cerium in coal combustion fly ashes: Implications for recovery of lanthanides. *Coal Combustion and Gasification Products*, *5*, 73–78. doi:10.4177/CCGP-D13-00007.1

Hower, J. C., Groppo, J. G., Joshi, P., Preda, D. V., Gamliel, D. P., Mohler, D. T., et al. (2020). Distribution of Lanthanides, Yttrium, and Scandium in the pilot-scale beneficiation of fly ashes derived from eastern Kentucky coals. *Minerals*, *10*, 105. https://doi.org/10.3390/min10020105

Hower, J. C., Henke, K. R., Dai, S., Ward. C. R., French, D., Liu, S., & Graham, U. (2017). The generation and nature of coal fly ash and bottom ash. In T.L. Robl, et al. (Eds.), *Coal Combustion Products* (Chapter 2, pp. 21–65). Cambridge, MA: Woodhead Publishing (Elsevier). http://dx.doi.org/10.1016/B978-0-08-100945-1.00002-2

Hower, J. C., Ruppert, L. F., & Eble, C. F. (1999). Lanthanide, Yttrium, and Zirconium anomalies in the Fire Clay coal bed, Eastern Kentucky. *International Journal of Coal Geology*, *39*, 141–153.

Hower, J. C., Qian, D., Briot, N., Henke, K. R., Hood, M. M., Taggart, R. K., & Hsu-Kim, H. (2018). Rare earth element associations in the Kentucky State University stoker ash. *International Journal of Coal Geology*, *189*, 75–82. 10.1016j.coal2018.02.022

Hower, J. C., Qian, D., Briot, N. J., Hood, M. M., & Eble, C. F. (2020). Nano-scale mineralogy of a Rare earth element-rich Manchester coal lithotype, Clay County, *Kentucky. International Journal of Coal Geology*, *220*, 103413. https://doi.org/10.1016/j.coal.2019.103413

Hu, Y., Florek, J., Larivière, D., Fontaine, F.-G., & Kleitz, F. (2018). Recent advances in the separation of rare earth elements using mesoporous hybrid materials. *The Chemical Record, 18*, 1261–1276.

Huang, Q., Noble, A., Herbst, J., & Honaker, R. (2018). Liberation and release of rare earth minerals from Middle Kittanning, Fire Clay, and West Kentucky No. 13 coal sources. *Powder Technology, 332*, 242–252.

Ice, G. E., Budai, J. D., & Pang, J. W. L. (2011). The race to X-ray microbeam and nanobeam science. *Science, 334*, 1234–1239.

International Energy Agency (2018). *World Energy Outlook 2018*. https://www.iea.org/reports/world-energy-outlook-2018 (accessed 7 March 2024).

Ireland, T. R. (2014). Ion microscopes and microprobes, section 15.21. *Treatise in Geochemistry* (2nd ed., vol. *15*, pp. 385–409). Amsterdam: Elsevier.

Ireland, T. R., Clement, S., Compston, W., Foster, J. J., Holden, P., Jenkins, B., et al. (2008). Development of SHRIMP. *Australian Journal of Earth Sciences, 55*, 937–954.

Joshi, P. B., Preda, D. V., Skyler, D. A., Tsinberg, A., Green, B. D., & Marinelli, W. J. (2013). *U.S. Patent Application Number 13/864677, Recovery of Rare Earth Elements and Compounds from Coal Ash*. In: Office, U.S.P.a.T. (Ed.).

Kaplan, S. (2020). Why coal lost—and can it recover? Power, https://www.powermag.com/why-coal-lost-and-can-it-recover/ (accessed 7 March 2024).

Kelmers, A., Canon, R., Egan, B., Felker, L., Gilliam, T., Jones, G., et al. (1982). Chemistry of the direct acid leach, calsinter, and pressure digestion-acid leach methods for the recovery of alumina from fly ash. *Resources and Conservation, 9*, 271–279.

Keystone Coal Industry Manual (2009). *2010 Keystone Coal Industry Manual*. Jacksonville, FL, Mining Media International, 631 p.

Kim, D., Powell, L., Delmau, L., Peterson, E., Herchenroeder, J., & Bhave, R. (2015). Selective extraction of rare earth elements from permanent magnet scraps with membrane solvent extraction, *Environmental Science & Technology, 49*, 9452–9459.

King, J. F., Taggart, R. K., Smith, R. C., Hower, J. C., & Hsu-Kim, H. (2018). Aqueous acid and alkaline extraction of rare earth elements from coal combustion ash. *International Journal of Coal Geology, 195*, 75–83.

Kolker, A. (2018). *Topics in Coal Geochemistry—Short course*. U.S. Geological Survey Open File Report 2018-1145. https://pubs.er.usgs.gov/publication/ofr20181145, https://doi.org/10.3133/ofr20181145 (accessed 7 March 2024)

Kolker, A., Scott, C., Hower, J. C., Vazquez, J. A., Lopano, C. L., & Dai, S. (2017). Distribution of rare elements in coal combustion fly ash, determined by SHRIMP-RG ion microprobe. *International Journal of Coal Geology, 184*, 1–10. https://doi.org/10.1016/j.coal.2017.10.002

Korotev, R. L. (2009). *"Rare Earth Plots" and the concentrations of Rare Earth Elements (REE) in Chondritic Meteorites*. http://meteorites.wustl.edu/goodstuff/ree-chon.htm (accessed 7 March 2024).

Kortright, J. B., & Thompson, A. C. (2009). X-ray emission energies. In A. C. Thompson, & D. Vaughan (Eds.), *X-ray Data Booklet* (Pp. 1–8–1–27). Berkeley, CA: Lawrence Berkeley National Laboratory.

Kose-Mutlu, B., Hsu-Kim, H., & Wiesner, M. R. (2020). Separation of rare earth elements from mixed-metal feedstocks by micelle enhanced ultrafiltration with sodium dodecyl sulfate. *Environmental Technology 43*, 1013-1025. https://doi.org/10.1080/09593330.2020.1812732

Kostova, I., Vassileva, C., Dai, S., & Hower, J. C. (2016). Mineralogy, geochemistry and mercury content characterization of fly ashes from the Maritza 3 and Varna thermoelectric power plants, Bulgaria. *Fuel, 186*, 674–684.

Kutchko, B. G., & Kim, A. G., (2006). Fly ash characterization by SEM-EDS. *Fuel, 85*, 2537–2544. doi:10.1016/j.fuel.2006.05.016

Lichte, F. E., Meier, A. L., & Crock, J. G. (1987). Determination of the rare-earth elements in geological materials by inductively coupled plasma mass spectrometry. *Analytical Chemistry, 59*, 1150–1157.

Lin, R., Bank, T. L., Roth, E. A., Granite, E. J., & Soong, Y. (2017). Organic and inorganic associations of rare earth elements in central Appalachian coal. *International Journal of Coal Geology, 179*, 295–301.

Lin, R., Howard, B. H., Roth, E. A., Bank, T. L., Granite, E. J., & Soong, Y. (2017). Enrichment of rare earth elements from coal and coal by-products by physical separations. *Fuel, 200*, 506–520.

Lin, R., Soong, Y., & Granite, E. J. (2018). Evaluation of trace elements in U.S. coals using the USGS COALQUAL database version 3.0. Part I: Rare earth elements and yttrium (REY). *International Journal of Coal Geology, 192*, 1–13.

Lin, R., Stuckman, M., Howard, B. H., Bank, T. L., Roth, E. A., Macala, M. K., et al. (2018). Application of sequential extraction and hydrothermal treatment for characterization and enrichment of rare earth elements from coal fly ash. *Fuel, 232*, 124–133.

Liu, J., Dai, S., He, X., Hower, J. C., & Sakulpitakphon, T. (2017). Size-dependent variations in fly ash trace-element chemistry: Examples from a Kentucky power plant and with emphasis on rare earth elements. *Energy & Fuels, 31*, 438–447. doi: 10.1021/acs.energyfuels.6b02644

Liu, P., Huang, R., & Tang, Y. (2019). Comprehensive understandings of Rare Earth Element (REE) speciation in coal fly ashes and implication for REE extractability. *Environmental Science & Technology, 53*(9), 5369–5377.

Liu, T., Hower, J.C. & Huang, C.-H. (2023). Recovery of Rare Earth Elements from Coal Fly Ash with Betainium Bis(trifluoromethylsulfonyl)imide: Different Ash Types and Broad Elemental Survey. *Minerals, 13*, 952.

Lo, Y.-C., Cheng, C.-L., Han, Y.-L., Chen, B.-Y., & Chang, J.-S. (2014). Recovery of high-value metals from geothermal sites by biosorption and bioaccumulation. *Bioresource Technology, 160*, 182–190.

Longerich, H. (2009). Laser ablation-inductively coupled plasma-mass spectrometry (LA-ICP-MS): An introduction. In P. Sylvester (Ed.), *Laser ablation ICP-MS in the Earth Sciences: Current practices and outstanding issues* (Chapter 1). Mineralogical Association of Canada Short Course 40, 1–18.

Makarova, I., Ryl, J., Sun, Z., Kurilo, I., Gornicka, K., Laatikainen, M., & Repo, E. (2020). One-step recovery of REE oxalates in electro-leaching of spent NdFeB magnets. *Separation and Purification Technology, 251*, 117362.

Mardon, S. M., & Hower, J. C. (2004). Impact of coal properties on coal combustion by-product quality: examples from a Kentucky power plant. *International Journal of Coal Geology*, *59*, 153–169.

McDowell, W. J., & Seeley, F. G. (1981). *Salt-soda sinter process for recovering aluminum from fly ash*. US Patent 4254088A. Sandia National Laboratories (SNL), Albuquerque, NM, and Livermore, CA.

Meawad, A. S., Bojinova, D. Y., & Pelovski, Y. G. (2010). An overview of metals recovery from thermal power plant solid wastes. *Waste Management*, *30*, 2548–2559.

Merriman, D., & Backeberg, N. (2018). *Rare earths: Is new non-Chinese supply diversifying the industry?* Roskill Information Services, https://roskill.com/news/rare-earths-is-new-non-chinese-supply-diversifying-the-industry/ (accessed 22 June 2020; no longer available).

Middleton, A., Park, D., Jiao, Y., & Hsu-Kim, H. (2020). Elemental composition controls rare earth element solubility during leaching of coal fly ash and coal by-products. *International Journal of Coal Geology*, *227*, 103532.

Miller, J. W. (2015). Molycorp to Suspend Production at California Mine; Firm, amid a price collapse for rare earths, to lay off almost 500 workers at Mountain Pass facility. *Wall Street Journal (Online)*, 26 August 2015. https://www.wsj.com/articles/molycorp-to-suspend-production-at-california-mine-1440616302 (accessed 7 March 2024).

Molycorp Minerals LLC (2010). *Revised mine and reclamation plan for the Mountain Pass mine; Draft subsequent mitigated negative declaration*. Submitted to County of San Bernardino, p. 97. https://www.sbcounty.gov/Uploads/lus/Desert/2682MolycorpDraftMND.pdf (accessed 7 March 2024)

Mutlu, B. K., Cantoni, B., Turolla, A., Antonelli, M., Hsu-Kim, H., & Wiesner, M. R. (2018). Application of nanofiltration for Rare Earth Elements recovery from coal fly ash leachate: Performance and cost evaluation. *Chemical Engineering Journal*, *349*, 309–317.

Neuendorf, K. K. E., Mehl, J. P., Jr., & Jackson, J. A. (2005). *Glossary of geology* (5th ed.). Alexandria, VA: American Geological Institute (now American Geosciences Institute), 779 p.

O'Connor, M. P., Coulthard, R. M., & Plata, D. L. (2018). Electrochemical deposition for the separation and recovery of metals using carbon nanotube-enabled filters. *Environmental Science: Water Research and Technology*, *4*(1), 58–66.

Oddo, G. (1914). Die Molekularstruktur der radioaktiven Atome. *Zeitschrift für anorganische Chemie*, *87*, 253. doi:10.1002/zaac.19140870118

Osborn, E. F., & Muan, A. (1960). *Phase equilibrium diagrams of oxide systems. The system CaO-Al2O3-SiO2. Plate 1*. The American Ceramic Society and the Edward Orton Jr. Ceramic Foundation, Columbus, Ohio.

Padilla, R., & Sohn, H. Y. (1985). Sintering kinetics and alumina yield in lime-soda sinter process for alumina from coal wastes. *Metallurgical Transactions B*, *16*, 385–395.

Palmer, C. A., Oman, C. L., Park, A. J., & Luppens, J. A. (2015). *The U.S. Geological Survey coal quality (COALQUAL) database version 3.0*. U.S. Geological Survey Data Series *975*, 43 p. with appendixes, http://dx.doi.org/10.3133/ds975

Park, D. M., Brewer, A., Reed, D. W., Lammers, L. N., & Jiao, Y. (2017). Recovery of rare earth elements from low-grade feedstock leachates using engineered bacteria. *Environmental Science & Technology*, *51*, 13471–13480.

Park, D., Middleton, A., Smith, R., Deblonde, G., Laudal, D., Theaker, N., et al. (2020). A biosorption-based approach for selective extraction of rare earth elements from coal byproducts. *Separation and Purification Technology*, *241*, 116726.

Peak Resources Limited (2017). *Ngualla Rare Earth Project—Updated Ore Reserve*. https://wcsecure.weblink.com.au/pdf/PEK/01847311.pdf. (accessed 7 March 2024).

Piispanen, M. H., Arvilommi, S. A., Van Den Broeck, B., Nuutinen, L. H., Tiainen, M. S., Perämäki, P. J., & Laitinen, R. S. (2009). A comparative study of fly ash characterization by LA-ICP-MS and SEM-EDS. *Energy and Fuels*, *23*(7), 3451–3456. doi:10.1021/ef801037a

Quercia, G., Perera, Y., Tovar, H., & Rodriguez, E. (2007). Thermal degradation of zirconium silicate ($ZrSiO_4$) ferrules. *Acta Microscopica*, *16*, 205–206.

Querol, X., & Chenery, S. (1995). *Determination of trace element affinities in coal by laser ablation microprobe-inductively coupled plasma mass spectrometry*. In M. K. G. Whateley & D. A. Spears (Eds.), *European Coal Geology* (pp. 147–155). London: Geological Society Special Publication 82.

Querol, X., Moreno, N., Umaña, J. C., Alastuey, A., Hernández, E., López-Soler, A., & Plana, F. (2002). Synthesis of zeolites from coal fly ash: An overview. *International Journal of Coal Geology*, *50*, 413–423.

Rahman, M. M., Awual, M. R., & Asiri, A. M. (2020). Preparation and evaluation of composite hybrid nanomaterials for rare-earth elements separation and recovery. *Separation and Purification Technology*, *253*, 117515.

Ratafia-Brown, J. A. (1994). Overview of trace element partitioning in flames and furnaces of utility coal-fired boilers. *Fuel Processing Technology*, *39*, 139–157.

Rayzman, V. L., Shcherban, S. A., & Dworkin, R. S. (1997). Technology for chemical–metallurgical coal ash utilization. *Energy & Fuels*, *11*, 761–773.

Roskill Information Services (2017). *Rare earths: What's the coal story?* Roskill Information Services, https://roskill.com/news/rare-earths-whats-coal-story/ (accessed 22 June 2020; no longer available).

Roskill Information Services (2018). *Rare earths: African ramp-up and Australian advances for rare earth supply.* Roskill Information Services, https://roskill.com/news/rare-earths-african-ramp-up-and-australian-advances-for-rare-earth-supply/ (accessed 22 June 2020; no longer available).

Rybak, A., & Rybak, A. (2021). Characteristics of some selected methods of rare earth elements recovery from coal fly ashes. *Minerals*, *11*, 142.

Scott, C., & Kolker, A. (2019). Rare earth elements in coal and coal fly ash. *U.S. Geological Survey Fact Sheet 2019-3048*, 4 p., https://doi.org/10.3133/fs20193048 (accessed 6 March 2024).

Seredin, V. V. (1996). Rare earth element-bearing coals from the Russian Far East deposits. *International Journal of Coal Geology*, *30*, 101–129.

Seredin, V. V. (2010). A new method for primary evaluation of the outlook for rare earth element ores. *Geology of Ore Deposits*, *52*, 428–433.

Seredin, V. V., & Dai, S. (2012). Coal deposits as potential alternative sources for lanthanides and yttrium. *International Journal of Coal Geology, 94*, 67–93.

Seydoux-Guillaume, A. M., Wirth, R., Nasdala, L., Gottschalk, M., Montel, J. M., & Heinrich, W. (2002). An XRD, TEM and Raman study of experimentally annealed natural monazite. *Physics and Chemistry of Minerals, 29*, 240–253.

Silva, L. F. O., Jasper, A., Andrade, M. L., Sampaio, C. H., Dai, S., Li., X., et al. (2012). Applied investigation on the interaction of hazardous elements binding on ultrafine and nanoparticles in Chinese anthracite-derived fly ash. *The Science of the Total Environment, 419*, 250–264.

Sloss, L. (2019). *Technology readiness of advanced coal-based power generation systems*. International Energy Agency, IEA Clean Coal Centre, CCC/ 292, 113 p.

Smith, R. C., Taggart, R. K., Hower, J. C., Wiesner, M. R., & Hsu-Kim, H. (2019). Selective recovery of rare earth elements from coal fly ash leachates using liquid membrane processes. *Environmental Science & Technology, 58*(8), 4490–4499.

Smolka-Danielowska, D. (2010). Rare earth elements in fly ashes created during the coal burning process in certain coal-fired power plants operating in Poland-Upper Silesian Industrial Region. *Journal of Environmental Radioactivity 101*, 965–968 doi:10.1016/j.jenvrad.2010.07.001

Spears, D. A. (2004). The use of laser ablation inductively coupled plasma-mass spectrometry (LA ICP-MS) for the analysis of fly ash. *Fuel, 83*, 1765–1770. doi:10.1016/j.fuel.2004.02.018

Stern, R. A. (2009). An introduction to secondary ion mass spectrometry (SIMS) in geology, In M. Fayek (Ed.), *Secondary mass spectrometry in the Earth Sciences: Gleaning the big picture from a small spot (Chapter 1)*. Mineralogical Association of Canada Short Course *41*, 1–18.

Stuckman, M. Y., Lopano, C. L., & Granite, E. J. (2018). Distribution and speciation of rare earth elements in coal combustion by-products via synchrotron microscopy and spectroscopy. *International Journal of Coal Geology, 195*, 125–138.

Suárez-Ruiz, I., Valentim, B., Borrego, A. G., Bouzinos, A., Flores, D., Kalaitzidis, S., et al. (2017). Development of a petrographic classification of fly-ash components from coal combustion and co-combustion. (An ICCP Classification System, Fly-Ash Working Group—Commission III.). *International Journal of Coal Geology, 183*, 188–203. doi: 10.1016/j.coal.2017.06.004

Sylvester, P. (Ed.). (2001). *Laser-ablation ICP-MS in the Earth Sciences: Principles and applications*. Mineralogical Association of Canada Short Course Series, *29*, 243 p.

Sylvester P. (Ed.). (2008). *Laser ablation ICP-MS in the Earth Sciences: Current practices and outstanding issues*. Mineralogical Association of Canada Short Course Series, *40*, 356 p

Taggart, R. K., Hower, J. C., Dwyer, G. S., & Hsu-Kim, H. (2016). Trends in the rare earth element content of U.S.-based coal combustion fly ashes. *Environmental Science & Technology, 50*, 5919–5926.

Taggart, R. K., Hower, J. C., & Hsu-Kim, H. (2018). Effects of roasting additives and leaching parameters on extraction of rare earth elements from coal fly ash. *International Journal of Coal Geology, 196*, 106–114.

Taggart, R. K., Rivera, N. A., Levard, C., Ambrosi, J.-P., Borschneck, D., Hower, J. C., & Hsu-Kim, H. (2018). Differences in bulk and microscale yttrium speciation in coal combustion fly ash. *Environmental Science: Processes & Impacts, 20*, 1390–1403.

Taylor, S. R., & McLennan, S. M. (1985). *The Continental Crust—Its composition and evolution*. Boston: Blackwell Scientific Publishers, 312 p.

Thomas, R. (2013). *Practical guide to ICP-MS; A tutorial for beginners* (3rd ed.). Boca Raton, FL: CRC Press.

Thompson, R. L., Bank, T., Montross, S., Roth, E., Howard, B., Verba, C., & Granite, E. (2018). Analysis of rare earth elements in coal fly ash using laser ablation inductively coupled plasma mass spectrometry and scanning electron microscopy. *Spectrochimica Acta—Part B Atomic Spectroscopy, 143*, 1–11. doi:10.1016/j.sab.2018.02.009

Thompson, R. L., Bank, T., Roth, E., & Granite, E. (2016). Resolution of rare earth element interferences in fossil energy by-product samples using sector-field ICP-MS. *Fuel, 185*, 94–101.

U.S. Department of Energy, National Energy Technology Laboratory. (2018). *Rare earth elements from coal and coal byproducts*. https://www.netl.doe.gov/research/coal/rare-earth-elements

Valentim, B. (2020). Petrography of coal combustion char: A review. *Fuel, 277*, 118271.

Valentim, B., Hower, J. C., Soares, S., Guedes, A., Garcia, C., Flores, D., & Oliveira, Á. (2009). Petrographic characterization of economizer fly ash. *Minerals and Metallurgical Processing, 26*, 208–216.

Van Gosen, B.S., Verplanck, P.L., Long, K.R., Gambogi, J., Seal, R.R., II,. (2014). *The rare-earth elements—Vital to modern technologies and lifestyles*. U.S. Geological Survey Fact Sheet 2014-3078, 4 p. https://pubs.usgs.gov/fs/2014/3078/. (accessed 7 May 2024).

Wang, Z., Dai, S., Zou, J., French, D., & Graham, I. T. (2019). Rare earth elements and yttrium in coal ash from the Luzhou power plant in Sichuan, Southwest China: Concentration, characterization and optimized extraction. *International Journal of Coal Geology, 203*, 1–14.

Ward, C. R., & French, D. (2006). Determination of glass content and estimation of glass composition in fly ash using quantitative X-ray diffractometry. *Fuel, 85*, 2268–2277. doi:10.1016/j.fuel.2005.12.026

Watson, B. (2018). Rare earth hunting in US coal country. *Defense One Radio*, 18 October 2018. https://www.defenseone.com/ideas/2018/10/ep-25-rare-earth-hunting-us-coal-country-chinas-military-might-uighurs-xinjiang-and-more/152167/ (starts at 28:50), (accessed 7 May 2024).

Wilcox, J., Wang, B., Rupp, E., Taggart, R., Hsu-Kim, H., Oliveira, M. L. S., et al. (2015). Observations and assessment of fly ashes from high-sulfur bituminous coals and blends of high-sulfur bituminous and subbituminous coals: Environmental processes recorded at the macro and nanometer scale. *Energy & Fuels, 29*, 7168–7177. 10.1021/acs.energyfuels.5b02033

Williams, G. P. (2009). Electron binding energies. In A. C. Thompson & D. Vaughan (Eds.), *X-ray Data Booklet* (pp. 1–1–1–7). Berkeley, CA: Lawrence Berkeley National Laboratory.

World Energy Council. (2016). *World Energy Resources: Coal.* London, World Energy Council, 61–63. https://www.worldenergy.org/publications/entry/world-energy-resources-2016 (accessed 7 March 2024)

Wraich, A. (2017). The feasibility of scandium as an expanding commodity. *Geological Society of America Annual Meeting, Seattle, Washington, paper no. 65-20*, https://gsa.confex.com/gsa/2017AM/webprogram/Paper293861.html (accessed 7 March 2024).

Wright, W. (2018). *See 30 years of strip mining spread across Eastern Kentucky in mere seconds.* Lexington Herald-Leader, 30 July 2018, https://www.kentucky.com/news/state/article215487870.html (accessed 7 March 2024).

Yakaboylu, G. A., Baker, D., Wayda, B., Sabolsky, K., Zondlo, J. W., Shekhawat, D., et al. (2019). Microwave-assisted pretreatment of coal fly ash for enrichment and enhanced extraction of rare-earth elements. *Energy & Fuels*, *33*(11), 12083–12095.

Yan, X., Dai, S., Graham, I. T., He, X., Shan, K., & Liu, X. (2018). Determination of Eu concentrations in coal, fly ash and sedimentary rocks using a cation exchange resin and inductively coupled plasma mass spectrometry (ICP-MS). *International Journal of Coal Geology*, *191*, 152–156.

Yang, J., Zhao, Y., Zyryanov, V., Zhang, J., & Zheng, C. (2014). Physical-chemical characteristics and elements enrichment of magnetospheres from coal fly ashes. *Fuel*, *135*, 15–26.

Zhang, W., Groppo, J. G., & Honaker, R. Q. (2015). *Ash beneficiation for REE recovery.* World of Coal Ash, 5–7 May 2015, Nashville, TN, paper 194-Groppo-2015, http://www.flyash.info/ (accessed 7 March 2024).

Zhang, W., & Honaker, R. Q. (2018). Rare earth elements recovery using staged precipitation from a leachate generated from coarse coal refuse. *International Journal of Coal Geology*, *195*, 189–199. doi: 10.1016/j.coal.2018.06.008

Zhang, W., Noble, A., Yang, X., & Honaker, R. (2020). A comprehensive review of rare earth elements recovery from coal-related materials. *Minerals*, *10*(5), 451.

Zhang, W., Yang, X., & Honaker, R. Q. (2018). Association characteristic study and preliminary recovery investigation of rare earth elements from Fire Clay seam coal middlings. *Fuel*, *215*, 551–560. doi: 10.1016/j.fuel.2017.11.075

Zhao, L., Dai, S., Graham, I. T., Li, X., Liu, H., Song, X., et al. (2017). Cryptic sediment-hosted critical element mineralization from eastern Yunnan Province, southwestern China: Mineralogy, geochemistry, relationship to Emeishan alkaline magmatism and possible origin. *Ore Geology Reviews*, *80*, 116–140. doi: 10.1016/j.oregeorev.2016.06.014

3

RECOVERING OF RARE EARTH ELEMENTS FROM UNCONVENTIONAL RESOURCES: BAUXITE RESIDUE

Efthymios Balomenos, Panagiotis Davris, Dimitrios Panias, and Ioannis Paspaliaris

Laboratory of Metallurgy, National Technical University of Athens, Athens, Greece

Abstract

A metallurgical by-product of the aluminum oxide production from bauxite ore, the bauxite residue (BR) has been known to contain exploitable amounts of rare earth elements (REEs), including scandium. However, there is significant complexity in extracting REEs from BR, as upon direct leaching, solutions dilute in REEs and rich in base metals (g/L) such as Fe, Ca, Na, Al, etc., are produced, effectively hampering downstream refining of the REEs. Several leaching agents have been proposed including inorganic acids and bases, organic acids, and ionic liquids. An initial comparison of such methods is presented. Finally, other methods are reviewed combining pyrometallurgical treatments such as roasting and smelting and subsequent leaching. Such methods achieve the extraction of base metals like iron and in some cases

Rare Earth Elements: Sustainable Recovery, Processing, and Purification, Special Publications 79, First Edition. Edited by Athanasios K. Karamalidis and Roderick Eggert.

DOI: 10.1002/9781119515005.ch3

aluminum, but suffer in complexity. Still, if applied, they can achieve a more holistic reuse of the BR by-product.

3.1. Bauxite Residue

Bauxite residue (BR) refers to the insoluble solid material generated during the extraction of alumina (Al_2O_3) from bauxite ore using the Bayer process. When bauxite ore is treated with caustic soda, the aluminum hydroxides/oxides contained within are solubilized, with approximately 50% of the bauxite mass being transferred to the liquid phase, while the remaining solid fraction constitutes the BR.

Active lime is usually added to control and reduce caustic soda and alumina losses during digestion occurring with the formation of the desilication products. The solid–liquid separation after ore digestion takes place in thickeners and washers, resulting in the formation of a red-colored BR slurry (approximately 50% solids) which was termed "red mud" in the past. Nowadays, many plants use, as a final step of slurry treatment, high-pressure filtration (the most efficient way regarding alkali recovery), in which the BR slurry is pressed to remove the maximum of remaining liquor and produce a compact filter cake with a relative humidity of 25–30%.

It is estimated that for each ton of alumina produced, 0.9–1.5 tons of solid residue (on a dry basis) is generated depending on the initial bauxite ore grade and alumina extraction efficiency (Power et al., 2011). BR consists of various metal oxides of Fe, Al, Ti, Si, Ca, Na, V, and Ga (depending on the initial chemical composition of the bauxite ore) along with inclusions of unwashed sodium aluminate solution.

As the global demand for primary aluminum metal increases, so does the BR production, currently in excess of 150 million tons per year (worldwide). This is generated at some 60 active alumina refining plants. In addition, there are at least 50 other closed legacy sites, so the combined stockpile of BR at active and legacy sites is estimated at between 3,000 and 4,000 million tons.

The primary aluminum industry has always focused on discovering potential applications for BR utilization. The vast amount of research and studies on BR utilization is justified by more than 734 patents from 1964. Possible applications can broadly be broken down into various categories, such as cement and building materials production, iron production, trace element (Ga, REEs, V, etc.) recovery, use as soil amelioration, landfill capping, acid mine drainage treatment, and others. The recent rare earth elements (REEs) crisis fueled significant research effort in recovering the REEs that are found in some BRs in concentrations between 1 and 2 kg of total REEs per ton of BR. Given the large quantity of the annual BR production, the total amount of contained REEs becomes significant and could cover part of the global REEs demand (Balomenos, Davris, Deady, et al., 2017; Davris et al., 2017). Furthermore, while the treatment of BR for the recovery of REEs does not solve the BR deposition problem, as the volume of the waste remains practically unaffected, it does help in the economic viability of holistic processing flow sheet seeking to achieve near-zero waste through multiple processing steps (Balomenos, Davris, Pontikes, & Panias, 2017).

3.1.1. Rare Earths in Bauxite Residue

The bauxite ore is one of the factors that affect the concentration of REEs in BR. Bauxites are classified in three categories, lateritic bauxites (88%), karstic bauxites (11.5%), and Tikhvin-type bauxites (0.5%) (Bardossy, 1982; Bárdossy & Aleva, 1990). Karstic bauxites are mainly found in Europe, Jamaica, Russia, and China. The karstic bauxites contain higher concentrations of rare earths than the lateritic bauxites. Rare earths are detected in bauxite ore as fluorocarbon or phosphate minerals which are very similar to the main industrial minerals of REEs (bastnaesite/monazite) (Li et al., 2013; Mouchos et al., 2017; Ochsenkühn-Petropulu & Ochsenkühn, 1995; Vind, Malfliet, Blanpain, et al., 2018). It has also been reported that in the Bayer process REEs end up in the BR in 2:1 ratio, compared to the initial bauxite ore (Derevyankin et al., 1981; Ochsenkühn-Petropulu et al., 1994; Wagh & Pinnock, 1987).

The worldwide typical concentration of REEs in BR is 800–2,500 mg/kg and is related to the initial bauxite ore and the operation conditions of the Bayer process (Deady et al., 2018). Recent research shows that REEs in BR can be found in secondary mineral phases produced by the Bayer process, known as the desilication product (DSP). DSP is the result of the silicon removal from the aluminate solution during the leaching of the bauxite ore, as silicon is the major pollutant for the final alumina product. Presence of REEs in the DSP can be attributed to REEs being initially hosted at alumina-bearing minerals that are dissolved in the Bayer process; these REEs are incorporated into the newly formed DSP mineral matrix that contains a mixture of Fe, Ti, Si, Al, Ca, and Na ions (Vind, Malfliet, Blanpain, et al., 2018).

Scandium often differs from the other REEs' mineral behavior. Especially in lateritic bauxites and their corresponding BR, it is often correlated with iron and titanium and zircon minerals (Liu et al., 2018; Vind, Malfliet, Blanpain, et al., 2018; Zhang et al., 2017), which for the most part are unaffected through the Bayer process. This is also confirmed by the laterite deposits in Australia and the Greek BR (Chassé et al., 2016) where main mineral, with high concentration of Sc, is goethite (Vind, Malfliet, Bonomi, et al., 2018). However, there are cases of BR where scandium is found to be related to a larger extent to the soluble Al-bearing minerals, as is reported by Russian researchers (Suss et al., 2018).

Published chemical analysis and leaching studies of BR focus on concentration of scandium, because scandium represents the 95% of REEs' financial values. Table 3.1 presents the Sc concentrations in different BRs worldwide as reported in literature.

It has been estimated that 70% of the world's Sc resources might be found in bauxite minerals and BR (Petrakova et al., 2016). Scandium is a very rarely concentrated element, making commercially usable deposits of this critical material very rare. As a result, it can be extracted only as byproduct from the metallurgical residues of the primary metallurgy W, REEs, Zr, Ni, Ti, and U (Wang et al., 2011). For this reason, scandium concentrations of 50–60 mg/kg are considered as exploitable (Wang et al., 2011). The main application of scandium is in the production of

Table 3.1 Concentrations of scandium in different bauxite residues.

Bauxite residue	Scandium concentration (ppm)	Reference
Guangxi, China	115	(Gu et al., 2018)
Henan, China	66	(Gu et al., 2018)
Shandong, China	54	(Gu et al., 2018)
Shandong, China	80	(Liu et al., 2017)
Shandong, China	76	(Liu et al., 2018)
Chinalco, China	158	(Qu & Lian, 2013)
China	41–93	(Wang et al., 2011)
Australia	54	(Wang et al., 2013)
Australia	68	(Rayzman, 1998)
India	50	(Abhilash et al., 2014)
Jamaica	130–254	(Wagh & Pinnock, 1987)
Jamaica	230	(Rayzman, 1998)
Jamaica	55	(Narayanan et al., 2018)
Kazakhstan	55	(Rayzman, 1998)
Russia	100–150	(Yatsenko & Pyagai, 2010)
Russia	110	(Rayzman, 1998)
Russia	240	(Petrakova et al., 2016)
Ural, Russia	90	(Petrakova et al., 2016)
Mykolaiv, Ukraine	60	(Smirnov & Molchanova, 1997)
Agios Nikolaos, Greece	100–135	(Borra et al., 2015; Davris et al., 2016b; Ochsenkühn-Petropulu et al., 1994)
Ajka, Hungary	80	(Ujaczki et al., 2017)
Hamburg, Germany	84	–
Canada	47	(Reid et al., 2017)

high-strength aluminum alloys suitable for aerospace applications (Ahmad, 2003). Due to the large Al alloy market, the production of scandium from the BR is easily connected with the primary production of aluminum to produce Al–Sc alloy.

There have been many studies on the extraction of REEs, and particularly scandium, from BR. Several authors have reported different methods based mainly on hydrometallurgical or combined pyrometallurgical and hydrometallurgical methods (Wang et al., 2011).

3.1.2. Direct Leaching with Inorganic Acids

REEs are a small fraction of the BR. Direct leaching of BR with mineral acids requires significant acid consumption, as BR is by nature highly alkaline, and results in leach solutions with low REE concentrations (<1,000 mg/L), while the major BR metals such as Fe, Al, Ti, Ca, Na, and Si dissolve extensively and are thus found in concentrations of several grams per liter. This complicates the REE extraction and refining from such solutions, as established methods (solvent extraction/ion exchange) start with solutions of several grams per liter of REEs. Hence, most researchers focus on the selective solubilization of REEs against major elements such as Fe, looking to produce leach solutions that can be further processed economically.

Initially, selective solubilization of REEs with weak acids was performed in BR from Jamaica (Fulford et al., 1991a). Sulfurous acid (H_2SO_3) produced by gas SO_2 blowing in BR pulp reduces gradually the pH of the pulp, allowing the selective leaching of REEs against iron. The leaching takes place in two to three stages and the recoveries of REEs reach 85% expressed by the recovery of Y, while the recovery of Sc is not reported. The same team, also, separated the REEs directly from the pregnant solution using di(2-ethylhexyl)phosphoric acid (DEHPA) organic solvent (Fulford et al., 1991b). Selectivity is achieved for pH values in the range of 1.5–3. In this range, Fe remains insoluble, while other metals such as Ca and Na are completely dissolved.

The use of inorganic acids (H_2SO_4, HCl, and HNO_3) results in selective leaching of REEs from Greek BR, but the recovery levels are quite low (Borra et al., 2016a). For instance, Sc selective leaching ranges always between 30% and 50% even at 150 °C (Sugita et al., 2016). It has been found that HCl addition in the Greek BR leads to a maximum Sc leaching of 50% against Fe (5%), while the increased recovery levels correspond to high Fe solubilization indicating indirect correlation between Sc and mineral Fe phases (Borra et al., 2015). The same team reports that during solubilization the complete Ca and Na dissolution takes place, while Al,

Ti, and Si are recovered in the range of 30%–40%. REE recoveries varied between 30% and 60% after 24 h treatment in a 2% w/v pulp density with 0.5–1.5 mol/L HCl. Increasing the HCl concentration to 3–6 mol/L increases REE and Fe recovery above 70%. At the same time, 30%–50% of Al, Ti, and Si are dissolved. Increasing the pulp density during leaching (at the same acid concentration) reduces the recovery of REEs. Finally, the increase of temperature during the HCl leaching does not affect the recoveries of rare earths (Borra et al., 2015). The absence of significant effect of factors such as temperature and pressure during the leaching of Greek BR is also reported in previous studies (Ochsenkühn-Petropulu et al., 1996). On the other hand, there are studies that report opposite results (Sugita et al., 2016). This happens because the leaching stage takes place under different conditions and there is heterogeneity between different BRs per country. Heterogeneity also occurs even from the same BR supplier in samples taken at different times due to possible alterations in Bayer process or the bauxite ore used in the plant. This fact does not always allow the direct comparison of results of acid leaching between different BRs.

Greek BR leaching with nitric acid (HNO_3) results in selective recovery of rare earths against Fe (Ochsenkühn-Petropulu et al., 1996). The selectivity of rare earths from BR based on the consumed acid increases with the following order $H_2SO_4 < HCl < HNO_3$. During the HNO_3 leaching, Y recovery was about 90%, heavy lanthanides (Dy, Er, and Yb) recovery was up to 70%, medium lanthanides (Nd, Sm, Eu, and Gd) recovery was up to 50%, and light lanthanides (La, Ce, and Pr) recovery was up to 30% (Ochsenkühn-Petropulu et al., 1996). Leaching with diluted HNO_3 was applied and optimized on a pilot scale (Ochsenkühn-Petropoulou et al., 2002), where scandium recovery reaches 70% after three steps of consecutive HNO_3 leaching. Furthermore, pH must be kept between 0 and 0.2 for the highest Sc/Fe ratio in the pregnant solution. The final solution undergoes purification stages to recover REEs using ion-exchange resin Dowex 50W and then solvent extraction with DEHPA (Ochsenkühn-Petropulu et al., 1995; Ochsenkühn-Petropoulou et al., 2002). Finally, Sc extraction from the organic solvent is performed with NaOH, while REEs are separated in a subsequent step with liquid chromatography (Tsakanika et al., 2004). The main drawback of this method is the large volume of liquid wastes containing nitrate anions when produced by the solid residue washing after leaching (Petrakova et al., 2014).

H_2SO_4 was selected as the most suitable extracting agent for the Australian BR leaching. Sc recovery is 47% against Fe 7%, while Y recovery is 5% at 10% w/v pulp, 1 mol/L H_2SO_4 at 50 °C for 2 h (Wang et al., 2013). Recent leaching studies in Greek BR using H_2SO_4 didn't show significant improvement in REEs recovery. Nevertheless, efforts have been made to reduce acid consumption by neutralizing BR with CO_2 (Rivera et al., 2017), they also tried to avoid the formation of silica gel by the method of digestion with concentrated acids and final water dilution (Rivera et al., 2018). Scandium recoveries are between 30% and 40% with high Fe dissolution due to the use of concentrated HCl and H_2SO_4. Furthermore, the avoidance of silica gel and recovery of Sc (68%) and Ti (91%) can also be achieved with 2.5 M H_2O_2 solution in 2.5 M H_2SO_4, but leaching is not selective (Alkan et al., 2018). The main scandium leaching mechanism from BR selectively using H_2SO_4 is described by the shrinking core model "thin-film diffusion" (Hatzilyberis et al., 2018), while the reaction rate is first class with the scandium recovery up to 50% selectively at 2% w/v pulp density, 1–3 mol/L for 1 h.

Correspondingly, in Canadian BRs, three different acids were examined. H_2SO_4 was selected as the most appropriate, based on the leaching efficiency, cost, and practicality (Reid et al., 2017). In the kinetic model, the process of BR's REEs leaching is described by the expansion of acid through the boundary layer at the boundary of grain. With the pretreatment of BR in direct microwave exposure before leaching, nano-sized pores were grown on the surface of grains, allowing the increase in rare earths' leaching efficiency, which means 40.0%–64.2% Sc and 54.3%–78.7% Nd, while no results on recoveries for other main metals are given (Reid et al., 2017).

The relatively moderate recoveries of REEs, during the direct selective leaching, led many teams to use more concentrated acids in higher temperatures to achieve higher scandium recovery. This BR process with concentrated acids provides the extraction of other critical metals with no selectivity, and is mainly examined by research teams in China. BR leaching with HCl in 6 mol/L at 50 °C for 1 h and 20% w/v pulp density can leach up to 80% Sc with 21.2 mol HCl consumption per kg of BR (Wang et al., 2010). Correspondingly, in a BR digestion process with concentrated H_2SO_4 at high temperatures followed by leaching with water and addition of H_2SO_4, the optimum conditions were determined for an average particle size of 65–80 μm, 90 °C, 3 h, and solid-to-liquid ratio 3:1 recovering 85% of Sc (Xue et al., 2010). In a similar process like this (Zhang et al., 2005), 6 mol/L HCl with solid-to-liquid ratio 4:1 is leached at 50 °C for 1 h recovering 80% of Sc, and then the residue is treated with 92% wt H_2SO_4, with solid-to-liquid ratio 3:1, at 200 °C, for 1.5 h to leach 97% of Ti. In these studies, the recoveries of other metals from BR and the subsequent recovery of Sc from the solutions produced are not reported. Nonselective leaching of Sc with concentrated sulfuric acid (3–4 mol/L) at 95 °C at 10%–20% pulp densities has also been reported for the Greek BR, leading to

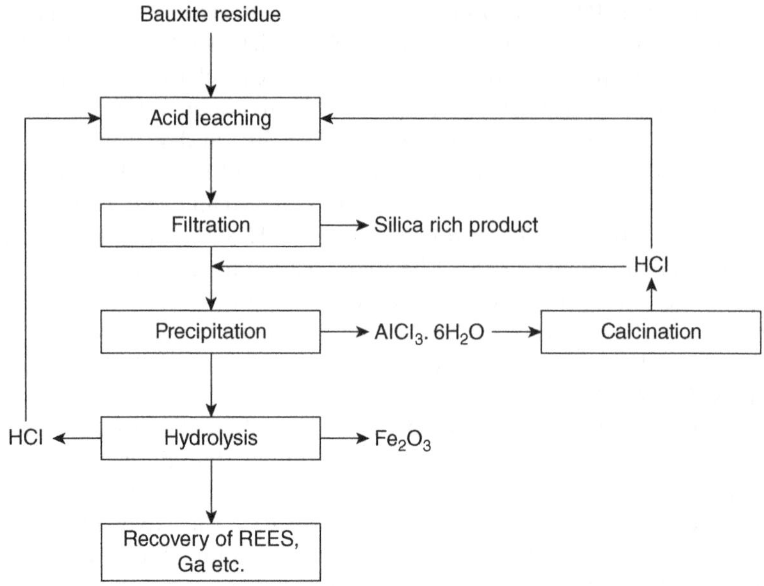

Figure 3.1 Leaching process with concentrated HCl developed by Orbite Technology Inc. (Borra et al., 2016a).

90% Sc and 95% Fe extraction. The produced pregnant liquor solution (PLS) was reported having 7 g/L Al, 5 g/L Ti, 42 g/L Fe, and 15 mg/L of Sc (Davris, Marino, et al., 2018).

For the nonselective recovery of metals from BR, Orbite Technologies Inc. (Boudreault et al., 2017) developed a hydrometallurgical process in which the BR is treated with concentrated HCl (18–33% wt.) solution in autoclave at 140–170 °C. During the leaching, all metals (except Ti) are dissolved. Aluminum, iron, and magnesium are recovered by treating the solution with HCl gas and subsequently removing steps (Figure 3.1). REEs are separated from the pregnant solution using organic solvents. The use of corrosive HCl, which requires glass-lined reactors as well as valves and tubes of high-chemical-resistance polymers, results in high cost and maintenance demand. In addition, the handling and storage of the produced insoluble acidic solid are of high importance, rendering the process as less effective.

3.1.3. Direct Leaching with Organic Acids

BR leaching using organic acids has been studied in the same way as inorganic acids. Greek BR treatment with citric, methanesulfonic, and acetic acids can correspond to inefficient rare earths' leaching, resulting in lower recoveries compared to the same done with inorganic acids. Exceptionally high REE recoveries, which are comparable to inorganic acids, can be achieved after citric acid treatment under the rising temperature (Borra et al., 2015). In contrast to these results, BR studies from Russia revealed that a mixture of formic and acetic acids at 80 °C and pH 1–3.6 obtained 63%–74% Sc recovery and similar Al recovery (yield) (Bogomazov & Senyuta, 2017), while the recoveries of other REEs are not reported. Further increase of pH during the leaching led to silica gel formation with filtration disorder. The main drawback of organic acids selection during leaching is the ineffective recovery which is attributed to the need of distillation in order to be reusable after the leaching stage. In addition, they are more expensive solvents than the one done with sulfuric acid. Bioleaching process is based on a more indirect use of BR leaching with organic acids. During bioleaching, metals are extracted with microorganisms (bacteria or fungi) which produce organic acids. Fungi are mainly used for the BR leaching because they can survive at high pH and produce organic acids, amino acids, and proteins in the presence of an organic agent forming complexes with metal ions. The produced organic acids are usually a mixture of citric, oxalic, and gluconic acids. REEs were leached from BR with the nematode fungi RM-10 in the presence of a sucrose agent (Qu & Lian, 2013). The authors found that leaching in two steps (pre-fungi culture followed by leaching) was better than the one-step leaching (culture in the presence of BR). Similarly, *Aspergillus niger* fungi were used (Qu et al., 2013) both in batch and continuous leaching. During the batch leaching at 2% w/v pulp density after 10 days, recovery of Sc is 44%, while the increase of pulp density to 10% w/v reduces the Sc recovery to 30% under continuous operation.

3.1.4. Leaching with Ionic Liquids

Ionic liquids (ILs) are versatile solvents consisting solely of ions having superior properties against classic organic solvents in terms of low vapor pressure and nonflammability. Ionic liquids have been applied as lixiviants to selectively dissolve REEs from BR.

A new leaching process that developed is based on a hydrophobic ionic liquid betainium bis(trifluoromethylsulfonyl)imide ([Hbet][Tf2N]) (Davris et al., 2016a, Davris et al., 2016b, Davris, Balomenos, et al., 2018). Leaching BR with [Hbet][Tf2N] at autoclave conditions of 150 °C for 4 h and water ratio of 40% v/v at 10% pulp density (w/v) can selectively dissolve REEs (>70%) against Fe, Al, Ti, and Si in the Greek BR. Scandium followed again a different leaching behavior from the rest REEs with maximum 45% recovery upon leaching, whereas Ca and Na contents are totally dissolved. The metals ions present in [Hbet][Tf2N] leachate obtained after filtration are stripped with an aqueous acidic solution, whereas hydrophobic [Hbet][Tf2N] is regenerated for reuse. During stripping and upon pH adjustment, a preconcentrated Sc solution with less impurities is generated. Where the concentration of the Sc in the organic phase before the stripping was 5–6 mg/L (directly comparable with the mineral acid selective leaching presented in Section 3.1.2), in the aqueous strip solution concentrations of 100 mg/L Sc have been achieved. This significant increase in concentration and the complete avoidance of silica gel formation are the most promising factors of this novel approach.

The proposed IL process (Figure 3.2) provides selectivity, recyclability of the leaching agent, and more efficient acid input in the process. Yet the high price of the ionic liquid in conjunction with the unavoidable IL losses in each contact with aqueous phases (in total, 3%–10% IL losses can be expected per cycle) makes this process financially unattractive. Future studies are focused on minimizing IL losses while maintaining the advantages of this newly developed process.

3.1.5. Leaching with $NaHCO_3$

Leaching BR studies at Russia indicated that selective Sc leaching can be achieved avoiding acidification stage through the alkaline leaching using $NaHCO_3$ as the leaching agent and utilizing CO_2 to regulate pH. It has been found that Sc has higher solubility in $NaHCO_3$ (16.7 g/L) than in Na_2CO_3 (0.43 g/L) solutions (Petrakova et al., 2016; Suss et al., 2018; Yatsenko & Pyagai, 2010). During alkaline leaching, part of $NaHCO_3$ is converted into Na_2CO_3, reducing Sc solubility. With the addition of CO_2 gas to the pulp, the $NaHCO_3$ content is increased allowing for higher Sc recovery. The flow sheet and the pilot plant developed by RUSAL are shown in Figure 3.3. During leaching, 26% of Sc is selectively recovered at 60 °C with pCO_2 = 6 atm and liquid-to-solid ratio 1:4 (Petrakova et al., 2016). About 30% Na, 68% Zr, and 6% Ti are dissolved and extracted during the purification of pregnant solution. Efforts to avoid Sc losses during the leaching are made by introducing resins directly into the pulp (Suss et al., 2018). Until now, this process is the only Sc extraction application from BR at a semi-industrial scale, allowing RUSAL to produce 2 tons of Sc_2O_3 per year.

The low recovery of Sc (approximately 25%) and the zero recovery of the other REEs are the main drawbacks in this case. The advantage of this process is the production of Sc with no acid addition, while the final pH 8 of BR after the process leads to a less environmentally aggressive waste to be disposed. In addition to this, it is one of the few methods that allow for actual CO_2 sequestration. It should be noted that this technology is

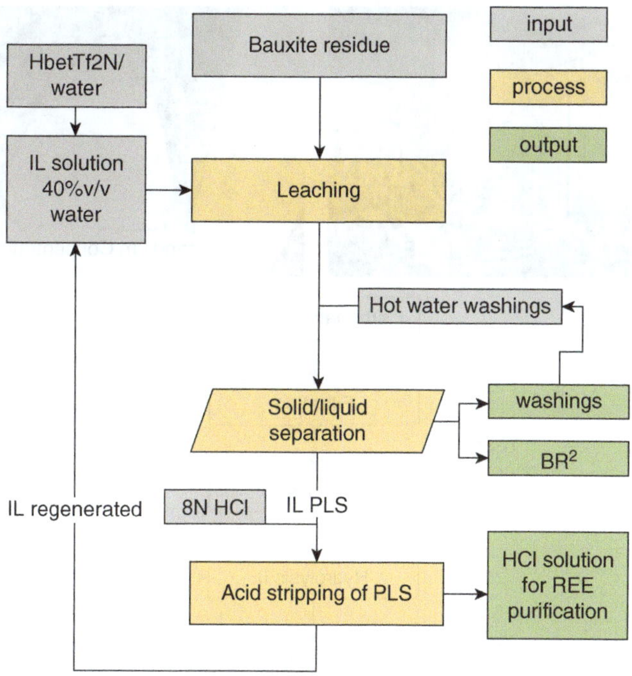

Figure 3.2 Bauxite residue leaching process flowsheet (top) and bench scale testring (bottom) with Ionic liquid HbetTf2N (Davris, Balomenos, et al., 2018).

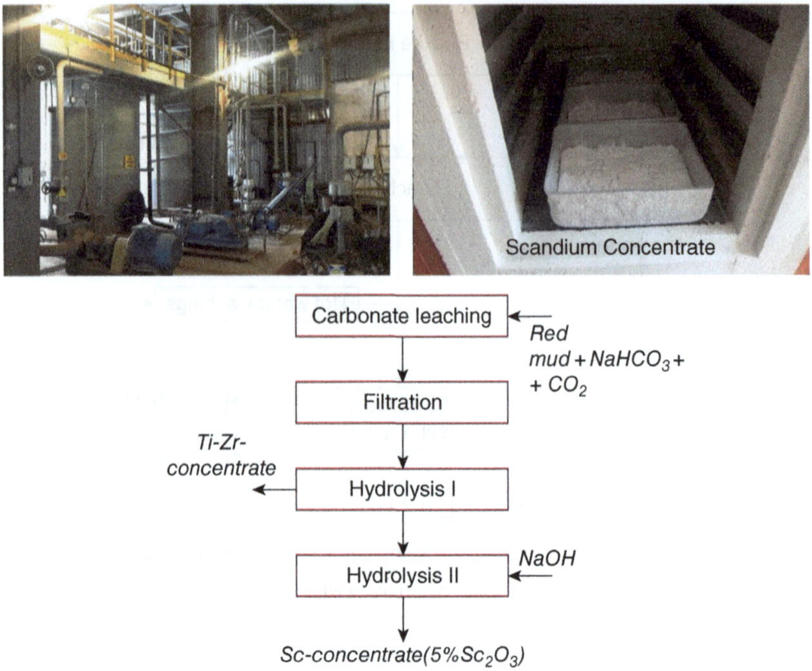

Figure 3.3 Sc pilot production at Rusal facilities (Petrakova et al., 2016).

effective when Sc is associated with aluminum mineral phases. Tests in the Greek BR (where Sc is in the mineral phases of Fe) did not result in significant Sc dissolution (<3%).

3.1.6. Comparison of Leaching Methods

Table 3.2 presents a comparison study regarding the direct BR leaching. In most REE studies, it is reported that the direct BR leaching by inorganic acids (HCl, HNO_3, and H_2SO_4) is more advanced than the use of organic acids such as citric acid, acetic acid, etc. (Borra et al., 2015; Ujaczki et al., 2017; Yatsenko & Pyagai, 2010). However, the recovery of REEs in some of the organic acids is comparable or even better to those of inorganic acids specially under a high-temperature leaching process. Nevertheless, the use of H_2SO_4 is considered to be most appropriate for BR leaching in terms of low cost and easy-to-handle process (Petrakova et al., 2014).

Considering the the chemical cost, it seems that citric acid leaching presents a double price compared to bioleaching, while H_2SO_4 leaching is

Table 3.2 Comparison of direct BR leaching methods.

Leaching means	Main advantages	Main disadvantages
H_2SO_4, HCl, HNO_3	20–80% REEs' extraction. Possibility to modify the recovery ratio of REEs and Fe depending on acid concentration and leaching steps. Byproducts production (Orbite Patent and HNO_3 Pilot NTUA Chemical Engineering Manual)	Low REEs' recovery. Large amounts of effluents and acid wastes. As acid concentration increases, selectivity decreases. Special equipment for corrosion resistance are needed in case of HCl. Silica gel formation at low acid concentrations
H_2SO_3	Digestion in two to three stages of selective dissolution of REEs, up to 85% Y recovery (Patent Alcan)	The results are reported in Y extraction and not in REEs separately. No data available for the Sc extraction
Carboxylic acids such as formic, citric, acetic, methanesulfonic acids, etc.	Selective REEs' leaching (similar to mineral acids); Ca and Al versus Fe. Bioleaching (RUSAL Patent)	Difficulty in recycling organic acids by distillation, producing large amounts of calcium sulfate. In some cases, low REE extraction. More expensive solvents than inorganic acids
CO_2, $NaHCO_3$	Selective metals recovery such as Sc, Ti, and Zr producing carbonate complexes in the caustic solution. Large intake of CO_2 in pulp (RUSAL pilot plant for production of 2 tons per year Sc_2O_3)	Low Sc recovery (<26%). Zero REEs recovery
Ionic liquid ([Hbet][Tf2N])	High selectivity of REEs against Fe, Si, and Ti; similar to mineral acid recoveries, low acid consumption (regeneration of IL), and purification and upconcentration of REE solution	High cost of IL (not commercial commodity-synthesized on demand), IL losses in aqueous phases

the most economically viable solution (Qu et al., 2015). The main disadvantages of acid extraction are the consumption of large acid quantities for the BR's neutralization, the waste manipulation, and the acidic residue's reuse after leaching (Table 3.2) (Petrakova et al., 2014). In addition, acid leaching dissolves large amounts of Fe, Ti, and Al by reducing selectivity (Wang & Cheng, 2011; Wang et al., 2011). The high concentration of these metals in the pregnant solution reduces the recovery efficiency during refining with ion-exchange resins, while increasing the acid consumption during the washing-regeneration phase of the resins. Unlike the acidic route, alkali extraction such as $NaHCO_3$ is advantageous due to the lack of neutralization and the production of less caustic BR, and while the scandium recovery in this process is quite low, it seems to be profitable as Sc_2O_3 can be produced at cost of well below $900/kg. However, this process is only effective in certain BRs from Russia. The patent of Orbite Technologies Inc. is also very promising, as it promotes metals recovery and acid's reuse, resulting in the minimization of the waste production. However, the use of concentrated HCl requires corrosion-resistance equipment, increasing the total budget as well. The IL leaching process is also very promising in delivering a concentrated Sc solution with low Fe, Ti, and Al impurities and overall low acid consumption, but it is uneconomic given the reported IL losses in each processing cycle and the current cost of the IL reagents.

3.1.7. Sulfuric Acid Baking

BR baking process with H_2SO_4 was developed after extensive research as a suitable method to increase the selective solubilization of REEs against base metals in BR, reducing the consumption of large quantities of chemical reagents (Borra et al., 2016c; Liu et al., 2017; Narayanan et al., 2018; Onghena et al., 2017). Based on this process, BR is digested at 100–120 °C with concentrated H_2SO_4 to form Fe, Al, Ti, and REE sulfates, and baking at 650–750 °C to break down unstable sulfates such as Fe, Al, and Ti in the form of oxides. REEs' sulfate salts remain in the form of sulfates even at 750 °C, allowing their selective solubilization in a subsequent step using water. The solid residue from the process contains mainly oxides Al, Ti, Fe, Si, and $CaSO_4$. This process results in the selective extraction of REEs with 60% Sc–90% REE recoveries (Borra et al., 2016c). In a similar study, 75% Sc–88% REE recoveries have been reported (Narayanan et al., 2018). The factors influencing the selective extraction are the amount of acid in the BR, the baking temperature, and the baking time. A big advantage of the method is the reduction in acid consumption and the high selectivity of REEs over Fe, Al, Si, and Ti. The slow rate of REEs' water leaching after the sulfate baking stage, which is reported between 2 and 7 days, and the SO_2 emissions are the main drawbacks of the process.

3.1.8. Combined Pyrometallurgical-Hydrometallurgical Leaching Techniques

The pyrometallurgical recovery of Fe from BR is a stepping stone for the development of holistic exploitation, valorizing about half the amount of BR and adjusting the properties of the slag to be further processed for critical metals extraction or to be used as building material. Iron recovery through the pyrometallurgical BR treatment can be achieved by reductive smelting or by reductive roasting and magnetic separation; REE oxides cannot be reduced by carbon and therefore will never follow the metal phase produced in such processes, and will be found in enhanced concentrations in the produced slag or from the nonmagnetic fraction, respectively.

Reductive smelting processes can be applied by several technologies (Corex, Finex, HIsmelt, Romelt, AusIron, and electric arc furnace [EAF]) to BR for the production of pig iron (Panov et al., 2012; Xenidis et al., 2016). So far, two methods have been applied on a large scale for the BR reductive smelting: the Romelt method (Mishra & Bagchi, 2002) and the EAF (Balomenos et al., 2011; Grzymek et al., 1982; Guccione, 1971; Logomerac, 1979). The Moscow Institute of Steel and Alloys (MISA), with National Aluminium Company Limited (NALCO) and Steel Authority of India Ltd. (SAIL) studied the pyrometallurgical process of BRs using the Romelt method (Mishra & Bagchi, 2002). The advantage of this method is that it approves materials with moisture levels of up to 10% wt. The main disadvantage is the high energy consumption and the poor quality of pig iron with a high concentration of S and P (Panov et al., 2012). In the EAF reductive smelting, mixture of BR, carbon, and fluxes is put at 1500–1700 °C to form pig iron with >95% iron recovery (Balomenos et al., 2011; Balomenos et al., 2014; Balomnenos et al., 2016). Recovery of residual iron can be further improved by later magnetic separation in slag dust (Borra et al., 2016b; Erçağ & Apak, 1997). Post-melting slag can be used to produce rockwool or building materials (Balomnenos et al., 2016; Raspopov et al., 2013), and for the recovery of nonferrous metals and REEs (Alkan et al., 2018; Borra et al., 2016b; Udy, 1958). REEs are extracted from the produced slag using strong inorganic acids (Sargic & Logomerac, 1974; Shaoquan & Suqing, 1996); in particular, slag extraction with 60% wt. H_2SO_4 dissolves virtually all of the Al and Ti contents residing predominantly in the silicon oxide (Udy, 1958), while extraction with inorganic acids HCl and HNO_3 results in high REE recovery yields (60%–100%) in solids to liquid (1/50) at 90 °C for 1 h at 2–3 mol/L acid concentration (Borra et al., 2016b). The use of H_2SO_4 in the same conditions showed similar recovery for Sc but lower for the remaining rare earths probably due to the formation of calcium sulfate. It is clear that a high acid concentration is required to dissolve REEs from the slag; however, the number of studies presented in the literature is limited in relation to the extraction of slag for Sc recovery using sulfuric acid as well as the description of extraction problems (Rivera et al., 2019). Interferences such as silicon gel formation, secondary $CaSO_4$ precipitation, and secondary aluminum sulfate precipitation have been reported and should be taken into account during the extraction of the slag (Davris, Gelestathi, et al., 2018).

BR sintering with Na and Ca carbonates (soda roasting) leads to the destruction of the aluminosilicate matrixes and the recovery of alumina. In such process, BR is mixed with sodium carbonate (soda [Na_2CO_3]) and the mixture is treated to temperatures between 800 and 1200 °C by converting the alumina into soluble phases of sodium aluminate, followed by extraction with water or mild alkaline solutions. Many research groups have applied this method for the recovery of alumina and sodium by BR (Bruckard et al., 2010; Chun et al., 2014; Kaußen & Friedrich, 2016; Li et al., 2009; Liu & Wu, 2012; Zhu et al., 2012). The resulting residue can be used for iron reduction by reducing smelting, while the produced slag is enriched in REEs and Ti where they can be extracted with strong inorganic acids (Borra et al., 2017). Additionally, the sequence of sintering and reductive smelting process implementation does not affect the REEs extraction at the final stage. Alternatively, a combined carbothermic reduction and soda roasting can also be applied to produce a magnetic iron fraction, soluble alumina phases, and a residue with upgraded REE content (Cardenia et al., 2018).

A process was investigated during the gradual extraction of Fe, Al, Ti, and REEs (Borra et al., 2017), where initially the BR sintering takes place with Na_2CO_3 at 950 °C for 4 h followed by water leaching at 80 °C to recover Al (75%) with further treatment of the insoluble residue by reductive smelting to recover Fe (98%) and extraction of the produced slag to recover Sc (80%) while recovery of the remaining REE is <5%. The produced slag is converted into amorphous by fast cooling, while more than 80% of the Ti content and REEs are recovered by leaching with 1 M HCl at 25 °C and 2% w/v pulp density. However, the presence of silica gel during the leaching process results in defective liquid–solids separation. According to BR sintering process with Na_2CO_3 and lignite (Li et al., 2014), the reduction of the hematite content to magnetite is achieved and is removed after breakage and magnetic separation. The nonmagnetic fraction is extracted in two steps (Deng et al., 2017), first with H_3PO_4 to recover SiO_2 and second with NaOH to recover Al, while the final insoluble solid is enriched in Sc and Ti obtaining the fourfold concentration compared to the original BR.

If CaO is added to the BR reductive smelting, an alkaline slag can be formed suitable for leaching alumina in alkaline environments (variation of the Pedersen process). The residue of the alkaline leaching can then be treated with acid leaching to recover REEs as well as Ti (Alkan et al., 2018; Vafeias et al., 2018).

An indicative mapping of various processing options aiming at extracting Fe, Al, and REEs is presented in Figure 3.4.

Such flow sheets result in more efficient leaching of the REEs, as the complex mineral matrix of the BR has been destroyed and significant amounts of base metals have been removed. At the same time, such flow sheets allow for significant reduction of the final residue that needs to be disposed and may even achieve financially a near break-even status (Balomenos et al., 2016). On the other hand, the numerous processing steps and the back and forth between wet and dry materials make for very energy- and cost-intensive processes. To achieve such processing in a sustainable (environmentally and financially) manner, a right balance between bulk applications with low-value products (i.e., pig iron production) and low-volume niche products (i.e., scandium) needs to be found.

3.1.9. Conclusions

Bauxite residue constitutes a significant resource for REEs and especially scandium production. However, there is significant complexity in extracting these metal values from BR, as upon direct leaching, solutions dilute in REEs and rich in base metals (g/L) such as Fe, Ca, Na, Al, etc., are produced, effectively hampering downstream refining of the REEs. Combined pyrometallurgical and hydrometallurgical routes that aim to recover base metals such as Fe, Al, and Ti and at the same time upconcentrate the REEs in slags are more attractive, as they would produce richer and cleaner REE leach solutions, while at the same time valorizing significant volumes of the BR itself. The goal of a near-zero waste and near break-even flow sheet for BR processing would be ideal for both the alumina and the REE industry. The realization of such a flow sheet, however, requires a high degree of industrial symbiosis and favorable logistics between different industries.

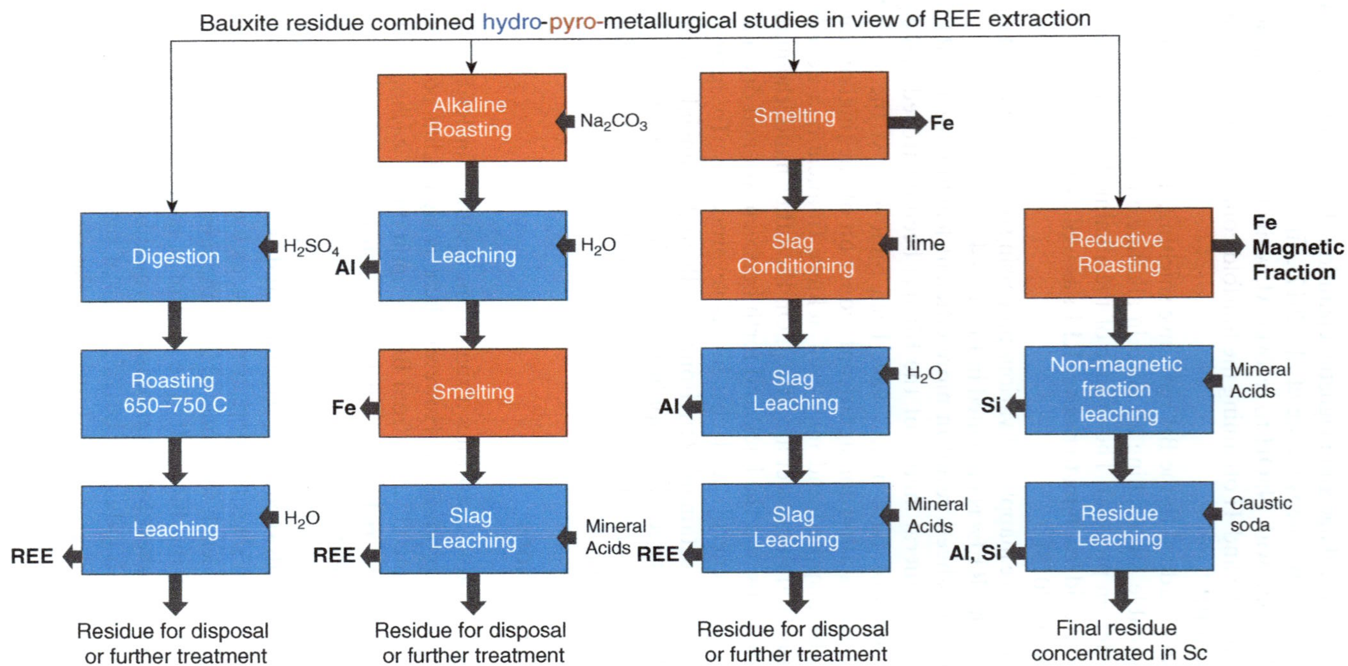

Figure 3.4 Main methods for major metals & REE extraction from BR by combining pyro- and hydro-metallurgical processes.

Acknowledgments

The research leading to these results has received funding from the European Community's FP7 and Horizon 2020 programmes (H2020/2017–2021) under grant agreements numbers 309373 (EURARE), 636876 (MSCA-ETN REDMUD), 730105 (SCALE), 767533 (ENSUREAL), and 776469 (REMOVAL).

References

Abhilash, Sinha, S., Sinha, M. K., & Pandey, B. D. (2014). Extraction of lanthanum and cerium from Indian red mud. *International Journal of Mineral Processing*, *127*, 70–73.

Ahmad, Z. (2003). The properties and application of scandium-reinforced aluminum. *Journal of the Minerals, Metals and Materials Society*, *55*(2), 35–39.

Alkan, G., Yagmurlu, B., Cakmakoglu, S., Hertel, T., Kaya, S., Gronen, L., et al. (2018). Novel approach for enhanced scandium and titanium leaching efficiency from bauxite residue with suppressed silica gel formation. *Scientific Reports*, *8*(1), 5676.

Balomenos, E., Davris, P., Deady, E., Yang, J., Panias, D., Friedrich, B., et al. (2017). The EURARE Project: Development of a sustainable exploitation scheme for Europe's rare earth ore deposits. *Johnson Matthey Technology Review*, *61*(2), 142–153.

Balomenos, E., Davris, P., Pontikes, Y., & Panias, D. (2017). Mud2Metal: Lessons learned on the path for complete utilization of bauxite residue through industrial symbiosis. *Journal of Sustainable Metallurgy*, *3*(3), 551–560.

Balomenos, E., Giannopoulou, I., Gerogiorgis, D., Panias, D., & Paspaliaris, I. (2014). Resource-efficient and economically viable pyrometallurgical processing of industrial ferrous by-products. *Waste and Biomass Valorization*, *5*(3), 333–342.

Balomenos, E., Giannopoulou, I., Panias, D., & Paspaliaris, I. (2011). A novel red mud treatment process: process design and preliminary results. *Travaux ICSOBA*, *36*(40), 255–266.

Balomenos, E., Kastritis, D., Panias, D., Paspaliaris, I., & Boufounos, D. (2016). The enexal bauxite residue treatment process: Industrial scale pilot plant results. In J. Grandfield (Ed.), *Light Metals 2014* (pp. 143–147). USA: Springer International Publishing, Cham.

Bárdossy, G., & Aleva, G. J. J. (1990). *Lateritic bauxites, 27*. Amsterdamn, The Netherlands: Elsevier Scientific Publishing Company.

Bárdossy, G. (1982). *Karst Bauxites (Bauxite deposits on carbonate rocks)*. Amsterdam, The Netherlands: Elsevier Scientific Publishing Company.

Bogomazov, A. V., & Senyuta, A. S. (2017). *Method for the acid treatment of red mud*. Google Patents.

Borra, C. R., Blanpain, B., Pontikes, Y., Binnemans, K., & Van Gerven, T. (2016a). Recovery of rare earths and other valuable metals from bauxite residue (red mud): A review. *Journal of Sustainable Metallurgy*, *2*(4), 365–386.

Borra, C. R., Blanpain, B., Pontikes, Y., Binnemans, K., & Van Gerven, T. (2016b). Smelting of bauxite residue (red mud) in view of iron and selective rare earths recovery. *Journal of Sustainable Metallurgy*, *2*(1), 28–37.

Borra, C. R., Blanpain, B., Pontikes, Y., Binnemans, K., & Van Gerven, T. (2017). Recovery of rare earths and major metals from bauxite residue (red mud) by alkali roasting, smelting, and leaching. *Journal of Sustainable Metallurgy*, *3*(2), 393–404.

Borra, C.R., Mermans, J., Blanpain, B., Pontikes ,Y., and Binnemans, K. (2016c). Selective recovery of rare earths from bauxite residue by combination of sulfation, roasting and leaching, *J. Minerals Engineering*, *92*, 151–159. https://doi.org/10.1016/j.mineng.2016.03.002

Borra, C. R., Pontikes, Y., Binnemans, K., & Van Gerven, T. (2015). Leaching of rare earths from bauxite residue (red mud). *Minerals Engineering*, *76*, 20–27.

Boudreault, R., Fournier, J., Primeau, D., & Labrecque-Gilbert, M.-M. (2017). *Processes for treating red mud*. Google Patents.

Bruckard, W. J., Calle, C. M., & Davidson, R. H. (2010). Smelting of bauxite residue to form a soluble sodium aluminium silicate phase to recover alumina and soda. *Mineral Processing and Extractive Metallurgy*, *119*(1), 18–26.

Cardenia, C., Balomenos, E., & Panias, D. (2018). Iron recovery from bauxite residue through reductive roasting and wet magnetic separation. *Journal of Sustainable Metallurgy*, *5*, 9–19. https://doi.org/10.1007/s40831-018-0181-5.

Chassé, M., Griffin, W. L., O'Reilly, S. Y., & Calas, G. (2016). Scandium speciation in a world-class lateritic deposit. *Geochemical Perspectives Letters*, *3*(2), 105–114.

Chun, T., Zhu, D., Pan, J., & He, Z. (2014). Recovery of alumina from magnetic separation tailings of red mud by Na2CO3 solution leaching. *Metallurgical and Materials Transactions B*, *45*(3), 827–832.

Davris, P., Balomenos, E., Panias, D., & Paspaliaris, I. (2016a). Chapter 12—Leaching rare earth elements from bauxite residue using brønsted acidic ionic liquids A2—Lima, Ismar Borges De. In W. L. Filho (Ed.), *Rare earths industry* (pp. 183–197). Boston: Elsevier.

Davris, P., Balomenos, E., Panias, D., & Paspaliaris, I. (2016b). Selective leaching of rare earth elements from bauxite residue (red mud), using a functionalized hydrophobic ionic liquid. *Hydrometallurgy*, *164*, 125–135.

Davris, P., Balomenos, E., Panias, D., & Paspaliaris, I. (2018a). Developing new process for selective extraction of rare earth elements from bauxite residue based on functionalized ionic liquids. *Light Metals 2018* (pp. 149–156). Springer International Publishing, Cham. USA.

Davris, P., Balomenos, E., Taxiarchou, M., Panias, D., & Paspaliaris, I. (2017). Current and alternative routes in the production of rare earth elements. *BHM Berg- und Hüttenmännische Monatshefte*, *162*(7), 245–251.

Davris, P., Gelestathi, N., Balomenos, E., Panias, D., & Paspaliaris, I. (2018b). Bauxite residue slag leaching for Al, Ti and Sc recovery. *Proceedings of the 2nd International Bauxite Residue Valorisation and Best Practices Conference*, Athens 7-10/05/2018, ISBN-number: 9789082825923, pp. 361–366.

Davris, P., Marino, D., Balomenos, E., Panias, D., & Paspaliaris, I. (2018c). Hydrometallurgical extraction of scandium from bauxite residue based on sulfuric acid process. *Proceedings of the 2nd International Bauxite Residue Valorisation and Best Practices Conference*, Athens 7-10/05/2018, ISBN-number: 9789082825923, pp. 449–454.

Deady, É. A., Mouchos, E., Goodenough, K., Williamson, B. J., & Wall, F. (2018). A review of the potential for rare-earth element resources from European red muds: Examples from Seydişehir, Turkey and Parnassus-Giona, *Greece. Mineralogical Magazine*, *80*(01), 43–61.

Deng, B., Li, G., Luo, J., & Ye, Q. (2017). Enrichment of Sc2O3 and TiO2 from bauxite ore residues. *Journal of Hazardous Materials*, *331*, 71–80.

Derevyankin, V., Porotnikova, T., Kocherova, E., Yumasheva, I., & Moiseev, V. (1981). Behaviour of scandium and lanthanum in the production of alumina from bauxite. *Izvestiya Vysshikh Uchebnykh Zavedenii, Tsvetnaya Metallurgiya*, 86–89.

Erçağ, E., & Apak, R. (1997). Furnace smelting and extractive metallurgy of red mud: Recovery of TiO2, Al2O3 and pig iron. *Journal of Chemical Technology and Biotechnology*, *70*(3), 241–246.

Fulford, G. D., Lever, G., & Sato, T. (1991a). Recovery of rare earth elements from Bayer process red mud. Google Patents.

Fulford, G. D., Lever, G., & Sato, T. (1991b). Recovery of rare earth elements from sulphurous acid solution by solvent extraction. Google Patents.

Grzymek, J., Derdacka-Grzymek, A., Konik, Z., & Grzymek, W. (1982). Methods for obtaining Iron, Alumina, Titania and binders from metallurgical slags and from "red mud" remaining in the Bayer method. *Light Metals*, 143–155.

Gu, H., Wang, N., & Hargreaves, J. S. J. (2018). Sequential extraction of valuable trace elements from bayer process-derived waste red mud samples. *Journal of Sustainable Metallurgy*, *4*(1), 147–154.

Guccione, E. (1971). Red mud, a solid waste, can now be converted to high-quality steel. *Engineering and Mining Journal*, *172*(9), 136–138.

Hatzilyberis, K., Lymberopoulou, T., Tsakanika, L. A., Ochsenkühn, K. M., Georgiou, P., Defteraios, N., et al. (2018). Process design aspects for scandium-selective leaching of bauxite residue with sulfuric acid. *Minerals*, *8*(3), 79.

Kaußen, F. M., & Friedrich, B. (2016). Methods for alkaline recovery of aluminum from bauxite residue. *Journal of Sustainable Metallurgy*, *2*(4), 353–364.

Li, G., Liu, M., Rao, M., Jiang, T., Zhuang, J., & Zhang, Y. (2014). Stepwise extraction of valuable components from red mud based on reductive roasting with sodium salts. *Journal of Hazardous Materials*, *280*, 774–780.

Li, X.-B., Xiao, W., Liu, G.-H., Peng, Z.-H., Zhaou Q.-S., & Qi, T.-G. (2009). Recovery of alumina and ferric oxide from Bayer red mud rich in iron by reduction sintering. *Transactions of Nonferrous Metals Society of China*, *19*(5), 1342–1347.

Li, Z., Din, J., Xu, J., & Liao, C. (2013). Discovery of the REE minerals in the Wulong–Nanchuan bauxite deposits, Chongqing, China: Insights on conditions of formation and processes. *Journal of Geochemical Exploration*, *133*, 88–102.

Liu, D.-Y., & Wu, C.-S. (2012). Stockpiling and comprehensive utilization of red mud research progress. *Materials*, *5*(7), 1232.

Liu, Z., Zong, Y., Li, H., Jia, D., & Zhao, Z. (2017). Selectively recovering scandium from high alkali Bayer red mud without impurities of iron, titanium and gallium. *Journal of Rare Earths*, *35*(9), 896–905.

Liu, Z., Zong, Y., Li, H., & Zhao, Z. (2018). Characterization of scandium and gallium in red mud with Time of Flight-Secondary Ion Mass Spectrometry (ToF-SIMS) and Electron Probe Micro-Analysis (EPMA). *Minerals Engineering*, *119*, 263–273.

Logomerac, V. (1979). Complex utilization of red mud by smelting and solvent extraction. *Trav. Com. Int. Etude Bauxites, Alumine Alum*, *15*, 279–285.

Mishra, S., & Bagchi, M. (2002). Mud to metal—Romelt is an answer. *Smelting reduction for iron making* (pp. 167–170). India: Allied Publishers Pvt. Ltd.

Mouchos, E., Wall, F., & Williamson, B. (2017). High-Ce REE minerals in the Parnassus-Giona bauxite deposits, *Greece. Applied Earth Science*, *126*(2): 82–83.

Narayanan, R. P., Kazantzis, N. K., & Emmert, M. H. (2018). Selective process steps for the recovery of scandium from Jamaican bauxite residue (red mud). *ACS Sustainable Chemistry & Engineering*, *6*(1), 1478–1488.

Ochsenkühn-Petropoulou, M. T., Hatzilyberis, K. S., Mendrinos, L. N., & Salmas, C. E. (2002). Pilot-plant investigation of the leaching process for the recovery of scandium from red mud. *Industrial & Engineering Chemistry Research*, *41*(23), 5794–5801.

Ochsenkühn-Petropulu, M., & Ochsenkühn, K. (1995). Rare earth minerals found in Greek bauxites by scanning electron microscopy and electron probe microanalysis. *European Microscopy and Analysis*, *37*, 33–34.

Ochsenkühn-Petropulu, M., Lyberopulu, T., & Parissakis, G. (1994). Direct determination of landthanides, yttrium and scandium in bauxites and red mud from alumina production. *Analytica Chimica Acta*, *296*(3), 305–313.

Ochsenkühn-Petropulu, M., Lyberopulu, T., & Parissakis, G. (1995). Selective separation and determination of scandium from yttrium and lanthanides in red mud by a combined ion exchange/solvent extraction method. *Analytica Chimica Acta*, *315*(1–2), 231–237.

Ochsenkühn-Petropulu, M., Lyberopulu, T., Ochsenkühn, K. M., & Parissakis, G. (1996). Recovery of lanthanides and yttrium from red mud by selective leaching. *Analytica Chimica Acta*, *319*(1–2), 249–254.

Onghena, B., Borra, C. R., Van Gerven, T., & Binnemans, K. (2017). Recovery of scandium from sulfation-roasted leachates of bauxite residue by solvent extraction with the ionic liquid betainium bis(trifluoromethylsulfonyl)imide. *Separation and Purification Technology*, *176*, 208–219.

Panov, A., Klimentenok, G., Podgorodetskiy, G., & Gorbunov, V. (2012). Directions for large scale utilization of bauxite residue. *Light Metals 2012* (pp. 93–98). USA: Springer Cham.

Petrakova, O., Klimentenok, G., Panov, A., & Gorbachev, S. (2014). Application of modern methods for red mud processing to produce rare earth elements. *Proceedings of the 1st European Rare Earth Resources Conference (ERES)*, Milos Greece, pp. 221–229.

Petrakova, O. V., Panov, A. V., Gorbachev S. N., & Klimentenok, G. N. (2016). Improved efficiency of red mud processing through scandium oxide recovery.

In M. Hyland (Ed.), *Light Metals 2015* (pp. 93–96). USA: Springer International Publishing, Cham.

Power, G., Gräfe, M., & Klauber, C. (2011). Bauxite residue issues: I. Current management, disposal and storage practices. *Hydrometallurgy*, *108*(1), 33–45.

Qu, Y., & Lian, B. (2013). Bioleaching of rare earth and radioactive elements from red mud using Penicillium tricolor RM-10. *Bioresource Technology*, *136*, 16–23.

Qu, Y., Lian, B., Mo, B., & Liu, C. (2013). Bioleaching of heavy metals from red mud using Aspergillus niger. *Hydrometallurgy*, *136*, 71–77.

Qu, Y., Li, H., Tian, W.,Wang, X., Wang, X., Jia, X., et al. (2015). Leaching of valuable metals from red mud via batch and continuous processes by using fungi. *Minerals Engineering*, *81*, 1–4.

Raspopov, N. A., Korneev, V., Avering, V. V., & Лайнер, Y.A. (2013). Reduction of iron oxides during the pyrometallurgical processing of red mud. *Russian Metallurgy (Metally)*, *2013*(1), 33–37.

Rayzman, V. L. (1998). Red mud revisited-special paper on scandium potential. *Aluminium International Today*, *10*(5), 64.

Reid, S., Tam, J., Yang, M., & Azimi, G. (2017). Technospheric mining of rare earth elements from bauxite residue (red mud): Process optimization, kinetic investigation, and microwave pretreatment. *Scientific Reports*, *7*(1), 15252.

Rivera, R. M., Ounoughene, G., Borra, C. R., Binnemans, K., & Van Gerven, T. (2017). Neutralisation of bauxite residue by carbon dioxide prior to acidic leaching for metal recovery. *Minerals Engineering*, *112*, 92–102.

Rivera, R. M., Ulenaers, B., Ounoughene, G., Binnemans, K., & Van Gerven, T. (2018). Extraction of rare earths from bauxite residue (red mud) by dry digestion followed by water leaching. *Minerals Engineering*, *119*, 82–92.

Rivera, R. M., Xahakalashe B. S., Ounoughene, G., Binnemans, K., Friedrich B., & Van Gerven, T. (2019). Selective rare earth element extraction using high-pressure acid leaching of slags arising from the smelting of bauxite residue. *Journal of Hydrometallurgy*, *184*. doi: 10.1016/j.hydromet.2019.01.005

Sargic, V., & Logomerac, V. (1974). Leaching and extraction in the complex processing of red mud. *Trav. Com. Int. Etude Bauxites, Alumine Alum*, *11*, 71.

Shaoquan, X., & Suqing, L. (1996). Review of the extractive metallurgy of scandium in China (1978 1991). *Hydrometallurgy*, *42*(3), 337–343.

Smirnov, D. I., & Molchanova, T. V. (1997). The investigation of sulphuric acid sorption recovery of scandium and uranium from the red mud of alumina production. *Hydrometallurgy*, *45*(3), 249–259

Sugita, K., Kobayashi, Y., Taguchi, Y., Wallen-Bryan, W., Lyew You, W., & Lyew-Ayee, P. A. (2016). Method of recovering rare-earth elements. Google Patents.

Suss, A., Panov, A., Kozyrev, A., Kuznetsova, N., & Gorbachev, S. (2018). Specific features of scandium behavior during sodium bicarbonate digestion of red mud. *Light Metals 2018* (pp. 165–173). USA: Springer International Publishing, Cham.

Tsakanika, L. V., Ochsenkühn-Petropoulou, M. T., & Mendrinos, L. N. (2004). Investigation of the separation of scandium and rare earth elements from red mud by use of reversed-phase HPLC. *Analytical and Bioanalytical Chemistry*, *379*(5), 796–802.

Udy, M. J. (1958). Process for the separation and recovery of Fe, Ti, and Al values from ores and wastematerials containing same. Google Patents.

Ujaczki, É., Zimmerman, Y. S., Gasser, C. A., Molnar M., Feigl V., & Lenz, M. (2017). Red mud as secondary source for critical raw materials–extraction study. *Journal of Chemical Technology and Biotechnology*, *92*(11), 2835–2844.

Vafeias, M., Marinos, D., Panias, D., Safarian, J., Van Der Eijk, C., Solheim, I., et al. (2018). From red to grey: Revisiting the Pedersen process to achieve holistic bauxite ore utilization. *Proceedings of the 2nd International Bauxite Residue Valorisation and Best Practices Conference*, Athens 7-10/05/2018, ISBN-number: 9789082825923, pp. 111–117.

Vind, J., Malfliet, A., Blanpain, B., Tsakiridis, P.E., Tkaczyk, A. H., Vassiliadou, V., & Panias D. (2018). Rare earth element phases in bauxite residue. *Minerals*, *8*(2), 77.

Vind, J. Malfliet, A., Bonomi, B., Paiste, P., Sajo, I. E., Blanpain, B., et al. (2018). Modes of occurrences of scandium in Greek bauxite and bauxite residue. *Minerals Engineering*, *123*, 35–48.

Wagh, A. S., & Pinnock, W. R. (1987). Occurrence of scandium and rare earth elements in Jamaican bauxite waste. *Economic Geology*, *82*(3), 757–761.

Wang, K.-Q., Yu, Y.-B., Wang, H., & Chen, J. (2010). experimental investigation on leaching scandium from red mud by hydrochloric acid. *Chinese Rare Earths*, *1*, 27.

Wang, W., & Cheng, C. Y. (2011). Separation and purification of scandium by solvent extraction and related technologies: A review. *Journal of Chemical Technology & Biotechnology*, *86*(10), 1237–1246.

Wang, W., Pranolo, Y., & Cheng, C. Y. (2011). Metallurgical processes for scandium recovery from various resources: A review. *Hydrometallurgy*, *108*(1), 100–108.

Wang, W., Pranolo, Y., & Cheng, C. Y. (2013). Recovery of scandium from synthetic red mud leach solutions by solvent extraction with D2EHPA. *Separation and Purification Technology*, *108*, 96–102.

Xenidis, A., Zografidis, C., Kotsis, I., & Boufounos, D. (2016). Reductive smelting of Greek bauxite residues for iron production. In: S. J. Lindsay (Ed.), *Light Metals 2011* (pp. 113–117). USA: Springer International Publishing, Cham.

Xue, A., Chen, X.-H., & Tang, X.-N. (2010). The technological study and leaching kinetics of scandium from red mud. *Nonferrous Metals Extractive Metallurgy*, *2*, 51–53.

Yatsenko, S. P., & Pyagai, I. N. (2010). Red mud pulp carbonization with scandium extraction during alumina production. *Theoretical Foundations of Chemical Engineering*, *44*(4), 563–568.

Zhang, J.-J., Deng, Z.-G., & Xu, T.-H. (2005). Experimental investigation on leaching metals from red mud. *Light Metals*, *2*, 13–15.

Zhang, N., Li, H.-X., Cheng, H.-J., & Liu, X.-M. (2017). Electron probe microanalysis for revealing occurrence mode of scandium in Bayer red mud. *Rare Metals*, *36*(4), 295–303.

Zhu, D.-Q., Chun, T.-J., Pan, J., & He, Z. (2012). Recovery of iron from high-iron red mud by reduction roasting with adding sodium salt. *Journal of Iron and Steel Research, International*, *19*(8): 1–5.

4

RARE EARTHS IN PHOSPHATE: CHARACTERIZATION AND EXTRACTION

Patrick Zhang[1], Haijun Liang[1], Zhen Jin[1], David DePaoli[2], Jan Miller[3], Chen-Luh Lin[3], and Raquel Crossman[4]

[1] *Florida Industrial and Phosphate Research Institute, Florida Polytechnic University, Lakeland, Florida, USA*
[2] *Oak Ridge National Laboratory, Oak Ridge, Tennessee, USA*
[3] *Department of Metallurgical Engineering, University of Utah, Salt Lake City, Utah, USA*
[4] *Freeport-McMoRan, Phoenix, Arizona, USA*

Abstract

Much of this chapter is based on results from a multi-year research project supported by the Critical Materials Innovation (CMI) Hub (formerly Critical Materials Institute) funded by the U.S. Department of Energy, Advanced Materials & Manufacturing Technologies Office. The Florida Industrial and Phosphate Research (FIPR) Institute, Florida Polytechnic University also provided substantial funding. Under the CMI project, FIPR Institute was engaged in three major activities: (1) chemical and mineralogical characterization of rare earth elements (REEs) in different phosphate-processing streams, (2) concentration of uranium- and REE-containing materials from the various processing streams, and

Rare Earth Elements: Sustainable Recovery, Processing, and Purification, Special Publications 79, First Edition. Edited by Athanasios K. Karamalidis and Roderick Eggert.

DOI: 10.1002/9781119515005.ch4

(3) extraction of REEs and uranium from the concentrated materials. This chapter covers the main findings of those three efforts, and is divided into four parts. Part I covers chemical analysis and basic properties of different samples. Part II is a detailed process-mineralogy study of the amine flotation tails. Part III focuses on isolation and characterization of rare earth mineral particles in three samples using dual energy (DE) rapid scan radiography and high-resolution X-ray microtomography (HRXMT). Part IV presents process development work on the concentration of REE-containing materials and extraction of REEs.

4.1. Introduction

4.1.1. Rare Earth Elements in Phosphate

Rare earth elements (REEs) have earned their reputation for being "rare" not because they are scarce in the Earth's crust but because they "rarely" exist in mineral forms that can be mined and extracted economically. For example, the average concentration of the REEs in the Earth's crust (ranging from 150 to 220 ppm) is much higher than that of copper (55 ppm).

Although there are over 200 minerals known to contain appreciable amounts of REEs, only three of them are economically significant, they include bastnaesite, monazite, and xenotime, with bastnaesite and monazite accounting for about 95% of the current sources for light rare earths. Some rare-earth-bearing clays are also significant sources for REEs. Xenotime is the primary mineral for heavy REEs and yttrium.

REEs may also be extracted as a byproduct from processing of minerals, such as copper, gold, uranium, and phosphate ores, with phosphate perhaps having the greatest potential. Certain phosphate deposits, specifically the fluorapatite ores, contain significant amounts of the rare earths Becker, 1983; Jorjani et al., 2008; Preston et al., 1996). Table 4.1 shows lanthanide content in some phosphate rock (Altschuler et al., 1967a, 1967b; Bliskoyskii et al., 1969; Watts, 2010).

Some phosphate ores or processing streams contain much higher REEs than what is shown in Table 4.1. A Canadian phosphate deposit near Quebec, for example, contains about 1,800 ppm of REEs. Another Canadian phosphate deposit in Ontario has an estimated content of 1.59% La_2O_3 + Ce_2O_3. In some recently discovered phosphate deposits in northern China (Xia, 2011), REE concentration (total R_2O_3) ranges from 1.5% to approximately 6.41%.

Table 4.1 Lanthanide content in selected phosphate rock.

Phosphate rock source	Ln_2O_3 (%)
Kola, Russia	0.8–1.0
Florida, USA	0.06–0.29
Algeria	0.13–0.18
Morocco	0.14–0.16
Tunisia	0.14

4.1.2. Rare Earth Elements in Florida Phosphate

The Florida phosphate ore (matrix) is mined in open pits using large draglines. Phosphate matrix is first transported to the beneficiation plant, and after several washing and separation steps, it is turned onto four streams, pebble product, flotation concentrate product, sand tailings, and waste clay. In the chemical-processing plant, the combined pebble/concentrate rock product is reacted with sulfuric acid, producing phosphoric acid and phosphogypsum byproduct.

Published analyses (Hill et al., 1932; Jarosinski et al., 1993) decades old, of trace elements in Florida phosphate rock have shown that many of these vital REEs are present. Though only present in trace amounts, because of the tonnage of phosphate produced, these elements are still significant in aggregate mass and have not been recovered. A relatively detailed investigation of REEs in Florida phosphate was conducted by Kremer and Chokshi (1989) of Mobil Research & Development Corporation. The total REEs in Florida phosphate matrix analyzed 282 ppm (88 ppm neodymium, 68 ppm cerium, 57 ppm yttrium, and 49 ppm lanthanum, accounting for 90%). Distributions of REEs in the mining and chemical-processing streams were also determined, showing 40% in waste clay, 37.5% in PG, 12.5% in phosphoric acid, and 10% in sand tailings. Data further indicated that REEs were concentrated in fine phosphate particles, as the pebble product analyzed 284 ppm REEs versus 575 ppm in the flotation concentrate. The high REE concentration (336 ppm) in waste clay is other evidence of REEs concentrating in fine phosphate particles. Another set of data by the United States Geological Survey (USGS) (Altschuler et al., 1967b) showed that Florida phosphate rock contained about 500 ppm REEs, with 150 ppm lanthanum, 120 ppm cerium, and 110 ppm yttrium. According to a report by the former United States Bureau of Mines (USBM) (May & Sweeney, 1983), Florida phosphogypsum contained 300 ppm REEs, with 130–170 ppm gadolinium, 49 ppm cerium, and 39 ppm lanthanum.

4.1.3. The CMI Efforts

As stated in the Abstract, the Critical Materials Institute (CMI) is an Energy Innovation Hub funded by the U.S. Department of Energy (DOE), Office of Energy Efficiency and Renewable Energy, Advanced Manufacturing Office. Phase 1 of CMI officially began 1st July 2013 and concluded 30th June 2018, with a total DOE funding of $120 million. CMI entered its Phase 2 on 1st July 2018, for another 5 years with slightly less funding than Phase 1. Figure 4.1 shows the logos of the CMI team members as of February 2020.

4.2. Part I: Chemical and Physical Characterizations of Phosphate-Processing Streams

Much of this chapter is based on analysis of development work for phosphate-processing streams from phosphate operations in central Florida, USA. In order to understand the fate of REEs during the phosphate mining and chemical-processing steps, a simplified processing flowchart is needed. Figure 4.2 shows a block diagram of the overall phosphate mining and processing flow sheet in Florida.

Figure 4.1 Logos for CMI team members, February 2020.

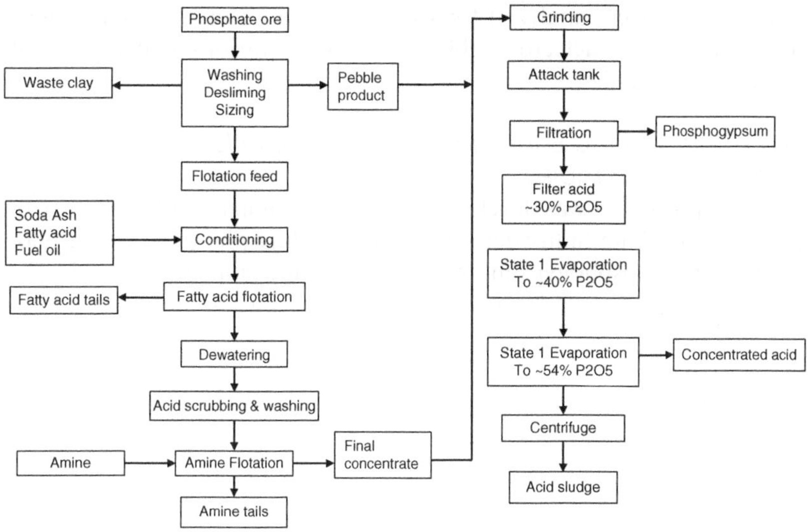

Figure 4.2 Flow sheet of phosphate mining and processing in central Florida, USA.

4.2.1. Sample Collection

The following samples were collected for this multi-year project:

- 2,000 pounds of amine flotation tails: in 3–55-gallon drums.
- 5 full 5-gallon buckets of primary slime (slurry): 10% thickened clay plus secondary clay.
- 100 pounds of acid plant rock feed (ground and wet): ball mill product.
- 2 full 5-gallon buckets of phosphoric acid: 30% clarified acid.
- 150 pounds of wet phosphogypsum: 3 5-gallon buckets of gypsum from an active stack.
- 95 pounds of phosphoric acid sludge: 3 barrels.

4.2.2. Chemical Analysis of Head Samples

Chemical analysis covered the following components: P_2O_5, CaO, Mg, Fe_2O_3, Al_2O_3, acid insoluble, REEs, U, Th, and Ra. The first six species were analyzed following a standard method developed by the Florida Association of Fertilizer and Phosphate Chemists. For REE analysis, contents were determined using a modified USGS procedure based on inductively coupled plasma atomic emission spectroscopy (ICP-AES). All head samples were analyzed for gamma-emitting isotopes using a high-purity germanium detector. Tables 4.2–4.4 show various analytical results for the head samples.

Procedure for Rare Earth Elements Determination in Phosphate Samples

This procedure was developed for the determination of REEs for the Florida phosphate industrial processes, including phosphate rock, phosphoric acid, phosphogypsum, sludge, and amine tailings. Besides lanthanide, the elements scandium, yttrium, thorium, and uranium are included in this procedure.

Inductively Coupled Plasma Optical Emission Spectroscopy (ICP-OES)

Standard Preparation

Reagents

A. Digestion acid (2 pts HNO_3, 2 pts deionized (DI) water).
B. Scandium stock (250 mL 1,000 ppm Fisher standard diluted to 10 L) (=25 mL 1,000 ppm standard diluted to 1 L).
C. Ca standard (50 g CaO digested to 1 L) (=10 g digested to 200 mL).
D. Mixed stock solution A: all Elements into a single 1000-mL volumetric flask.
   i. Group 1 metals
      1. 5 mL each of group 1 metals As and Cd.
         a. From 1,000 ppm Fisher standard solutions [5 ppm each in final volume].
   ii. Group 2 metals
      1. 20 mL each of group 2 metals Cr, Mn, Ti, and U.
         a. From 1,000 ppm Fisher standard solutions [20 ppm each in final volume].
      2. 5 mL each of group 2 metals Y, Ce, La, Th, and Nd.
         a. From 1,000 ppm Fisher standards [5 ppm each in final volume].
         b. [5,000 μg each or 5 μg/mL of final solution].
      3. 5 mL each of group 2 metals Gd, Ho, Eu, Pr, Er, and Dy.
         a. From 1,000 ppm Fisher standards [5 ppm each in final volume].
   iii. Group 3 metals
      1. 5 mL each of group 3 metals Sc, Sm Lu, Yb, Tm, Tb, and Nb.
         a. From 100 ppm standard [0.5 ppm each in final volume].
   iv. Diluted to 1 L using 1,000 ppm Fisher standards [20 ppm each in final volume].

E. Part A [53.70 g 85% phosphoric acid, 1.0579 g 96% sulfuric acid, 100 mL 10,000 ppm silicon standard and 80 mL digestion acid diluted to 1 L].
F. Part B [dilute 10,000 ppm standards of Fe (75 mL), Al (75 mL), Na(45 mL), K(10 mL), and Mg (20 mL) to 1 L].

Rock Standards

Blank (digest 20 mL digestion acid, 10 mL part A, 10 mL part B, 9 mL Ca standard in 100 mL volumetric and bring to volume).

Standard 1

- (Digest) 20 mL digestion acid, 10 mL part A, 10 mL part B, 9 mL Ca standard in 100 mL volumetric; cool, and add 2 mL of mixed stock solution A and bring to volume.
  - [Group 1: 0.1 ppm each element in final volume].
  - [Group 2: 0.4 ppm each element in final volume].
  - [Group 3: 0.01 ppm each element in final volume].

Standard 2

- (Digest) 20 mL digestion acid, 10 mL part A, 10 mL part B, 9 mL Ca standard in 100 mL volumetric; cool, and add 5 mL of mixed stock solution A and bring to volume.
  - [Group 1: 0.25 ppm each element in final volume].
  - [Group 2: 1.0 ppm each element in final volume].
  - [Group 3: 0.025 ppm each element in final volume].

Standard 3

- (Digest) 20 mL digestion acid, 10 mL part A, 10 mL part B, 9 mL Ca standard in 100 mL volumetric; cool, and add 10 mL of mixed stock solution A and bring to volume.
  - [Group 1: 0.5 ppm each element in final volume].
  - [Group 2: 2.0 ppm each element in final volume].
  - [Group 3: 0.05 ppm each element in final volume].

Standard 4

- (Digest) 20 mL digestion acid, 10 mL part A, 10 mL part B, 9 mL Ca standard in 100 mL volumetric; cool, and add 20 mL of mixed stock solution A and bring to volume.
  - [Group 1: 1.0 ppm each element in final volume].
  - [Group 2: 4.0 ppm each element in final volume].
  - [Group 3: 0.10 ppm each element in final volume].

Sample Preparation

A. Solid sample
   i. Dry sample in the furnace at 105 °C, and then grind the sample as fine as possible.
   ii. Weigh sample according to the desired concentration of the final product (usually 1 g diluted into a 100-mL flask, if sample concentration range is unknown).
   iii. Add 20 mL of 1:1 HNO_3.
   iv. Heat on hot plate at 250–300 °C until boiling gently and boil for 15 min.
   v. Cool and fill to the 100 mL mark with DI water.
   vi. Filter through a Whatman #40 paper into a bottle.

B. Liquid sample
   i. Weigh sample according to the desired concentration of the final product (usually 1 g diluted into a 100-ml flask, if sample concentration is unknown).
   ii. Add 20 mL of 1:1 HNO_3.
   iii. Heat on hot plate at 300 °C until boiling gently and boil for 15 min.
   iv. Cool and fill to the 100 mL mark.

C. If the concentrations of certain elements are too high for the high standard (a concentration between the standard concentration ranges), dilution of the 100 mL sample should be done. Also, initial weight and dilution of the sample can be adjusted so that a second dilution is not necessary.
   i. ICP-OES setup
      1. Run all the recommended wavelengths of the elements to be analyzed, and select the ones with the least interference and strongest intensity.
      2. Wavelengths (λ)

Element	Sc	La	Ce	Pr	Nd	Sm	Eu	Gd
Wavelength (λ)	424.7	408.7	413.8	414.3	401.2	363.4	382.0	342.2

Element	Tb	Dy	Ho	Er	Tm	Yb	Lu	Y
Wavelength (λ)	350.9	353.2	339.9	349.9	346.2	328.9	261.5	371.0

   ii. Quality assurance and quality control
      1. Run the 100% standard as a sample in each batch, since it has a known concentration, thus assuring accuracy of the ICP readings.
      2. A reference apatite concentrate with rare earth values (0.5000 g) and a duplicate of a random sample are prepared and run with each batch.
      3. The quality control (QC) range should be 2–5% depending on the sample nature.

Inductively Coupled Plasma Mass Spectrometry (ICP-MS)

To maintain total dissolved solids at less than 0.5% for ICP-MS, the standards with the same matrix as used for ICP-OES are diluted ten times, and the sample weight is adjusted to less than 0.5% total dissolved solids level. The preliminary results indicate that for the less sensitive elements, such as thorium and uranium, the results from ICP-MS are more stable and reproducible than those from ICP-OES. ICP-MS has a lot lower detection limits than ICP-OES, which are required for the low concentrate elements analyses.

4.2.3. Chemical Analysis Results

A routine chemical analysis of phosphate rock covers six (6) components. Table 4.2 shows chemical analysis results on the first set of samples.

Radioactivity analysis was conducted by the Department of Health of the State of Florida, see Table 4.3.

REE contents in various samples are shown in Table 4.4.

In this table, "dup" represents a duplicate digestion and analysis of the same sample. The PG samples from the three different buckets were sampled and analyzed separately in order to determine whether the samples were homogenized in the collection process. The three PG samples show small variations in chemical analysis, but can be considered to be representative of the same source.

Table 4.2 Routine chemical analysis of head samples (wt%).

Sample ID	P_2O_5	Insol	MgO	Fe_2O_3	Al_2O_3	CaO
Phosphate rock	27.36	15.84	0.50	1.19	1.19	38.84
Phosphate Rock dup*	27.39	16.30	0.48	1.18	1.17	38.80
Phosphoric acid	27.81	0.00	0.40	0.83	0.72	0.80
Phosphoric acid dup	27.70	0.02	0.41	0.83	0.78	0.82
Phosphogypsum 1	1.33	47.82	0.02	0.15	0.17	15.87
Phosphogypsum 1 dup	1.45	45.06	0.03	0.16	0.20	20.25
Phosphogypsum	1.47	45.03	0.02	0.14	0.18	19.71
Phosphogypsum 2 dup	1.37	44.06	0.02	0.13	0.15	19.48
Phosphogypsum 3	1.42	46.40	0.02	0.13	0.13	17.70
Phosphogypsum 3 dup	1.43	41.42	0.02	0.13	0.12	18.00
Amine tail	2.49	86.11	0.03	0.29	0.28	4.25
Amine tail dup	2.72	90.07	0.03	0.29	0.28	4.12
Waste clay	5.64	51.04	2.33	2.16	7.84	9.55
Waste clay dup	5.55	51.75	2.13	1.86	6.48	9.40

* Dup represents a duplicate digestion and analysis of the same sample.

Table 4.3 Radioactivity analysis of head samples (pCi/g).

Sample	Radium-226	Radium-228	U-235	U-238
Phosphogypsum 1	19.4 ± 0.8	0.5 ± 0.1	0.4 ± 0.1	2.8 ± 0.1
Phosphogypsum 2	21.7 ± 0.9	0.5 ± 0.1	0.4 ± 0.1	2.9 ± 0.2
Phosphogypsum 3	17.5 ± 0.6	0.5 ± 0.1	0.4 ± 0.1	2.8 ± 0.3
Phosphoric acid	0.2 ± 0.2	0.1 ± 0.1	2.7 ± 0.1	36.0 ± 1.4
Amine tails	5.5 ± 0.3	2.8 ± 0.1	0.3 ± 0.1	4.2 ± 0.3
Waste clay	14.9 ± 0.6	0.9 ± 0.1	0.8 ± 0.1	12.9 ± 0.6
Phosphate rock	27.9 ± 2.5	1.0 ± 0.1	2.0 ± 0.1	20.3 ± 0.7

4.2.4. Sizing Analysis of Phosphogypsum

Simple sizing and chemical analysis of PG revealed an extremely encouraging piece of information on REEs in PG. As shown in Table 4.5, about 65% of the REEs are concentrated in the −500 mesh (approximately 30 μm) fraction that represents 10% of the total PG. Another fact is that the finest fraction also contains most of the thorium. Distributions of other chemical species are shown in Table 4.6.

4.2.5. Sizing Analysis of Phosphate Clay

Phosphate beneficiation in Florida generates over 1 ton of phosphatic clay (slime) per ton of phosphate rock produced. Since the beginning of the large-scale washing and desliming practice for phosphate beneficiation, over two billion tons of such slime have been accumulated, which contain approximately 600 million tons of phosphate rock, 600,000 tons of REE, and 80 million kg of uranium. However, recovery of these valuable elements from the phosphatic clay may well be the most challenging subject in mineral processing, because the clay is very dilute (averaging 3% solids) and extremely fine in size (Zhang et al., 2012). Under the CMI project, two slime samples were analyzed using a Horiba laser scattering particle size distribution analyzer, LA-960, with a scale from 0.01 μm to 3,000 μm for dry and wet samples. The medium nozzle was set with data acquisition times of the sample as 50,000 and data acquisition times of the blank as 5,000. The air pressure is set at 0.3 MPa (3 bar, 45 psi). The instrument is set at the right standard refractive index. Particle size distribution default is the volume which shows the volume ratios of particles for each particle diameter with 100 representing the total volume of all particles. For analytical testing of the wet samples, the instrument initially was run on blank. The samples were stirred well and a very small amount (about 0.25 mL) of the stirred clay was then transferred to a recirculation system for measurements.

Table 4.4 Analysis of REEs, thorium, and uranium in head samples (ppm).

Sample ID	Pr	Eu	Tb	Dy	Ho	Er	Tm	Yb	Lu	Sc	Gd	Sm	Th	U	Ce	Y	La	Nd
Phosphate rock	7.92	3.48	2.17	13.81	3.25	9.62	1.22	8.69	1.34	4.90	16.46	12.04	9.15	87.77	131.16	132.06	77.58	85.26
Phosphate rock dup	8.23	3.39	2.13	13.48	2.91	9.28	1.10	8.51	1.37	4.80	16.34	11.45	9.69	89.25	129.21	129.11	76.38	83.92
Phosphoric acid	0.00	0.00	0.00	0.54	0.11	1.92	0.57	3.18	0.56	4.37	0.91	0.00	6.21	119.88	15.52	30.96	3.28	9.61
Phosphoric acid dup	0.00	0.00	0.00	0.49	0.06	1.81	0.60	3.19	0.58	4.36	1.03	0.00	5.26	123.33	15.75	30.99	3.23	9.72
Phosphogypsum 1	5.30	0.87	0.07	4.04	1.16	2.22	0.10	1.49	0.00	0.07	1.84	4.27	0.00	30.64	46.55	31.96	24.99	28.83
Phosphogypsum 1 dup	5.44	0.85	0.12	4.15	1.21	2.29	0.32	1.50	0.00	0.11	1.88	4.06	0.00	31.57	46.49	31.99	24.76	28.58
Phosphogypsum 2	5.41	0.84	0.03	4.12	0.99	2.41	0.35	1.47	0.00	0.04	1.73	4.10	0.00	30.17	46.23	31.92	24.72	28.34
Phosphogypsum 2 dup	5.78	0.96	0.10	4.53	1.07	2.51	0.14	1.60	0.00	0.06	2.32	4.46	0.00	31.79	49.41	34.65	27.23	30.81
Phosphogypsum 3	6.41	1.04	0.38	4.84	1.13	2.60	0.34	1.70	0.00	0.04	2.67	4.89	0.00	32.41	49.49	36.53	28.78	32.70
Phosphogypsum 3 dup	6.70	1.03	0.33	4.82	1.28	2.67	0.57	1.69	0.00	0.02	2.64	5.04	0.00	32.10	49.68	36.63	28.69	32.29
Amine tail	5.72	0.96	0.18	4.72	0.99	2.81	0.43	2.26	0.07	1.48	4.03	4.97	6.58	25.16	65.12	35.86	29.40	39.81
Amine tail	5.69	0.95	0.23	4.77	1.19	2.82	0.56	2.27	0.08	1.50	3.87	4.88	6.50	24.42	66.35	36.19	29.31	39.50
Waste clay	2.55	1.92	1.02	7.26	0.55	4.77	0.60	4.41	0.84	6.05	10.42	6.07	5.24	37.12	85.65	61.00	42.02	54.16
Waste clay dup	2.63	1.82	0.95	7.11	0.80	4.49	0.52	4.27	0.79	6.16	9.82	6.47	6.61	37.49	86.28	59.35	41.30	53.19

Table 4.5 Size distribution of rare earths in PG.

Screen size mesh	Major REEs (ppm)					Total REEs (ppm)	REE distribution (%)
	La	Ce	Pr	Nd	Y		
+200	23.02	41.22	6.83	28.53	27.52	141.69	26.08
−200 + 270	43.63	71.85	11.28	50.92	51.02	256.17	2.19
−270 + 325	55.09	87.13	12.90	59.30	60.49	306.47	3.37
−325 + 400	65.79	103.47	14.95	69.28	74.33	366.30	0.70
−400 + 500	88.81	135.32	18.07	90.21	90.74	467.96	2.28
−500	579.13	954.51	63.29	630.41	559.53	2,981.29	65.39

Table 4.6 Size distribution of major chemical components in PG.

Screen size mesh	Yield (wt%)	Major chemical components (wt%)					U (ppm)	Th (ppm)
		Insol	P_2O_5	Fe_2O_3	Al_2O_3	CaO		
+200	79.24	59.94	0.15	0.04	0.12	13.06	34.25	0.00
−200 + 270	3.68	19.23	0.60	0.09	0.19	29.83	35.87	0.89
−270 + 325	4.73	7.44	0.63	0.10	0.34	31.02	34.33	0.20
−325 + 400	0.82	6.13	0.72	0.11	0.41	33.82	35.44	0.37
−400 + 500	2.10	7.10	0.62	0.11	0.59	31.90	34.68	0.91
−500	9.44	6.40	3.52	2.04	3.84	27.62	0.00	10.93

Figure 4.3 shows a size distribution graph for a phosphate clay sample from Four Corners (FC) Mine. The mean size of the sample is 8.30 μm and the mode size is 8.23 μm. The size distribution follows a bimodal distribution with 10% of the particles having diameters averaging 0.32 μm, 50% averaging 6.96 μm, and 90% averaging 16.67 μm. Figure 4.4 shows a size distribution graph for phosphate clay sample from South Fort Meade (SFM) Mine. The mean size of the sample is 5.33 μm and the mode size is 4.79 μm. The size distribution follows a bimodal distribution with 10% of the particles having diameters averaging 0.13 μm, 50% averaging 3.45 μm, and 90% averaging 9.04 μm.

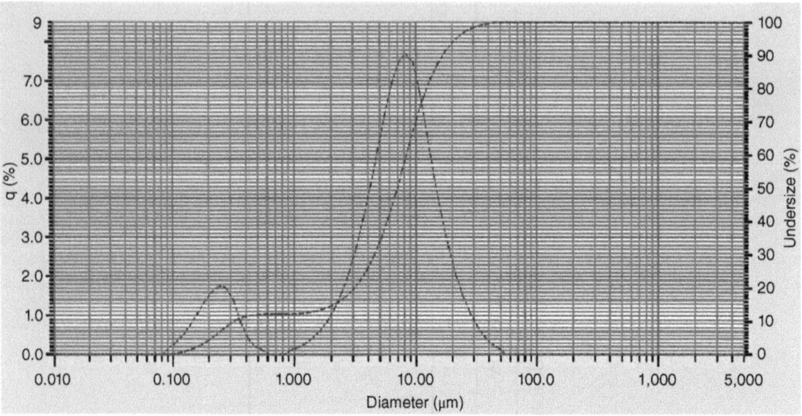

Figure 4.3 Particle size distribution of phosphate clay sample from FC.

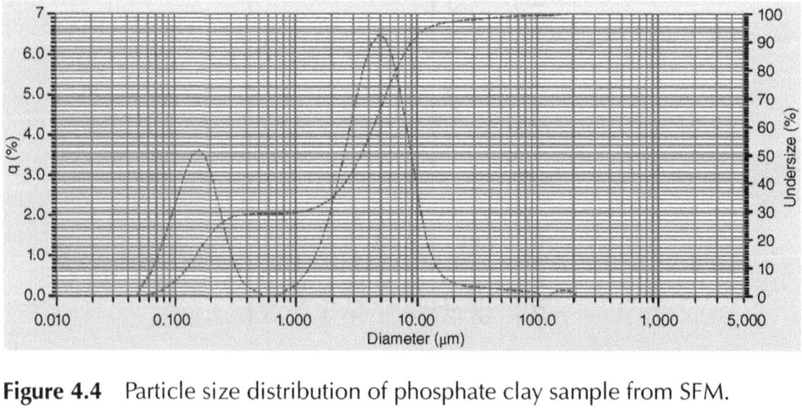

Figure 4.4 Particle size distribution of phosphate clay sample from SFM.

4.2.6. Rare Earths Mass Balance in Phosphate Mining and Processing Streams

Case 1, Plant A

The comprehensive investigation of REEs in Florida phosphate conducted by Kremer and Chokshi (1989) showed that about 50% of the REEs in the phosphate matrix reported to the final rock product, 40% to waste clay, and 10% to flotation tailings. Since the current waste clay contains significantly less phosphate than that in the past, and that the portion of the pebble product is lower in the current phosphate ore, the percentage of REE distribution in the various streams could be very different. FIPR therefore undertook a major effort to sample and analyze two sets of samples for REE mass balance calculations. Table 4.7 summarizes the total REE concentrations in various samples from plant A.

Based on weight distribution in the different size fractions shown in Table 4.8, REEs in the head samples of rougher tails and cleaner tails can be calculated as 39.68 and 152.78, respectively.

A typical plant's mass flow rates are shown in Table 4.9.

By simple calculations based on the data in Tables 4.8–4.10, REE distributions in different streams can be derived as follows:

$$\begin{aligned}&\text{Percentage REE distribution in the product}\\&\quad = 100\times(608\times350+229\times90+163\times160)/\\&\qquad(160\times1{,}850+102\times750+229\times90+163\times160)=61.9.\end{aligned}$$

$$\begin{aligned}&\text{Percentage REE distribution in the flotation tailings}\\&\quad = 100\times(39.68\times1{,}300+152.78\times200)/\\&\qquad(160\times1{,}850+102\times750+229\times90+163\times160)=19.6.\end{aligned}$$

Table 4.7 Total REE concentrations (ppm) in plant A samples.

Sample ID	Total REEs
Final concentrate	608
Intermediate pebble	229
Regular pebble	163
Flotation feed	160
Cleaner tail: +50	265
Cleaner tail: −50 + 100	168
Cleaner tail: −100 + 230	114
Rougher tail: +50a	61
Rougher tail: −50 + 100	19.6
Rougher tail: −100 + 230	18.2
Primary slime head	102
Phosphogypsum	119

Table 4.8 Sizing analysis of rougher and cleaner tails from plant A.

Sample	Size range (mesh)	Wt%
Rougher tail	+50	48.69
	−50 + 100	45.54
	−100 + 230	5.76
	−230	0.00
Cleaner tail	+50	2.91
	−50 + 100	63.62
	−100 + 230	33.38
	−230	0.09

Table 4.9 Plant mass flow rates (ton/hour) for plant A.

Stream	Flow rate (tons/hour)
Flotation feed	1,850
Flotation concentrate	350
Regular pebble	160
Intermediate pebble (to bin)	90
Rougher tails (cell underflow)	1,300
Amine tails (cell overflow)	200
Primary clay to Clay Settling Area (CSA)	750

$$\text{Percentage REE distribution in the waste clay}$$
$$= 100 \times (102 \times 750)/(160 \times 1{,}850 + 102 \times 750 + 229 \times 90 + 163 \times 160)$$
$$= 18.2.$$

Assuming that 5 tons of PG are generated for each ton of P_2O_5 in the final product, and that the average P_2O_5 content in the final product is 30%, REE distribution in PG can then be estimated as follows:

$$\text{Percentage REE distribution in PG}$$
$$= 100 \times 5 \times 0.3(350 + 160 + 90) \times 119/$$
$$(160 \times 1{,}850 + 102 \times 750 + 229 \times 90 + 163 \times 160) = 25.88.$$

The above mentioned total distribution of REEs in PG represents 42.15% of the REEs in the overall product, derived as follows:

$$\text{Percentage distribution of the product REE in PG}$$
$$= 100 \times 5 \times 0.3(350 + 160 + 90) \times 119/$$
$$(350 \times 608 + 160 \times 163 + 229 \times 90 + 163 \times 160) = 42.15.$$

The total REEs in the matrix can be back calculated, as follows:

$$\text{REE concentration in the matrix (ppm)} = (160\times1{,}850+102\times750+229\times90+163\times160)/(1{,}850+750+90+160)=145.$$

To better understand the REE mass balance numbers, one may refer to Figure 4.2 showing a flow sheet schematic of phosphate mining and processing.

Case 2, Plant B

Table 4.10 summarizes total REE concentrations in various samples from plant B.

A typical plant mass distribution for this plant is shown in Table 4.12.

By simple calculations based on the data in Tables 4.11–4.13, REE distributions in different streams can be estimated, as follows:

$$\text{Percentage REE distribution in the product} = 100 \times (901 \times 111{,}400 + 262 \times 40{,}100)/[217 \times (451{,}000 + 78{,}500) + 262 \times 40{,}100 + 346 \times 240{,}000] = 53.19.$$

$$\text{Percentage REE distribution in the flotation tailings} = 100 \times ([335 \times (22{,}700 + 7{,}011) + 20 \times (342{,}000 + 47{,}800)]/[217 \times (451{,}000 + 78{,}500) + 262 \times 40{,}100 + 346 \times 240{,}000] = 8.50.$$

$$\text{Percentage REE distribution in the waste clay} = 100 \times (346 \times 240{,}000)/[217 \times (451{,}000 + 78{,}500) + 262 \times 40{,}100 + 346 \times 240{,}000] = 39.83.$$

The above mentioned distributions can be normalized as 52.40% to the product, 39.23% to the waste clay, and 8.37% to the flotation tails.

Table 4.10 Total REE concentrations (ppm) in samples from plant B.

Sample	Total REEs
Concentrate	901
Feed averaged	217
Pebble	262
Rougher tail: +50	26.5
Rougher tail: −50+100	17.9
Rougher tail: −100+230	19.5
Rougher tail calculated	20.07
Amine tail: −50+100	290
Amine tail: −100+230	354
Amine tail calculated	335.0
Waste clay (CF clay)	346
Gypsum	112

Table 4.11 Sizing analysis of rougher and cleaner tails from plant B.

Sample	Size range (mesh)	Wt%
Rougher tail	+50	22.51
	−50 + 100	63.00
	−100 + 230	14.20
	−230	0.29
Cleaner tail	+50	2.00
	−50 + 100	33.81
	−100 + 230	61.25
	−230	2.94

Table 4.12 Plant B mass distribution (tons).

Stream	Tons
Fine flotation feed	451,000
Fine fatty acid tails	342,000
Fine amine tails	22,700
Coarse flotation feed	78,500
Coarse fatty acid tails	47,800
Coarse amine tails	7,011
Concentrate	111,400
Pebble	40,100
Waste clay	240,000

4.3. Part II: Detailed Process-Mineralogy Studies of Amine Flotation Tails

4.3.1. Amine Tails

Elemental and Chemical Components Analysis of Amine Tails

Table 4.13 shows concentrations of major compounds in the as-received amine tails, and Table 4.14 lists contents of the REEs.

Corresponding analyses for gravity separation concentrate and tails from the amine tails are shown in Tables 4.15–4.17. Gravity separation was conducted using a jig. The jigging concentrate was sized into three fractions, +70 mesh, 70 × 100 mesh, and −100 mesh. Results show that the −100 mesh fraction contains a higher concentration of REEs totaling about 618 ppm. Another interesting fact is that REE content has no obvious correlation with the P_2O_5 concentration, indicating that REEs in the amine tails are not associated with the phosphate minerals.

Sizing Analysis of Amine Tails

Table 4.18 shows detailed sizing analysis of the amine tails.

Table 4.13 Chemical compositions (wt%) of the amine flotation tails.

Compound	SiO_2	Al_2O_3	TFe_2O_3	MgO	CaO	Na_2O	K_2O
Content	86.36	2.51	0.88	0.10	3.56	0.11	0.81
Compound	TiO_2	P_2O_5	MnO	H_2O^-	LOI	Σ	
Content	1.07	2.31	0.039	0.08	1.44	99.269	

Table 4.14 Analysis (ppm) of rare earth elements in the amine tails.

Element	La	Ce	Pr	Nd	Sm	Eu	Gd	Tb
Content	29.36	65.77	5.71	39.65	4.92	0.95	3.93	0.20
Element	Dy	Ho	Er	Tm	Yb	Lu	Y	ΣREE
Content	4.75	1.10	0.50	0.50	2.26	0.08	36.02	195.70

Table 4.15 Chemical compositions (wt%) of jigging products from the amine tails.

Sample ID	P_2O_5	MgO	Fe_2O_3	Al_2O_3	CaO	Insol
Jig concentrate	3.84	0.03	0.32	0.4	4.99	88.38
Jig tails	3.12	0.03	0.4	0.31	3.98	90.42
+70 mesh jig concentrate	3.61	0.03	0.36	0.24	4.41	90.08
−70 + 100 jig concentrate	5.07	0.05	0.25	0.54	6.77	85.09
−100 jig concentrate	3.11	0.05	0.24	1.23	4.21	88.65

Table 4.16 Analysis (ppm) of REEs in the jigging products from the amine tails.

Sample ID	Pr	Eu	Tb	Dy	Ho	Er	Tm	Yb	Lu	Sc
Concentrate	5.62	0.99	0.65	4.7	0.88	3.55	0.91	2.51	0.27	2.09
Tails	6.09	0.94	0.61	4.62	0.83	3.19	0.83	1.96	0.09	2.44
+70 mesh concentrate	3.91	0.62	0.56	3.32	0.73	2.54	0.68	1.62	0.09	1.36
−70 + 100 concentrate	6.65	1.5	1.02	7.06	1.65	5.25	1.06	3.72	0.42	2.53
−100 concentrate	16.29	1.69	1.8	7.97	1.28	5.29	1.05	3.37	0.5	4.21

Table 4.17 Analysis (ppm) of U and REEs in the jigging products from the amine tails.

Sample ID	Gd	Sm	Th	Ce	Y	La	Nd	U	ΣREE
Concentrate	6.02	3.85	9.04	63.59	38.67	27.64	40.2	32.11	211.18
Tails	5.05	6.84	0	55.45	29.89	23.7	38.48	33.76	181.01
+70 mesh concentrate	3.27	5.2	0	36.92	24.86	15.29	26.2	33.48	127.17
−70 + 100 concentrate	7.78	7.36	0	66.16	53.42	32.44	46.65	25.96	244.67
−100 concentrate	13.21	11.44	18.09	209.9	48.14	86.02	187.6	6.95	617.85

Table 4.18 Size distribution of the amine tails head sample.

Mesh size	Wt%	Cumulative wt%
−200	7.01	7.01
−140 + 200	53.60	60.61
−100 + 140	14.09	74.7
−60 + 100	25.18	99.88
+60	0.12	100

Mineral Components and Their Concentrations

Quantitative mineral analysis was conducted using a MLA. The MLA is perhaps the most advanced analyzer in today's market for mineral identification, mineral quantification, mineral liberation, and grain size. It is an automated, X-ray based analyzer coupled with smart software modules, and analyzes mineral samples in polished sections of drill core, particulate, or lump materials. MLA can produce vast information, invaluable to geologists, mineralogists, and process engineers. Two rare earth minerals were identified by MLA—they are monazite and xenotime, both at very low concentrations. Major non-rare-earth minerals include quartz, fluorapatite, feldspar, zircon, and rutile (Table 4.19).

Size Distribution of Major Minerals

The size range of the mineral particles was also analyzed using MLA. Results show that the major rare earth minerals are relatively fine with 100% of the xenotime particles smaller than 40 μm and 88.93% of the monazite particles below 80 μm. The particle sizes of other minerals ranged from 0.02 to 0.16 mm. Table 4.20 shows the size analysis details.

Table 4.19 Quantitative mineral analysis of amine tails using MLA.

Mineral	Wt%	Mineral	Wt%	Mineral	Wt%
Monazite	0.053	Tourmaline	0.353	Pseudo-rutile	1.125
Xenotime	0.003	Garnet	0.477	Rutile	0.450
Zircon	0.614	Epidote	0.119	Leucoxene	0.148
Apatite	8.703	Kyanite	1.174	Sphene	0.337
Wavellite	0.512	Staurolite	0.329	Limonite	0.016
Woodhouseite	0.009	Kaolin	0.048	Siderite	0.143
Quartz	80.011	Pyrite	0.047	Zinc spinel	0.003
Feldspar	5.294	Sphalerite	0.009	Others	0.003
Biotite	0.004	Calcite	0.003	Total	100.000

Table 4.20 Size distribution analysis for the major minerals in amine tails.

Size range (mm)	Weight distribution (%)						
	Apatite	Monazite	Xenotime	Zircon	Rutile	Pseudo-rutile	Leucoxene
−0.32 + 0.16	3.32						
−0.16 + 0.08	44.68	11.07		24.96	30.10	21.56	24.46
−0.08 + 0.04	34.60	78.29		62.83	63.23	68.29	53.88
−0.04 + 0.02	12.13	8.54	86.71	10.21	5.95	9.13	15.94
−0.02 + 0.01	3.97	1.98	5.20	1.47	0.55	0.78	3.60
−0.01	1.30	0.12	8.09	0.53	0.17	0.24	2.12
Total	100.00	100.00	100.00	100.00	100.00	100.00	100.00

Dissemination Characteristics of Major Minerals

Monazite

Micro-area chemical analysis of monazite was performed using energy dispersive X-ray spectroscopy (EDS), and the results are shown in Table 4.21. The monazite particles contain the rare earth elements La, Ce, Nd, and Pr, with Ca, U, and Rh substitutions. Other impurities include Si and Al. Figure 4.5 shows a clear monomer monazite particle, while Figure 4.6 indicates inclusion of monazite within quartz.

Xenotime

Micro-area chemical analysis of xenotime was also performed using EDS, and the results are shown in Table 4.22. The xenotime particles contain the rare earth elements Y, Gd, Dy, Yb, Nd, Er, and Sm, with Ca

Table 4.21 Chemical components of monazite analyzed using EDS.

Spectrum	Chemical components (%)									
	CaO	La_2O_3	Ce_2O_3	Nd_2O_3	UO_2	ThO_2	Pr_2O_3	Al_2O_3	SiO_2	P_2O_5
1	1.38	13.69	28.62	12.00	1.35	5.26	3.65	0.01	1.75	32.28
2	1.49	12.73	29.41	12.65	0.26	8.06	3.81	0.05	2.06	29.48
3	1.26	17.69	31.45	9.09	0.24	5.58	2.47	0.99	1.33	29.91
4	1.45	9.93	26.81	15.33	0.13	9.35	3.92	0.92	1.85	30.30
5	1.14	13.40	28.10	12.59	0.78	6.35	3.65	0.48	2.03	31.48
6	1.67	13.81	28.76	11.76	1.40	6.50	3.57	0.10	1.44	31.00
7	1.22	14.24	29.50	11.27	0.57	6.28	3.32	0.01	2.01	31.59
8	0.90	13.84	29.76	12.05	0.06	5.88	3.76	1.10	1.61	31.04
9	0.95	13.95	30.81	13.38	0.46	3.62	3.94	0.83	1.10	30.96
10	1.25	14.04	29.57	12.43	1.05	5.22	3.40	0.13	1.71	31.20
Average	1.27	13.73	29.28	12.26	0.63	6.21	3.55	0.46	1.69	30.92

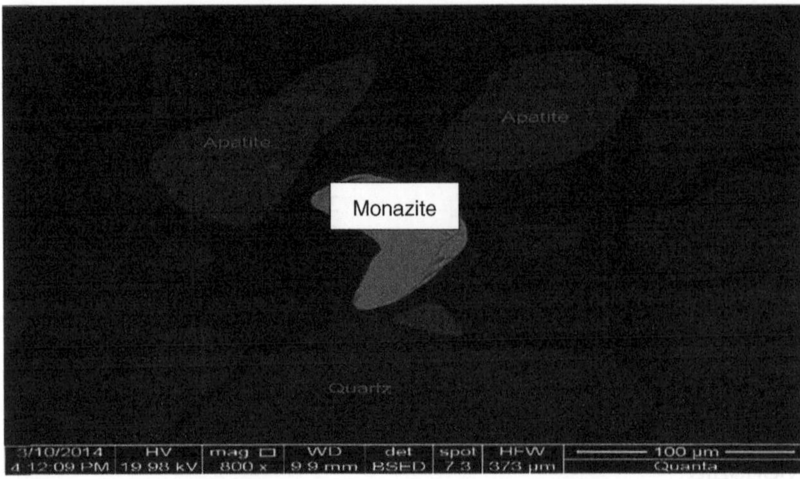

Figure 4.5 Scanning electron microscope-backscattered electrons (SEM-BSE) mage showing monazite monomer.

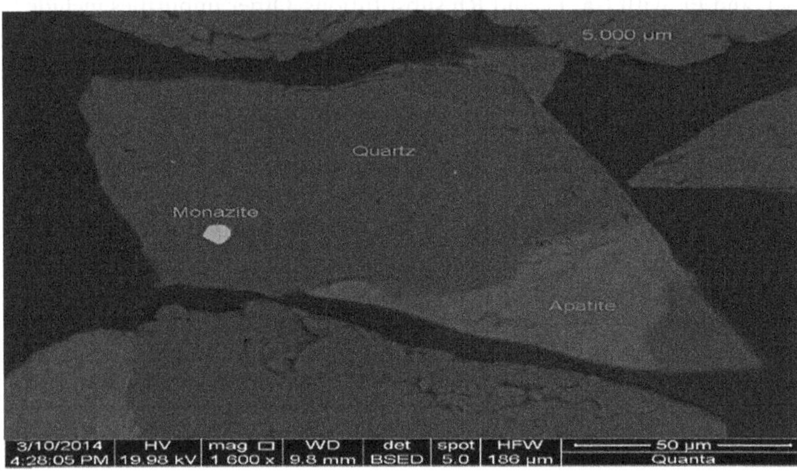

Figure 4.6 SEM-BSE image showing monazite inclusion in quartz.

Table 4.22 Chemical components of xenotime analyzed using EDS.

	Chemical components (%)									
Spectrum	Y_2O_3	Gd_2O_3	Dy_2O_3	Yb_2O_3	Nd_2O_3	Er_2O_3	Sm_2O_3	UO_2	CaO	P_2O_5
1	44.18	2.66	5.90	4.89	0.37	4.85	0.81	1.48	0.22	34.63
2	47.67	2.04	4.93	3.31	0.29	4.67	0.53	0.77	0.10	35.69
3	47.48	2.17	4.89	3.60	0.26	4.32	0.64	0.93	0.25	35.47
Average	46.44	2.29	5.24	3.93	0.31	4.61	0.66	1.06	0.19	35.26

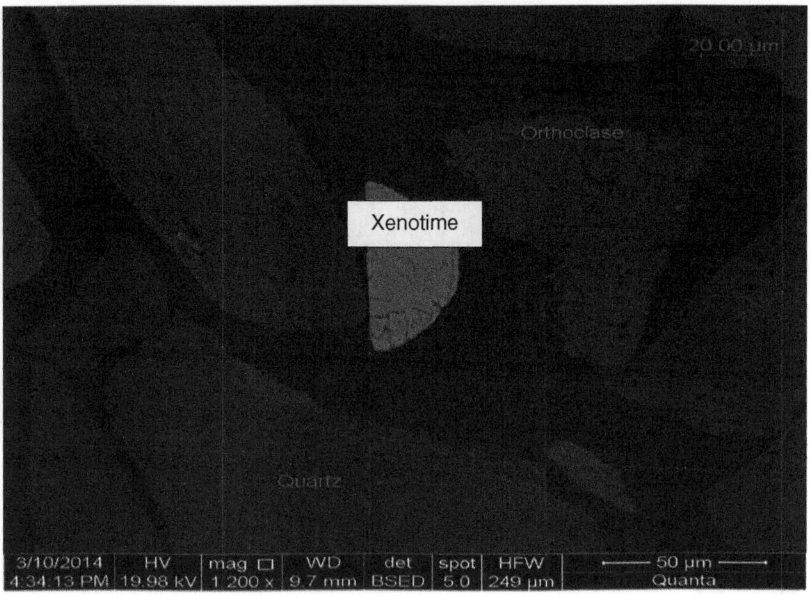

Figure 4.7 SEM-BSE image showing xenotime monomer.

and U substitutions. Figure 4.7 shows a clear monomer of a xenotime particle, while Figure 4.8 indicates inclusion of xenotime within apatite.

Zircon

Micro-area chemical analysis of zircon was also performed using EDS, and the results are shown in Table 4.23. The zircon particles do not contain many REEs, but some particles contain a minute amount of iron. Figure 4.9 shows clear zircon monomers.

Rutile, Pseudo-Rutile, and Leucoxene

Major titanium oxide minerals in the sample include rutile, pseudo-rutile, and leucoxene with pseudo-rutile being the most abundant. Again, micro-area chemical analysis of these three minerals was performed using EDS, and the results are shown in Figure 4.10 and Tables 4.24–4.26.

Phosphate Minerals

Francolite is the primary phosphate mineral with apatite being the minor one. However, MLA measurement identifies minerals based on chemical components, thus it is unable to distinguish francolite from apatite. Micro-area chemical analysis of the phosphate was performed using EDS, and the results are shown in Table 4.27. All phosphate particles contain fluorine and

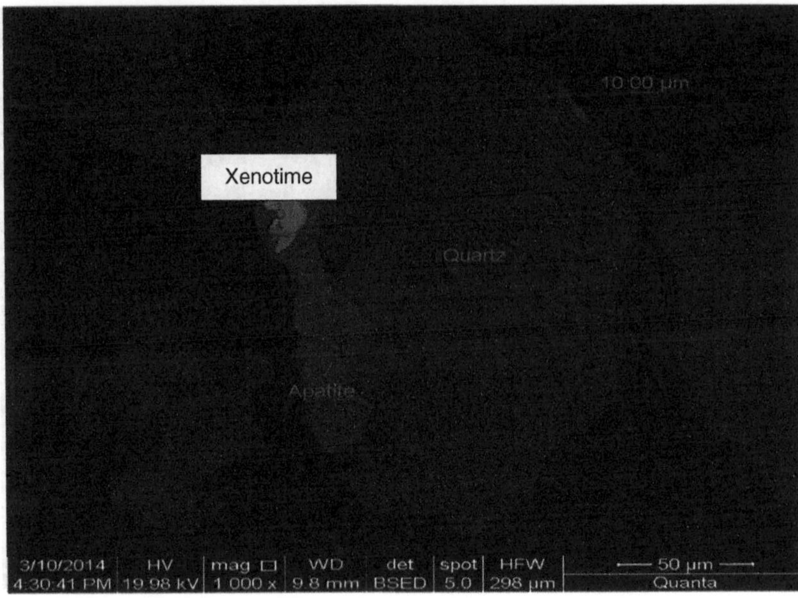

Figure 4.8 SEM-BSE image showing xenotime inclusion in quartz.

Table 4.23 Chemical components of zircon analyzed using EDS.

	Chemical components (%)			
Spectrum	ZrO_2	HfO_2*	FeO	SiO_2
1	66.20	1.15	0.00	32.65
2	66.40	1.25	0.00	32.35
3	65.93	1.14	0.03	32.90
4	66.15	1.40	0.06	32.39
5	66.17	1.42	0.00	32.40
6	65.11	1.74	0.00	33.15
7	66.45	1.05	0.00	32.50
8	66.35	1.22	0.00	32.43
9	66.37	1.27	0.00	32.35
10	65.98	1.59	0.00	32.44
Average	66.11	1.32	0.01	32.56

* Hafnium (Hf) analysis may have higher errors.

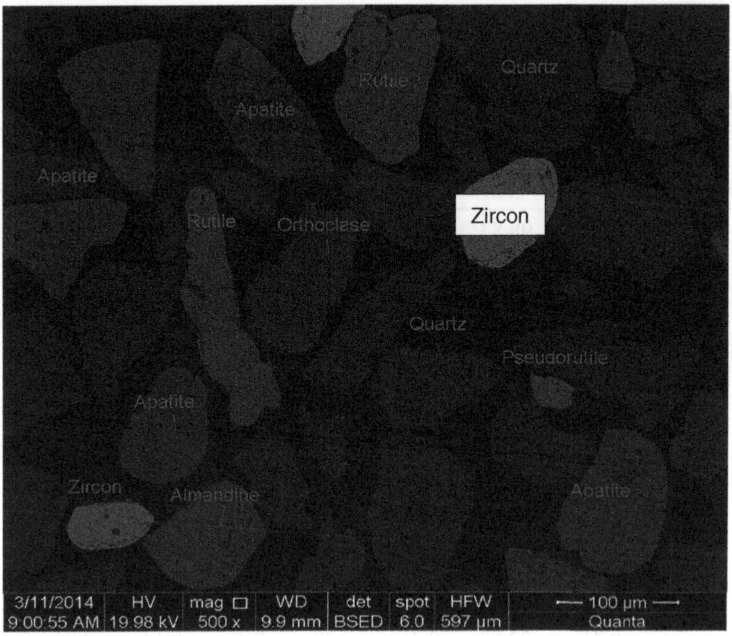

Figure 4.9 SEM-BSE image showing zircon monomer.

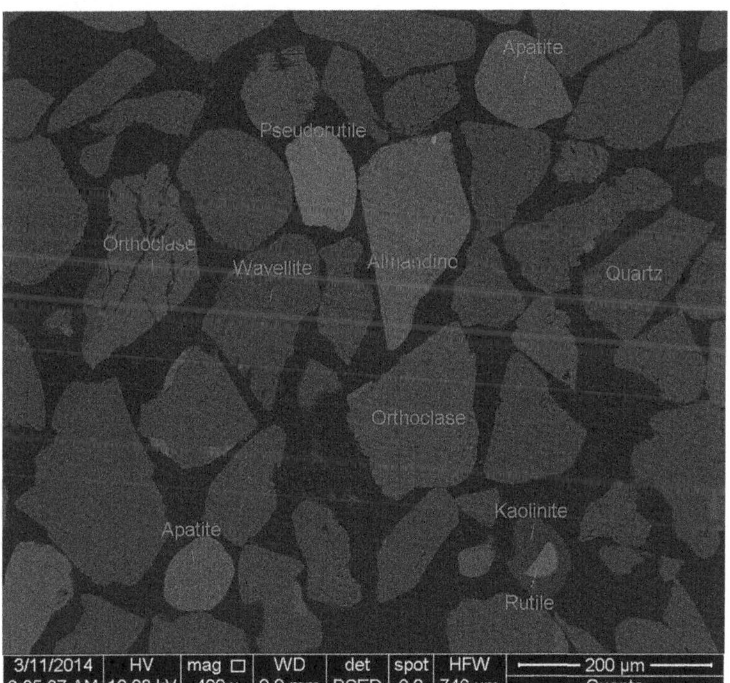

Figure 4.10 SEM-BSE image showing rutile and pseudo-rutile monomers.

Table 4.24 Chemical components of rutile analyzed using EDS.

	Chemical components (%)							
Spectrum	Nb_2O_5	TiO_2	FeO	Al_2O_3	SiO_2	CaO	Cr_2O_3	P_2O_5
1	0.47	98.83	0.13	0.08	0.49	0.00	0.00	0.00
2	0.33	98.63	0.46	0.13	0.43	0.02	0.00	0.00
3	0.08	98.58	0.54	0.16	0.46	0.19	0.00	0.00
4	0.44	98.85	0.17	0.08	0.37	0.10	0.00	0.00
5	1.08	97.71	0.53	0.16	0.50	0.02	0.00	0.00
6	0.00	99.05	0.42	0.08	0.38	0.06	0.00	0.00
7	0.00	94.58	1.26	2.05	0.48	0.16	0.78	0.69
8	0.23	95.82	0.53	0.78	0.72	1.05	0.09	0.79
9	0.21	98.40	0.12	0.54	0.26	0.12	0.23	0.13
10	0.28	98.95	0.11	0.20	0.44	0.01	0.00	0.00
11	0.24	98.90	0.12	0.38	0.31	0.06	0.00	0.00
12	1.08	97.10	0.95	0.24	0.53	0.11	0.00	0.00
13	0.77	98.03	0.58	0.22	0.37	0.03	0.00	0.00
14	0.51	98.42	0.51	0.20	0.34	0.02	0.00	0.00
Average	0.41	97.99	0.46	0.38	0.43	0.14	0.08	0.12

Table 4.25 Chemical components of pseudo-rutile analyzed using EDS.

	Chemical components (%)									
Spectrum	TiO_2	FeO	MgO	P_2O_5	MnO	CaO	ZnO	Al_2O_3	SiO_2	SO_3
1	60.59	37.50	0.00	0.01	0.31	0.05	0.77	0.26	0.51	0.00
2	67.29	30.82	0.07	0.11	0.94	0.11	0.00	0.24	0.41	0.01
3	75.16	21.27	0.12	0.66	0.92	0.60	0.00	0.76	0.50	0.02
4	61.90	33.58	1.05	0.00	0.92	0.25	0.00	0.52	1.53	0.26
5	62.94	33.55	0.04	0.08	2.38	0.15	0.00	0.30	0.57	0.00
6	69.09	28.64	0.03	0.07	1.13	0.20	0.00	0.20	0.64	0.00
7	63.58	32.64	0.14	0.02	2.02	0.12	0.00	0.26	0.58	0.64
8	62.49	34.56	0.37	0.00	1.48	0.10	0.00	0.24	0.64	0.12
9	71.38	25.70	0.18	0.19	1.10	0.16	0.24	0.47	0.56	0.01
10	64.28	32.39	0.07	0.06	2.26	0.09	0.00	0.26	0.59	0.00
11	59.56	37.94	0.38	0.10	0.76	0.07	0.00	0.44	0.70	0.05
12	65.94	31.25	0.07	0.00	1.95	0.12	0.00	0.28	0.40	0.00
13	66.89	29.96	0.14	0.10	2.00	0.09	0.00	0.28	0.49	0.06
Average	65.47	31.52	0.20	0.11	1.40	0.16	0.08	0.35	0.62	0.09

Table 4.26 Chemical components of leucoxene analyzed using EDS.

Species	Chemical components (%)											
	Nb_2O_5	TiO_2	MnO	FeO	CaO	MgO	Cr_2O_3	Cl	Al_2O_3	SiO_2	P_2O_5	SO_3
1	0.28	67.18	2.20	25.82	0.19	0.21	0.35	0.00	1.22	2.03	0.52	0.00
2	0.00	72.73	3.85	20.97	0.16	0.25	0.21	0.13	0.63	0.79	0.28	0.00
3	0.00	78.88	0.64	17.74	0.88	0.11	0.02	0.12	0.90	0.34	0.39	0.00
4	0.00	72.03	3.75	22.75	0.13	0.14	0.15	0.00	0.41	0.48	0.14	0.00
5	0.09	78.75	0.61	17.62	0.86	0.12	0.09	0.08	0.87	0.37	0.41	0.11
6	0.07	85.57	0.11	11.39	0.18	0.06	0.39	0.00	1.12	0.47	0.62	0.00
7	0.00	68.00	0.31	4.58	15.85	0.09	0.00	0.02	0.47	0.37	9.90	0.40
8	0.00	63.60	2.59	30.95	0.19	0.22	0.00	0.00	1.52	0.48	0.45	0.00
9	0.03	77.23	0.52	19.86	0.22	0.10	0.26	0.00	0.77	0.51	0.49	0.00
10	0.00	68.76	0.26	3.88	15.28	0.00	0.01	0.02	0.12	0.71	9.65	1.31
11	0.00	67.26	3.45	25.95	0.31	0.23	0.25	0.00	1.32	0.53	0.70	0.00
Average	0.04	72.73	1.66	18.32	3.11	0.14	0.16	0.03	0.85	0.64	2.14	0.17

Table 4.27 Chemical components of phosphate analyzed using EDS.*

Spectrum	Chemical components (%)													
	F	Cl	Na_2O	MgO	Al_2O_3	SiO_2	P_2O_5	SO_3	K_2O	CaO	FeO	MnO	SrO	Ce_2O_3	Nd_2O_3
1	2.21	0.00	0.44	0.27	0.94	3.42	38.91	2.29	0.50	49.56	1.47	0.00	0.00	0.00	0.00
2	2.41	0.00	0.09	0.04	0.62	2.22	41.56	0.40	0.36	51.63	0.67	0.00	0.00	0.00	0.00
3	2.65	0.00	0.44	0.24	1.50	3.18	39.41	0.74	0.38	50.74	0.45	0.27	0.00	0.00	0.00
4	2.37	0.00	0.37	0.13	1.15	0.83	42.11	0.47	0.22	51.73	0.63	0.00	0.00	0.00	0.00
5	2.47	0.00	0.39	0.24	1.69	1.94	37.50	4.68	0.36	48.10	2.64	0.00	0.00	0.00	0.00
6	2.57	0.00	0.59	0.29	1.36	3.12	39.22	1.14	0.15	50.96	0.60	0.00	0.00	0.00	0.00
7	2.50	0.00	0.27	0.19	1.94	3.20	40.76	0.64	0.13	49.89	0.47	0.00	0.00	0.00	0.00
8	2.17	0.00	0.85	0.27	1.52	1.04	40.62	1.04	0.16	48.91	0.94	0.05	2.44	0.00	0.00
9	2.34	0.00	1.03	0.70	1.10	1.45	40.72	2.14	0.19	49.56	0.76	0.00	0.00	0.00	0.00
10	2.56	0.00	0.42	0.16	0.44	1.39	40.61	0.91	0.27	52.40	0.84	0.00	0.00	0.00	0.00
11	2.51	0.10	0.87	0.17	0.59	0.81	39.98	2.10	0.30	50.41	0.31	0.00	1.84	0.00	0.00
12	2.53	0.00	0.47	0.18	1.74	1.37	40.94	1.07	0.29	50.91	0.50	0.00	0.00	0.00	0.00
13	2.33	0.00	0.30	0.26	2.14	4.03	37.89	2.90	0.46	48.02	1.66	0.00	0.00	0.00	0.00
14	2.08	0.00	0.29	0.13	0.77	1.75	40.10	2.47	0.39	48.21	1.50	0.07	1.81	0.23	0.20
15	2.47	0.00	0.40	0.26	1.46	3.69	39.77	0.68	0.22	50.49	0.55	0.00	0.00	0.00	0.00
16	2.30	0.00	0.42	0.30	1.29	2.36	36.32	6.25	0.41	46.48	3.88	0.00	0.00	0.00	0.00
17	2.44	0.00	0.61	0.43	2.19	5.17	37.19	1.14	0.62	49.54	0.67	0.00	0.00	0.00	0.00
18	2.45	0.00	0.34	0.19	0.36	1.43	41.59	0.69	0.09	52.43	0.43	0.00	0.00	0.00	0.00
19	2.44	0.00	0.62	0.21	0.58	1.40	39.06	2.54	0.32	51.23	1.60	0.00	0.00	0.00	0.00
20	2.45	0.00	0.81	0.20	0.62	2.90	38.33	2.57	0.07	50.81	1.25	0.00	0.00	0.00	0.00
21	2.42	0.00	1.32	1.29	0.98	5.16	35.62	2.80	0.39	47.97	2.06	0.00	0.00	0.00	0.00
22	2.32	0.00	0.15	0.09	0.67	2.34	41.66	0.43	0.04	51.66	0.66	0.00	0.00	0.00	0.00
23	2.75	0.00	0.58	0.25	1.19	1.35	40.30	1.10	0.17	51.97	0.33	0.00	0.00	0.00	0.00
24	2.44	0.00	0.28	0.10	1.44	1.00	42.26	0.97	0.05	51.01	0.45	0.00	0.00	0.00	0.00
Avg	2.42	0.00	0.51	0.27	1.18	2.36	39.68	1.76	0.27	50.19	1.06	0.02	0.25	0.01	0.01

* MLA could not tell the difference between francolite and apatite.

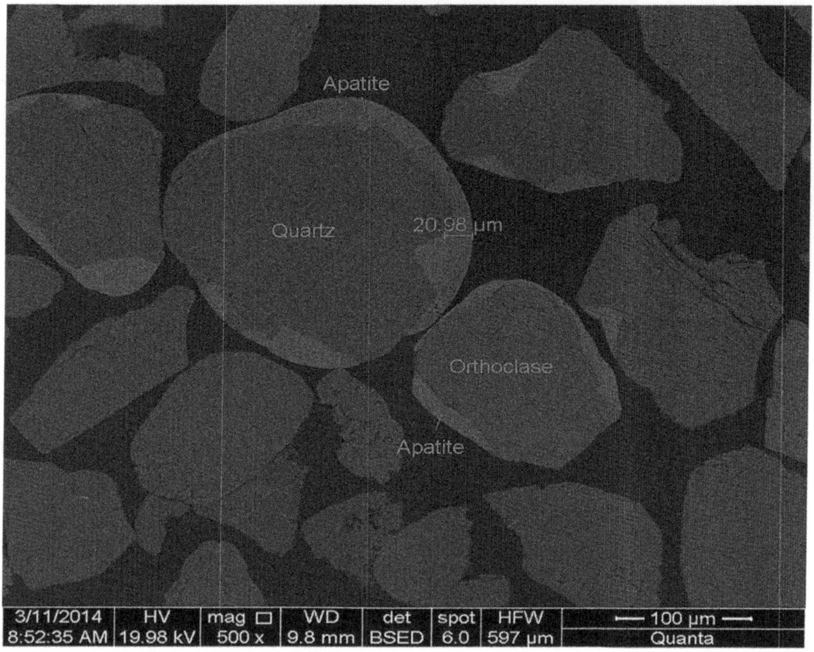

Figure 4.11 SEM-BSE image showing wrapping of quartz by phosphate.

substantial amounts of impurities such as silica, aluminum, and iron. A small number of francolite particles contain REEs.

In addition to francolite monomers, substantial amounts of francolite exist as shell-shaped particles covering quartz surfaces, as is shown in Figure 4.11.

4.3.2. Mineralogical Examination of Phosphate Clay

Phosphate mining in Florida has accumulated more than 2 Gt of phosphate clay (slime), containing approximately 600 Mt of phosphate rock, 600 kt of REEs, and 80 million kg of uranium (Zhang et al., 2017). However, the recovery of these valuable elements from the phosphate clay is not easy for the following reasons: (1) the waste clay coming from the beneficiation plant is very dilute, with an average solids concentration of 3%; (2) the material is extremely fine in size, with more than 50% having a particle size smaller than 2 µm; (3) about 50% of the material are clay minerals; and (4) large amounts of magnesium, iron, and aluminum are present in the phosphate clay.

Table 4.28 Mineralogical composition of a clay sample.

Phase	Chemical formula	Weight %
Quartz	SiO_2	10.7
Apatite	$Ca_5P_3O_{12}(OH, F)$	15.6
Wavellite	$Al_3P_2O_{16}H_{13}$	1.2
Dolomite	$CaMgC_2O_6$	5.4
Kaolinite	$Si_2Al_2O_9H_4$	8.8
Montmorillonite	$Al_4Si_8CaO_{24}H_2$	2.0
Illite	$Al_4Si_2KO_{12}H_3$	0.6
Amorphous content (mainly clay)		55.7

X-ray diffraction analysis shows that the slime sample contains over 60% clay minerals, making it extremely challenging to recover the rare earths and phosphorus values (see Table 4.28).

4.3.3. Mineralogical Examination of Phosphoric Acid Sludge

The solid fraction of phosphoric acid sludge is mainly gypsum with REEs existing as monazite and xenotime. Figure 4.12 shows an X-ray diffraction (XRD) graph of the sludge solids.

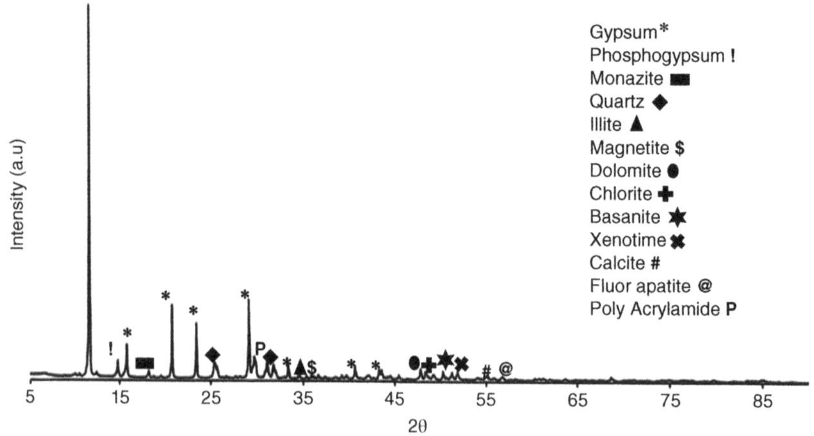

Figure 4.12 XRD peaks of phosphoric acid sludge solids.

4.4. Part III: Isolation and Characterization of Rare Earth Mineral Particles in Florida Phosphate Rock by DE Rapid Scan Radiography and HRXMT

4.4.1. Sample Collection and Characterization

Particle Size Separation

The first two samples shipped to University of Utah were the shaking table concentrate and the acid plant feed. There was about a kilogram of each sample that needed to be separated into size classes. To do this, three sieves were chosen: 140 mesh (106 μm), 200 mesh (75 μm), and 270 mesh (53 μm). A vibrating sieve shaker was used to do this separation of each sample, first as dry sieving, and then again as a wet sieving for thorough and effective separation. Four size classes were prepared: >106 μm, 75–106 μm, 53–75 μm, and <53 μm. Because the last class was so small, no further testing was done on this portion, which left three size classes to be examined for each sample.

Because of the length of time required for the double separation of dry sieving followed by wet sieving, when the third sample (phosphogypsum) arrived, a different method was applied. Coning-and–quartering method was used instead of dry separation, reducing the bulk sample in content from approximately 1 kg to a bit more than 100 g. The 100 g sample was then separated into size classes using the same wet sieving on the vibrating sieve shaker as was used for the other two samples.

Chemical Mineralogical Characterization

Before proceeding to DE radiography and HRXMT, an idea about what was present in the samples was necessary. A chemical analysis of the acid plant feed was provided, as can be seen in Table 4.29, and an approximate estimation for the shaking table concentrate, as in Table 4.30. To get a more exact concentration for the shaking table concentrate, a portion of the >106 μm sample was further separated using gravity separation by hand panning. The heavy particles were separated out from the lighter ones and taken for XRD and HRXMT. The minerals identified in the sample were zircon ($ZrSiO_4$), apatite ($Ca_5(PO_4)_3(OH,F,Cl)$), and monazite ($(Ce,La)PO_4$).

4.4.2. Dual Energy Calibration

Materials and Methods

Plastic plates, 3″ × 1″, and double-sided sticky tape were used to prepare the samples for DE radiography. Pure samples were placed on

Table 4.29 Chemical analysis of acid plant feed.

Main compounds, %		Rare earth elements and trace elements, ppm					
P_2O_5	27.36	Pr	Eu	Tb	Dy	Ho	Er
Insol	15.84	7.92	3.48	2.17	13.81	3.25	9.62
MgO	0.50	Tm	Yb	Lu	Sc	Gd	Sm
Fe_2O_3	1.19	1.22	8.69	1.34	4.90	16.46	12.04
Al_2O_3	1.19	Ce	Y	La	Nd	Th	U
CaO	38.84	131.16	132.06	77.58	85.26	9.15	87.77

Table 4.30 Provided estimated chemical analysis of shaking table concentrate.

Main compounds, %		Rare earth elements and trace elements, ppm					
P_2O_5	6–9	Pr	Eu	Tb	Dy	Ho	Er
Insol	70–80	10–16	2–3	0.5–0.7	5.0–7.5	2–3	6–11
MgO	0.03–0.08	Tm	Yb	Lu	Sc	Gd	Sm
Fe_2O_3	0.29–0.9	1.0–1.5	4.0–6.0	.10–.15	3.04.5	8–11	10–18
Al_2O_3	0.28–0.8	Ce	Y	La	Nd	Th	U
CaO	8–15	130–150	70–100	60–90	80–110	8.0–10.0	30–50

double-sided sticky tape, which was attached to a plastic plate, and then were stabilized with a second plate. For this case, two types of samples were prepared for mineral identification, which are shown in Figure 4.13. The image on the left shows the first sample with minerals used for calibration, and the image on the right shows the second sample with minerals used for calibration verification. Table 4.31 shows more detailed information about the minerals used for verification.

Reference Samples for Calibration

Reference samples were prepared for the estimation of reconstruction coefficients for effective atomic numbers using DE radiography measurements. In this case, aluminum, copper, and lead were selected as known materials. These three elements can represent the three different mineral phases usually present in samples from metal mining operations having low attenuation coefficients (gangue and quartz/silicates), medium attenuation coefficients (copper and most base metal minerals), and high attenuation coefficients (molybdenite, galena, lead, or other precious metals).

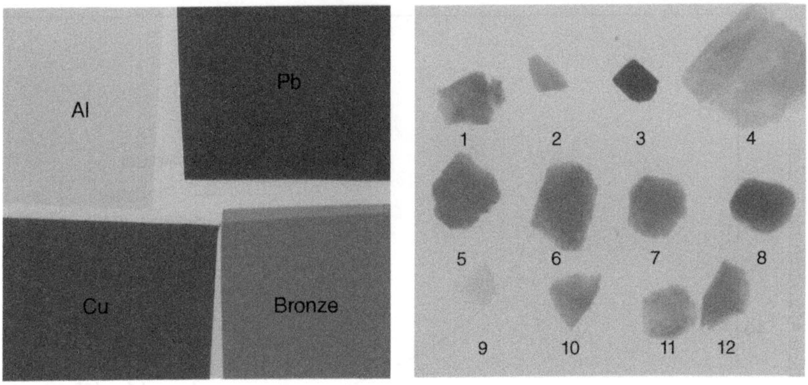

Figure 4.13 Radiographs of the pure reference minerals and minerals in Cu–Mo flotation tailing for DE calibration.

Table 4.31 Detailed information for main minerals in Cu–Mo flotation tailing.

	Mineral		Effective atomic number, (Z_{actual})	Density (g/cm³)
	Name	Formula		
1	Molybdenite	MoS_2	36.87	4.6–4.7
2	Galena	PbS_2	78.96	7.2–7.6
3	Chalcopyrite	$CuFeS_2$	25.00	4.1–4.3
4	Bornite	Cu_5FeS_4	26.58	4.9–5.3
5	Barite	$BaSO_4$	48.73	4.3–5.0
6	Malachite	$Cu_3(CO_3)_2(OH)_2$	24.78	3.6–4.0
7	Azurite	$Cu_2CO_3(OH)_2$	24.85	3.7–3.8
8	Chalcocite	Cu_2S	27.40	5.5–5.8
9	Quartz	SiO_2	11.85	2.6–2.7
10	Hematite	Fe_2O_3	23.56	5.1–5.2
11	Sphalerite	$(Zn, Fe)S_2$	27.18	3.9–4.2
12	Pyrite	FeS_2	22.06	4.9–5.1

Coefficients k_1, k_2, k_3, and k_4 were then calculated according to the equation (see Figure 4.14) using $p = 3.8$. Yielding $k_1 = -8.97 \times 10^{11}$, $k_2 = 7.95.97 \times 10^{11}$, $k_3 = 1.88 \times 10^{6}$, and $k_4 = -1.16 \times 10^{6}$.

Then, the calculated effective atomic number is compared to the actual effective atomic number to determine the accuracy in data correlation and verify calibration, shown in Figure 4.14.

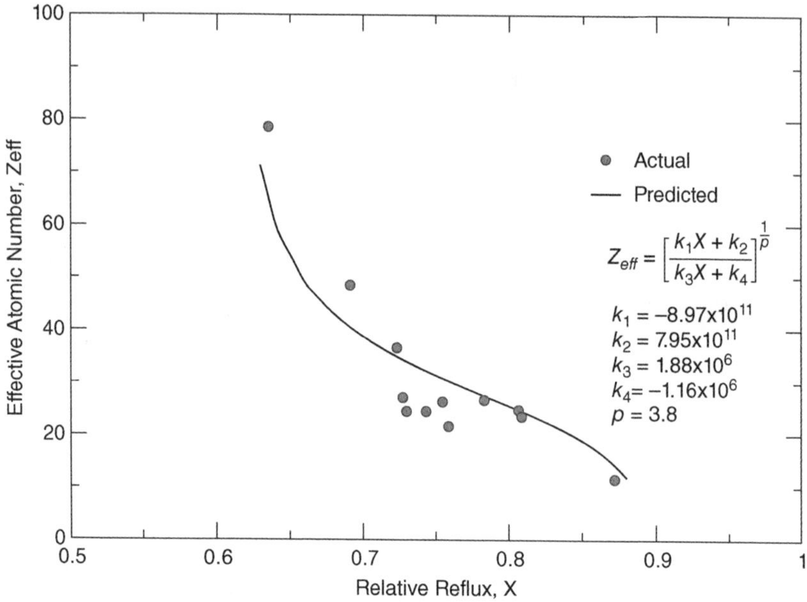

Figure 4.14 Comparison between calculated and actual effective atomic numbers.

4.4.3. High-Resolution X-Ray Microtomography Calibration

Estimation of Mineral Attenuation Coefficients

It is important to estimate the linear attenuation coefficient before the scan has been taken in order to use the proper settings for the scans. The mass attenuation coefficient has a linear relationship with energy according to Beer's law. Scans for heavy materials with high density and high atomic number require a high-level X-ray energy source in order to get enough X-ray photons to the detector. Therefore, it is very important to determine which level of X-ray energy is sufficient to pass through high-atomic-number and high-density rare earth samples. A preview of the linear relationship of X-ray mass attenuation coefficient and energy, using XMuDat software, is shown in Figure 4.15.

From the estimation of mineral attenuation coefficients, it is evident that the lower the photon energy, the greater the attenuation coefficient difference between minerals. This will make it easier to apply a threshold to radiograph scans and identify only potential rare earth particles. Additionally, the photon energy should be large enough to pass through the high-atomic-number and high-density minerals; for example, monazite

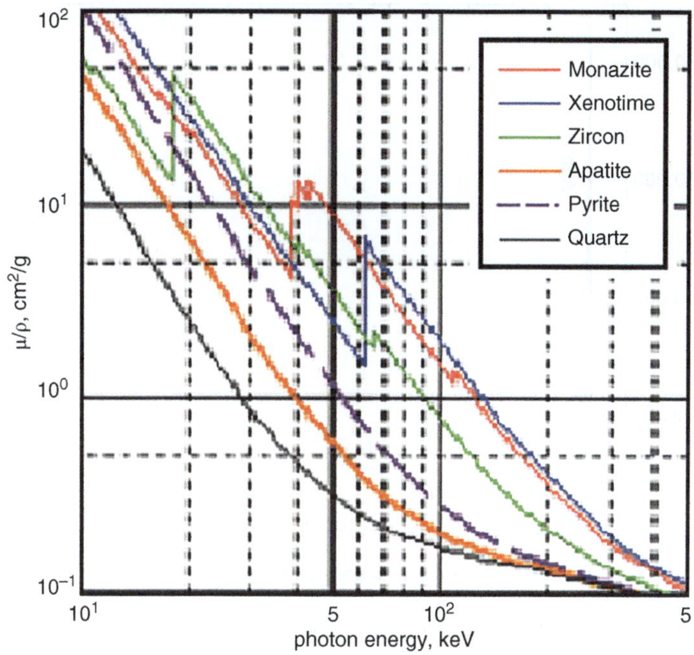

Figure 4.15 Mineral attenuation coefficients estimated using XMuDat.

and zircon. Taking both of those factors into consideration, an energy voltage 70 kV was selected to scan the samples. Other parameters are defined by considering the projection quality and include an exposure time of 10 s for each projection image and a magnification level of 4×.

4.4.4. Calibration of Mineral Standards

In order to obtain an accurate analysis of the rare earth sample, usually the computed tomography (CT) standard has to be set for the mineral characterization. Here, from the initial XRD analysis, monazite, zircon, apatite, and quartz minerals are expected to be present in the samples. Monazite is the mineral of interest as the rare earth mineral. Therefore, monazite and zircon standards were scanned using the same conditions as the rare earth sample, which are 4× magnification level, 70 kV energy voltage, and 10 s exposure time per projection.

After the reconstruction of CT standards, the CT number distribution map can be exported and transformed into an extension .txt file, which can then be plotted using EXCEL or XMGrace. The CT number

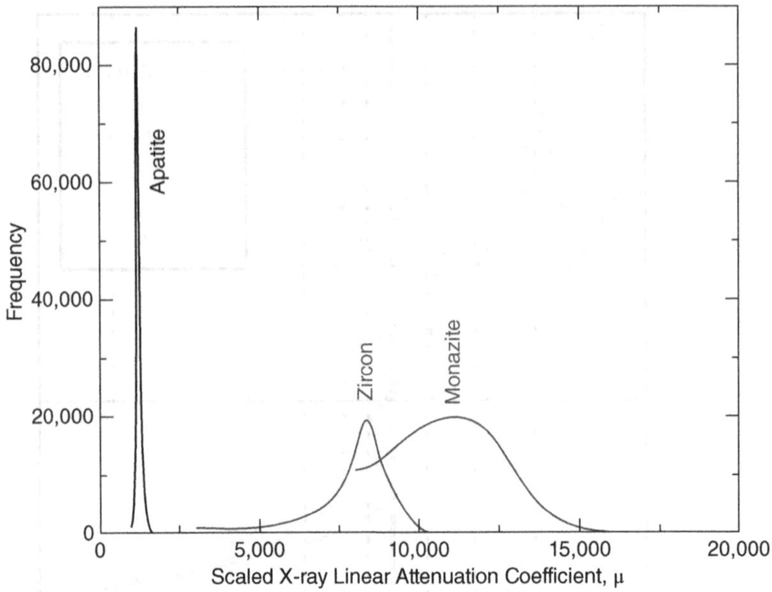

Figure 4.16 Scaled CT number distribution map for mineral phases.

distribution map of minerals is shown in Figure 4.16. From the CT number distribution map of CT standards, it can be seen that monazite has the highest attenuation coefficient and apatite has the lowest. There is a big difference between the high-density and high-atomic-number minerals and the gangue mineral (apatite). The minerals are clearly distinguished based on their attenuation coefficient. There is some overlap between monazite and zircon, but this is not a serious problem.

4.4.5. Calibration of Rare Earth Phosphate Sample

Since the CT standards of pure monazite, zircon, and apatite samples have been made, it is possible to generate CT-scaled data from the software XMController. This was done for the heavy fraction of the shaking table concentrate after scanning and the reconstruction with scaled data using the standards. The same scaling has been applied to the HRXMT samples of this project.

4.4.6. Duel Energy Radiography

In order to begin the DE radiography scans, a vibrating riffler was used to separate out a representative portion from each size class of each sample and the subsample was attached to a slide (glass plate) using double-sided sticky tape and secured using a second slide. Three slides were prepared from each size class for each sample, making a total of nine slides per sample and 27 slides in total. Each slide has roughly 34,000 particles attached to it, so a total of 102,000 particles from each size class for each sample were then imaged.

The DE radiography scans were completed by splitting the slide into 40 sections, making a 4×10 grid, and scanning each section at a low energy level (80 kV) and a high energy level (140 kV).

The attenuation coefficients recorded from the DE radiography were used to find the relative reflex, R, which then gave an effective atomic number using the calibration curve completed in Section 4.2.3. To do this, MATLAB was utilized to read each radiograph at both high and low energies and compare the two scans, which gave the relative reflex (X) pixel by pixel. Once the relative reflex was calculated, a threshold was applied to identify the potential rare earth particles. Threshold discards all the relative reflex values above a given point and accepts those below it. The threshold number used was $X = 0.70$, which corresponds to $Z_{eff} = 38$. This means that all of the values where $X < 0.70$ were accepted, giving particles with $Z_{eff} > 38$. This threshold was chosen based on the calibration curve (see Figure 4.16), and tested on a sample of pure bastnasite, seen in Figure 4.17.

Looking at the calibration curve, 0.70 might seem a little high when looking for rare earth particles. However, as the particle density increases, the 80 and 140 kV energies that were used might have some difficulties penetrating the particles. Due to this effect, the accuracy of the calibration

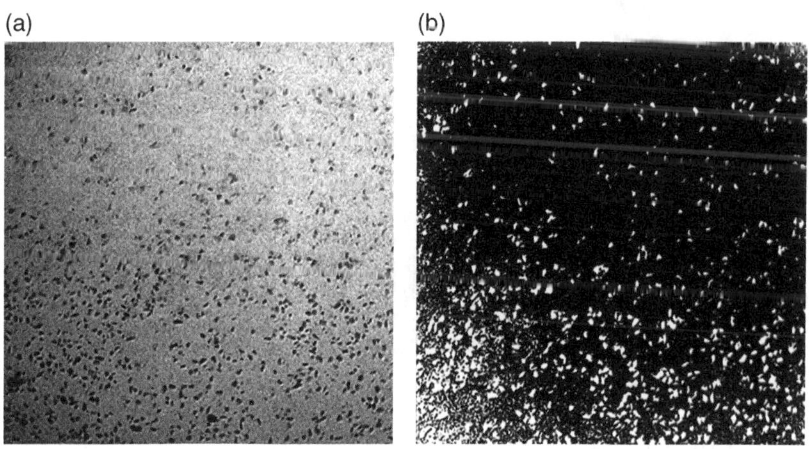

Figure 4.17 (a) Bastnasite radiograph and (b) resulting threshold image used for threshold verification.

curve as it nears higher atomic numbers decreases. Therefore, 0.70 was utilized to make sure that all the potential rare earth particles were gathered. Examples of the DE radiography scans taken at high and low energies, their corresponding relative reflex image, and the image after threshold can be seen in Figures 4.18–4.20. Note that the dark particles in the DE scans correspond to the white portions in the threshold images and are potential rare earth particles.

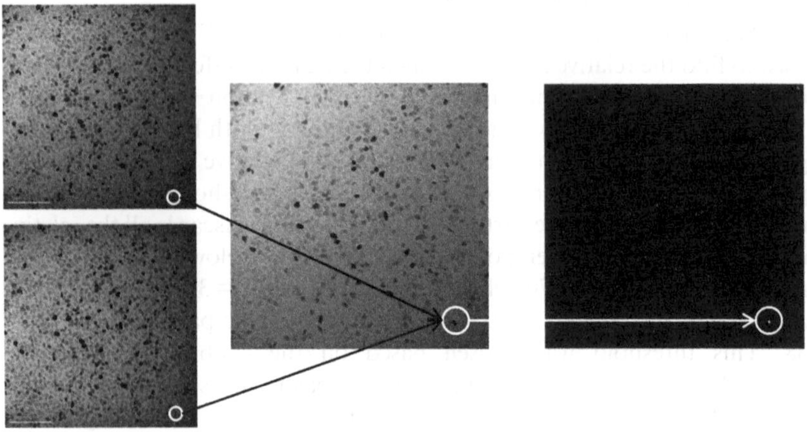

Figure 4.18 DE scans, relative reflex, and threshold image of a section from shaking table concentrate.

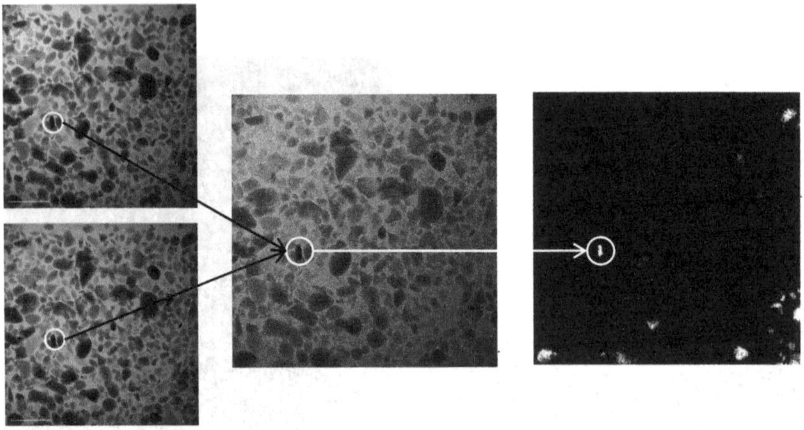

Figure 4.19 DE scans, relative reflex, and threshold image of a section from acid plant feed.

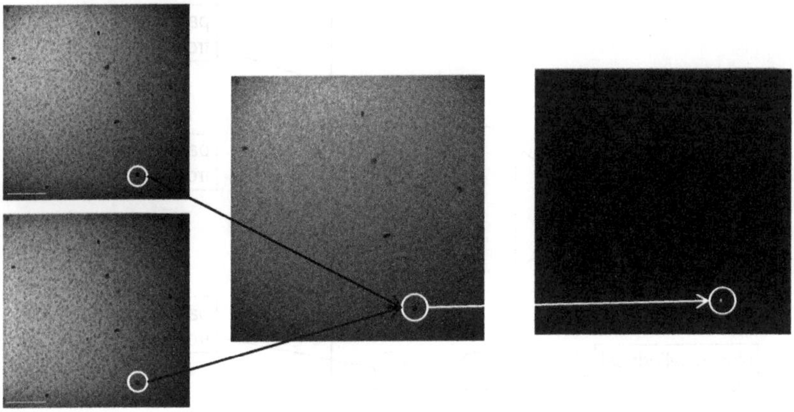

Figure 4.20 DE scans, relative reflex, and threshold image of a section from phosphogypsum.

While some of these threshold images appear to have a good number of potential rare earth particles, it is important to remember that each image is only one of 120 possible images for each size group of each sample, and the majority of the other images have no potential particles shown.

4.4.7. High-Resolution X-Ray Microtomography

The particles that were identified and isolated using DE radiography were removed and prepared for HRXMT scanning. For this portion of the test, the samples were secured in a plastic cylindrical tube. Samples from different slides were separated by a circular paper, as can be seen in Figure 4.21. The final prepared samples can be seen in Figure 4.22.

The samples were then scanned using the conditions determined in Section 4.5.4, that is, 70 kV energy and 10 s/image. 1000 projection images were taken while rotating the samples 360°. This created a series of projection images, several of which can be seen in Figure 4.23, which were then used to reconstruct a 3D digital copy of each sample using the phosphate rare earth standard created and used for calibration. These can be seen in the Results section.

4.4.8. DE Radiography Results

Out of the 27 slides scanned using DE radiography, 11 slides had no potential rare earth particles after threshold. The number of sections removed from each slide and which slides contained potential rare earth particles can be seen in Table 4.32.

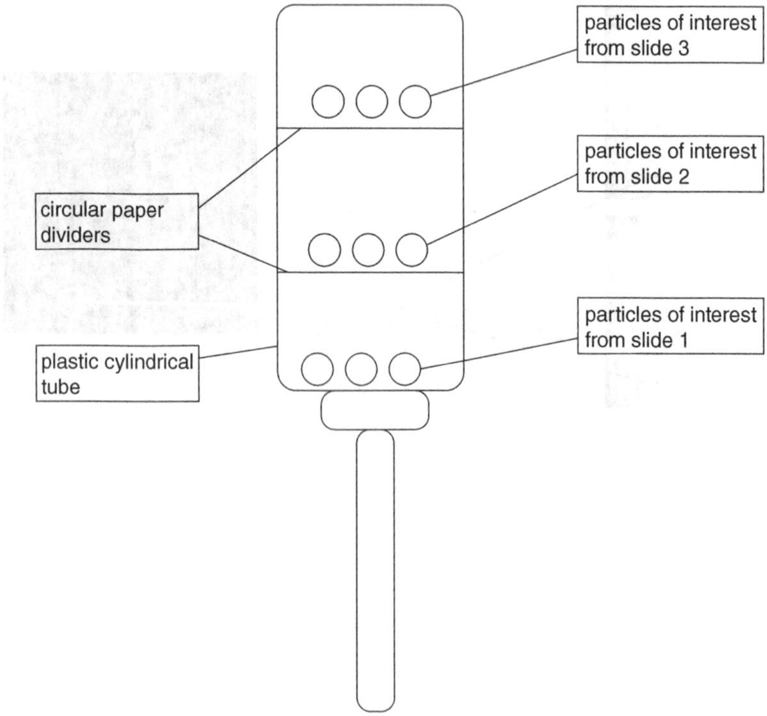

Figure 4.21 Sample preparation setup for HRXMT scanning.

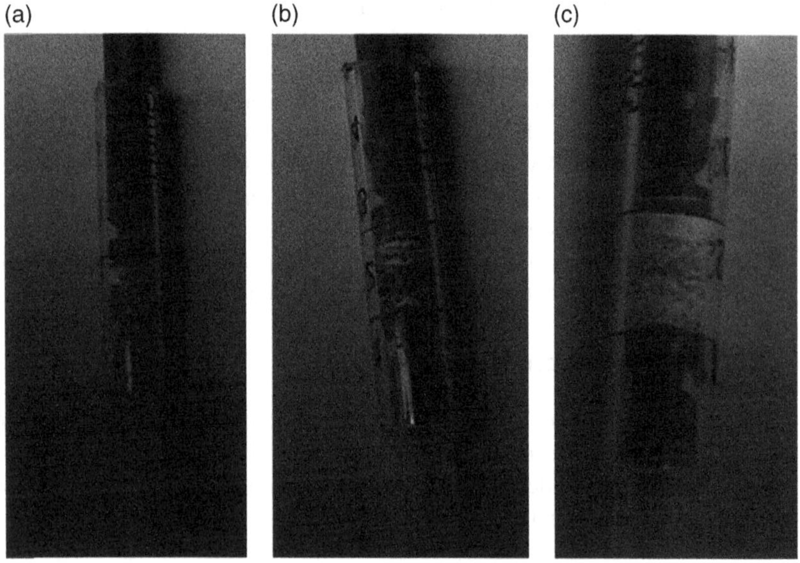

Figure 4.22 Final HRXMT samples for each (a) shaking table concentrate, (b) acid plant feed, and (c) phosphogypsum.

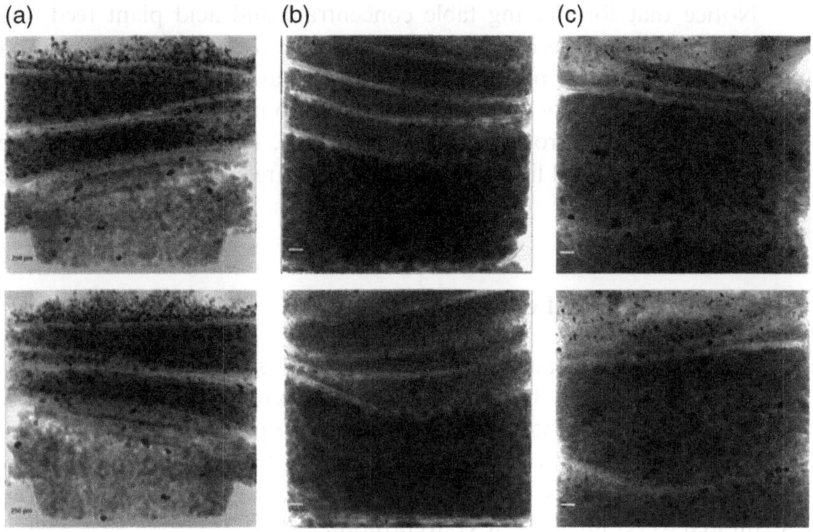

Figure 4.23 Projection images from (a) shaking table concentrate, (b) acid plant feed, and (c) phosphogypsum before reconstruction.

Table 4.32 Number of sections with potential rare earth particles on each slide.

Sample stream	Size range	Slide number	Number of sections with potential rare earth particles
Shaking table concentrate	>106 μm	2	3
Shaking table concentrate	>106 μm	3	3
Shaking table concentrate	75–106 μm	11	8
Shaking table concentrate	75–106 μm	13	5
Shaking table concentrate	53–75 μm	22	1
Acid plant feed	>106 μm	32	2
Acid plant feed	>106 μm	33	8
Acid plant feed	75–106 μm	41	8
Acid plant feed	75–106 μm	42	6
Acid plant feed	75–106 μm	43	3
Acid plant feed	53–75 μm	53	10
Phosphogypsum	>106 μm	61	11
Phosphogypsum	>106 μm	62	14
Phosphogypsum	75–106 μm	71	1
Phosphogypsum	75–106 μm	72	7
Phosphogypsum	53–75 μm	82	12

Notice that for shaking table concentrate and acid plant feed, the majority of the sections contain particles in the size range 75–106 μm and only a small amount of particles are in the size class 53–75 μm, and that the phosphogypsum had the most particles come from the size range >106 μm and the least amount from the 75–106 μm range. This is important for the possible size range and liberation analysis for rare earth particles in those sample streams.

4.4.9. HRXMT Scan Results

The 3D digital samples can be viewed as 2D slices, as seen in Figure 4.24, or as full 3D renderings, as in Figure 4.25. Using a digital viewing program, Drishti, the 3D images can be put on threshold to view their individual mineral breakdown, shown in Figures 4.26–4.28. When imported to ImageJ, the number of particles past a certain threshold can be counted to get an accurate concentration.

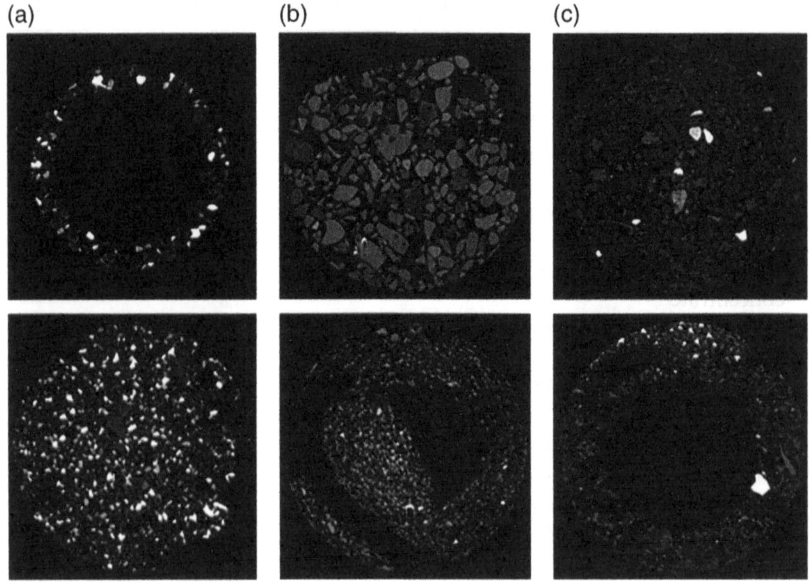

Figure 4.24 2D slice from (a) shaking table concentrate, (b) acid plant feed, and (c) phosphogypsum after reconstruction.

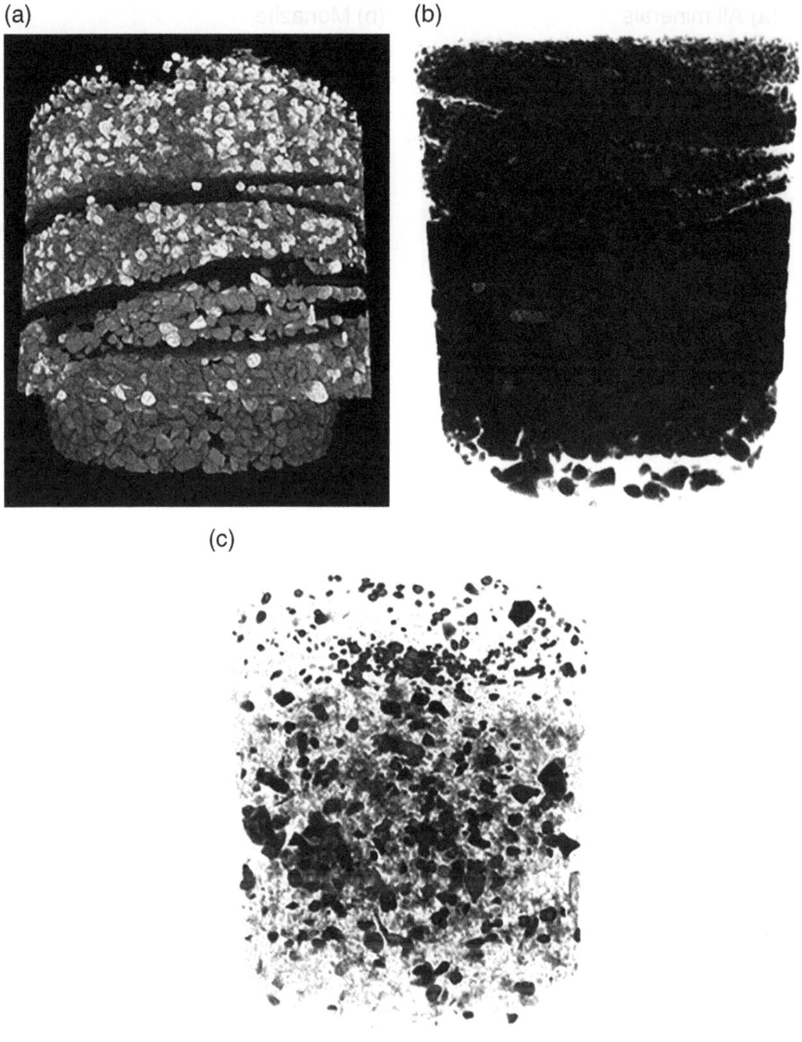

Figure 4.25 3D reconstruction of (a) shaking table concentrate, (b) acid plant feed, and (c) phosphogypsum.

4.4.10. Particle Analysis from HRXMT

The final mineral count per sample after reconstruction was found using the 3D Object Counter plugin through ImageJ and using the threshold values per mineral found in Figure 4.18. The final mineral count can be found in Table 4.33.

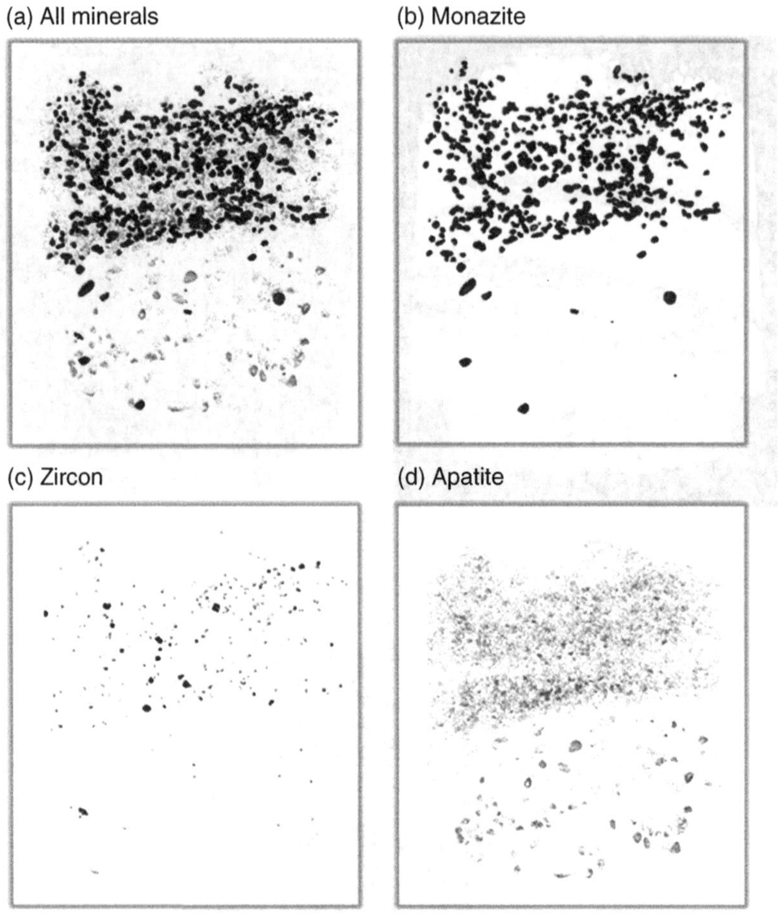

Figure 4.26 Shaking table concentrate 3D reconstruction by mineral composition.

During the final threshold of the particles, there was some inevitable overlap between the monazite CT numbers and the zircon CT numbers, as shown in Figure 4.18. There could be some error in the particles identified as monazite. Per sample, 18 particles from the shaking table concentrate, 5 particles from the acid plant feed, and 7 particles from phosphogypsum belong to this somewhat nebulous zone. However, as these are at the very edge of the zircon identification zone, where the CT number is between 9,000 and 10,000, it was ultimately decided that these particles are most likely monazite and so they were counted in the rare earth particle

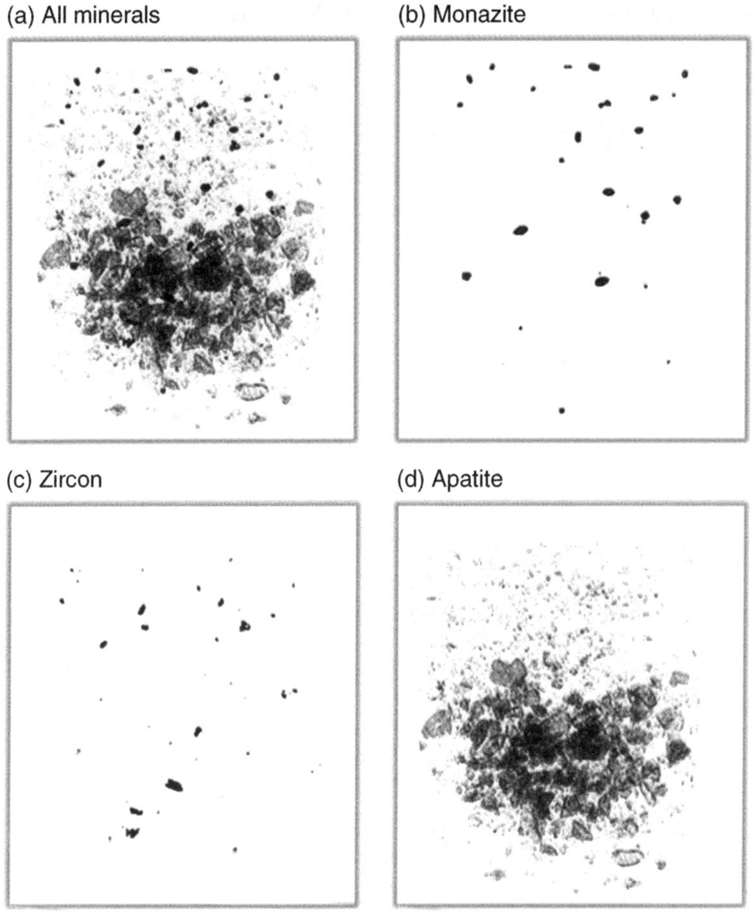

Figure 4.27 Acid plant feed 3D reconstruction by mineral composition.

category. Of those identified as zircon, 164 particles from the shaking table concentrate, 26 particles from the acid plant feed, and 28 particles from phosphogypsum also belong to the region belonging to both monazite and zircon. However, as the CT numbers in this range, between 8,000 and 9,000, are closer to the peak of zircon and only belong to the edge of monazite, the particles can be reasonably identified as zircon and not monazite. Therefore, the ppm count shown in Table 4.32 can be reasonably concluded as accurate.

Using ImageJ, a visual examination of each sample can determine the degree of liberation of the monazite particles. Because the DE slides were

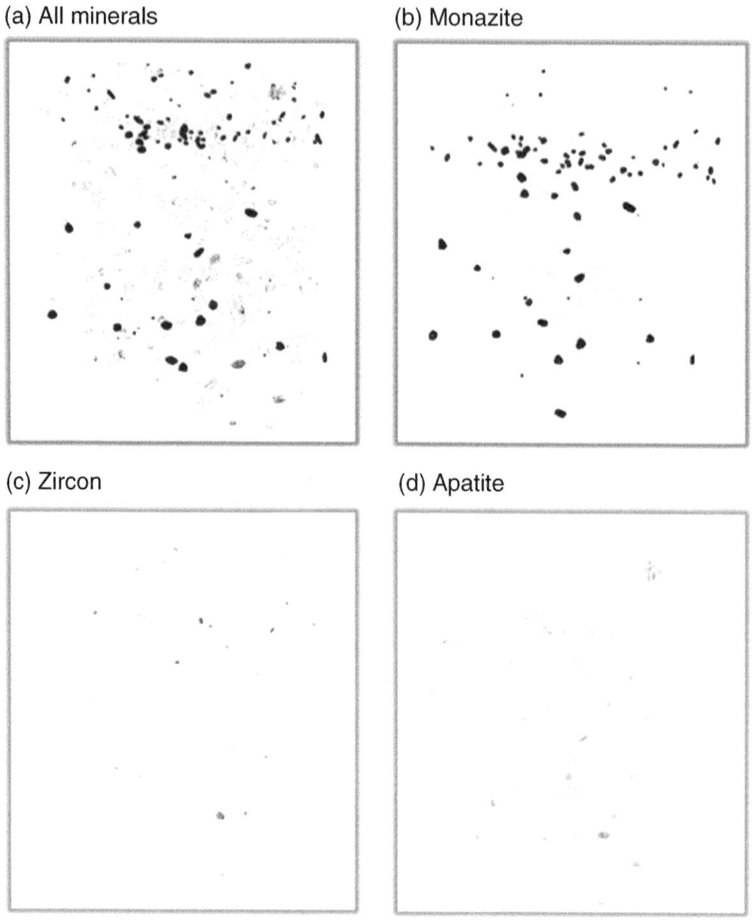

Figure 4.28 Phosphogypsum 3D reconstruction by mineral composition.

Table 4.33 Final monazite count per sample.

	Number of rare earth particles	Rare earth particle concentration (ppm)
Shaking table concentrate	660	2,157
Acid plant feed	32	104
Phosphogypsum	87	284

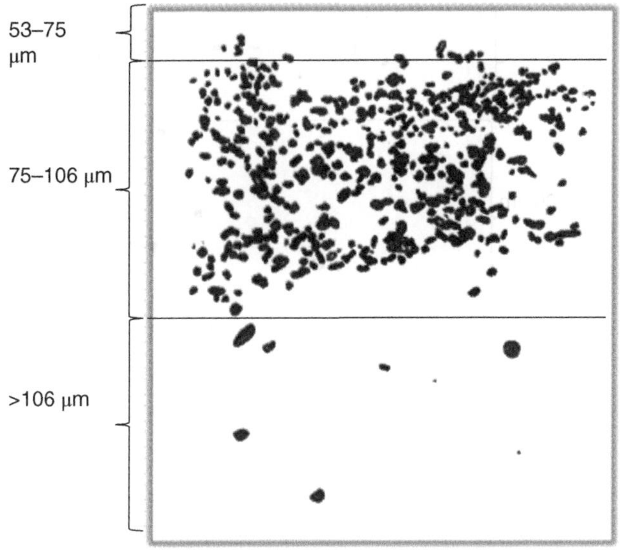

Figure 4.29 Monazite in 3D reconstruction of shaking table concentrate separated approximately by size class.

separated in the HRXMT samples by paper, it is possible to look at each size class individually to examine the degree of liberation for each size class of each sample. For the shaking table concentrate, the monazite particles in all three size classes are fully liberated. Additionally, reexamining the mineral breakdown by approximate size classes, found in Figures 4.29–4.31, notice that in the shaking table concentrate the majority of the monazite is found at the top of the sample. All three HRXMT samples were prepared so that the largest particles were found at the bottom and the smallest at the top. However, because the shaking table concentrate only contained one section of one slide in the size range 53–75 μm, most of these particles near the top of the sample were located on the slides with particle size 75–106 μm. This means that the majority of the monazite particles can be found in this size range and fully liberated, making it the optimum size for grinding and processing.

The phosphogypsum is liberated to a similar degree. In the size range >106 μm, the majority of the monazite particles are fully liberated, though with a few particles that are partially locked, making about a 95% degree of liberation. For the size classes 75–106 μm and 53–75 μm, all the monazite particles are fully liberated. The phosphogypsum sample (Figure 4.31) has a similar grouping as the shaking table concentrate further from the

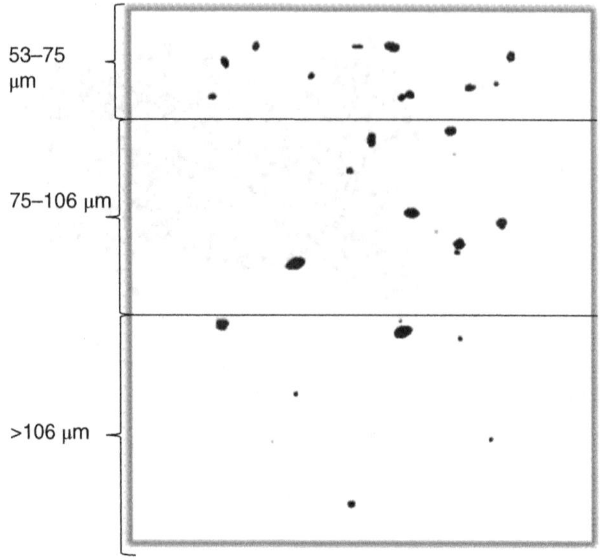

Figure 4.30 Monazite in 3D reconstruction of acid plant feed separated approximately by size class.

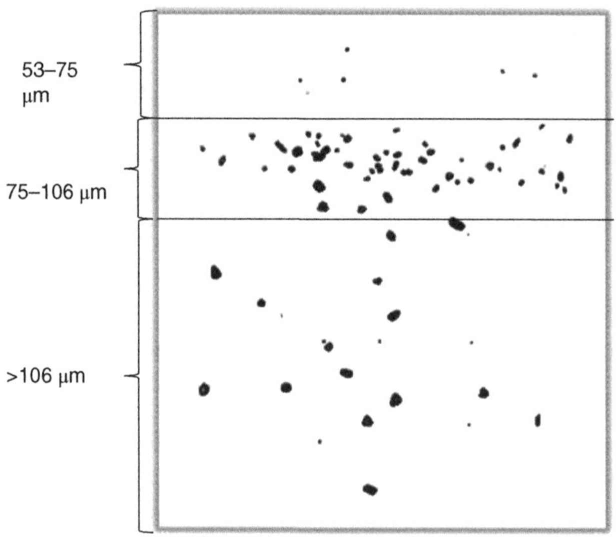

Figure 4.31 Monazite in 3D reconstruction of phosphogypsum separated approximately by size class.

top of the sample, where particles from the 75–106 μm were located as well. However, there are many particles grouped together throughout the rest of this sample as well, especially in the >106 μm range. Because the majority of the rare earth particles are liberated as well, the optimum size range cannot be stated as clearly as the shaking table concentrate, but should be in the approximate range of 75–106 μm, although a larger size could reasonably be considered.

The acid plant feed had a bit different liberation analysis than the other two samples. For the size class >106 μm, the majority of the monazite particles are not liberated at all. In fact, the rare earth particles are about 95% fully locked. An example of this can be seen in the 2D slice of the sample shown in Figure 4.32. However, in the smaller size ranges, the monazite is nearly fully liberated. As with the other two samples, only a small amount of the HRXMT sample came from the size range 53–75 μm, therefore making this size range not ideal for grinding, even though monazite particles in this size range are fully liberated. The acid plant feed (Figure 4.30) has a much lower concentration of monazite than the other two samples and it is concentrated relatively uniformly throughout the sample and throughout the size groups. Because the particles >106 μm are not liberated, but not many particles from 53–75 μm were potential rare earth minerals, the size class 75–106 μm should be considered as a good size range for monazite retrieval.

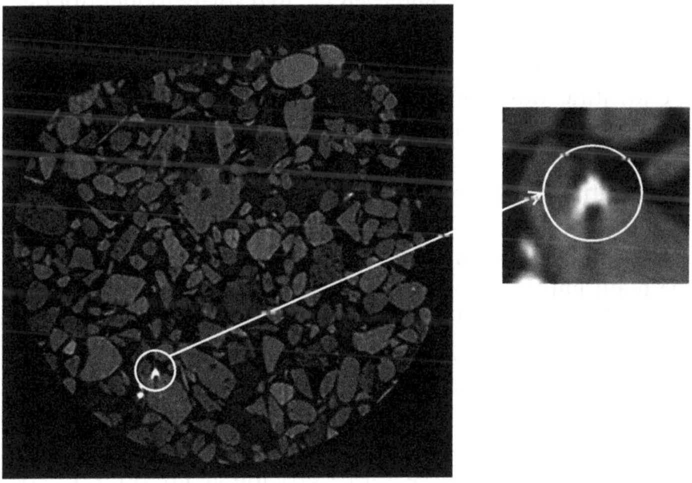

Figure 4.32 A locked particle of acid plant feed found in the size class >106 μm.

4.4.11. Conclusions and Recommendations

Using DE radiography and HRXMT, the concentrations of rare earth minerals in the samples provided were as follows: 2,157 ppm in the shaking table concentrate, 104 ppm in the acid plant feed, and 284 ppm in phosphogypsum. Judging from the degree of liberation for each sample and size range, the best particle size to find fully liberated monazite particles is 75–106 μm, although other sizes can be reasonably considered for the acid plant feed and phosphogypsum.

While the concentrations follow the trend from the chemical analysis, meaning that the shaking table concentrate has a higher concentration than the acid plant feed, the concentration for the acid plant feed is much lower than anticipated. Part of this error can be explained due to a portion of all the samples, the particles that were <53 μm, that were not being included in the analysis. However, some error could be caused by rare earth particles not being identified on the DE slides and so were not transferred to the HRXMT sample. For this to happen, there could have been a manual error in the physical process of removing the particles or an error due to an incorrect threshold value. The lower than expected concentration could be caused by any of the three described scenarios or all three possibilities compounded.

From these results, it can be concluded that DE radiography followed by HRXMT scanning is an effective and efficient method for resource identification, particularly for rare earth mineral identification. Based on the accuracy of DE radiography correctly identifying potential rare earth particles, DE radiography could even be used solely to identify rare earth particles given a machine that can produce X-rays at a sufficiently high energy level. Because of the amount of information this method of resource identification can provide about the individual particles in each sample, in addition to the level of accuracy HRXMT has due to its three-dimensional nature, it is recommended that DE radiography be used to semi-quantitatively identify minerals of interest followed by a thorough particle analysis from HRXMT scanning and reconstruction in order to isolate and characterize rare earth particles.

The main limitation of this method is the amount of time required for HRXMT scanning. This is reduced by using DE radiography to "pre-concentrate" the HRXMT samples with particles more likely to be rare earth minerals, but the time still required to scan enough particles to gain a reliable accuracy is not insignificant. However, if an X-ray source could produce X-rays of high enough energy level, identification confirmation by HRXMT would not be a necessary follow-up to DE radiography. Rare earth particle identification could be done in a matter of minutes instead of hours or days. In this sense, the project is further limited by the maximum energy level the X-ray machine can produce. As demonstrated in previous literature (Lin, Hsieh, Tserendagva, & Miller, 2013), the maximum effective atomic number that can be accurately identified using solely DE radiography with the machine used in this project is an atomic number of 40 and so HRXMT verification is necessary.

Despite these limitations, this project has demonstrated that DE radiography followed by HRXMT is still a more effective method for particle identification and liberation analysis than other methods available. Using DE radiography first reduces the time necessary for just HRXMT scanning, and using HRXMT for particle verification compensates for insufficient energy levels during DE radiography. It is recommended that this method be used in the future for rare earth identification and analysis.

4.5. Part IV: Process Development on the Concentration of REE-containing Materials and Extraction of REEs

4.5.1. Extraction of Rare Earths from PG

REE in Phosphogypsum from Wet Phosphoric Acid Process Production

The "wet acid" process usually refers to the manufacturing of phosphoric acid by reacting phosphate rock with sulfuric acid. The first phosphoric acid plant was built in Germany in about 1870 and in the USA in 1890 (Beaton, 2010). The past decade has seen many new phosphoric acid plants constructed in major phosphate-producing countries such as Morocco, China, and the Middle East. The primary chemical reaction in the "wet acid" process may be expressed in the following equation using fluorapatite to represent phosphate rock:

$$Ca_{10}F_2(PO_4)_6 + 10H_2SO_4 + 10nH_2O \rightarrow 10CaSO_4 \cdot nH_2O + 6H_3PO_4 + 2HF. \quad (1)$$

Depending on the value of n, the process is defined as the di-hydrate (DH) ($n = 2$) process, hemi-hydrate ($n = 1/2$) process, or anhydrate process. The term $CaSO_4 \cdot nH_2O$ in the equation is simply referred to as phosphogypsum.

The DH process is used most widely in the world, and is the main wet acid process practiced in the United States. It was reported that the DH process required a low capital cost, with a very low production cost and great flexibility in using various qualities of phosphate rock. One distinguishable advantage of the process is its capability of producing an acid from which uranium can be extracted easily. This process is generally designed for a 24–28% P_2O_5 acid.

Under the CMI project, FIPR collected and characterized numerous phosphogypsum and phosphoric acid samples. Results show that during the wet process, approximately 70% of the REEs in the phosphate rock feed report to phosphogypsum, with the remaining 30% mainly ending up in the filter acid.

Leaching of Phosphogypsum and Acid Sludge

Considering the easy integration of the sulfuric acid system with the existing phosphoric acid manufacturing process, phosphogypsum leaching focused on leaching with sulfuric acid solutions. Experimental results indicated that the REE leaching efficiency increased rapidly, reached a maximum, and then began to decrease with sulfuric acid concentrations ranging from 0% to 10% and at temperatures ranging from 20 °C to 70 °C. At a sulfuric acid concentration of 5% and leaching temperature of 50 °C, the REE leaching efficiency obtained a maximum value of approximately 43%. Increasing the leaching time or liquid/solid ratio further increased the leaching efficiency. Leaching recovery of over 80% was achieved by countercurrent leaching of the phosphogypsum. An intensive leaching study also showed that a 2.5% sulfuric acid solution was as effective as a 5% solution.

Three sets of leaching tests were conducted. The first set of tests was designed to study the effect of acid concentration using sulfuric acid solutions of 0%, 1%, 2.5%, 5%, and 10% sulfuric acid by weight. These tests were conducted in the abovementioned glass reactor, each with 500 g of phosphogypsum at two temperature levels (room temperature and 75 °C), respectively. Five samples of the leachate were taken at leaching time intervals of 5, 10, 30, 60, and 120 min, with each sample analyzed for REE content to calculate recovery and plot leaching kinetics curves.

Another set of tests was stage wise washing and leaching to maximize REE recovery by the following sequence:

5% (or 2.5% depending on results from the first set of tests) sulfuric acid solution; room temperature, washing for 10–20 minutes.

5% sulfuric acid solution; room temperature, leaching for about 60 min.

23% sulfuric acid solution; room temperature, leaching for about 60 min.

93% sulfuric acid solution; 230 °C, leaching for about 60 min.

The third set of tests was the development of a countercurrent leaching flowchart, as shown in Figure 4.33. The numbers in this flowchart represent phosphoric acid sludge, which will be discussed later, but the concept is the same for PG.

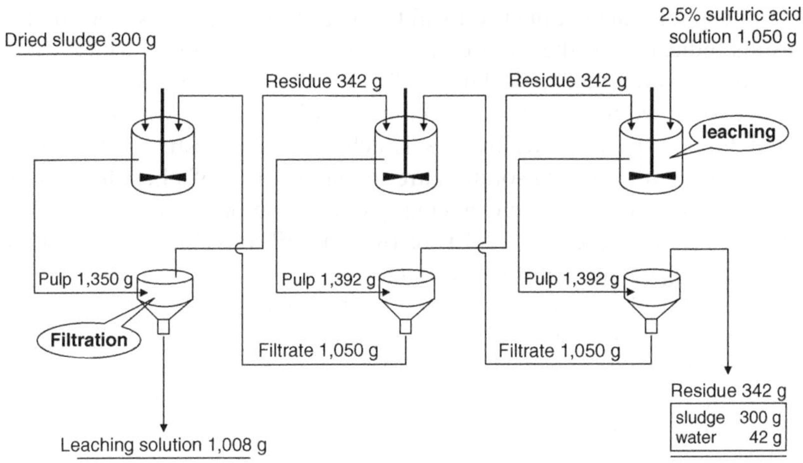

Figure 4.33 Conceptual countercurrent leaching flowchart.

Solvent Extraction Experiments on the Phosphogypsum Leachate

The optimal leachate from the abovementioned three sets of leaching tests was selected for solvent extraction experiments using the reagents system recommended by Oak Ridge National Laboratory (ORNL). These tests were conducted using laboratory shakers with each test consuming 250 mL of leachate. Most of the solvent extraction parameters were evaluated, including the O/A ratio, contacting time, temperature, solvent concentration, and acid strength. Extraction isothermals and kinetics curves were established.

4.5.2. Recovery of REEs from Phosphoric Acid Sludge

Pre-treatment of Phosphoric Acid Sludge

Over 80% of the world's phosphate rock is used for the production of wet process phosphoric acid. In many operations, the filter acid (about 28 30% P_2O_5) is evaporated in two stages, first to 40% P_2O_5 and then to 54% P_2O_5. Each evaporation stage produces a sludge, but the majority of the REEs in the filter acid eventually end up in the final sludge. Results show that REE concentration in the 54% P_2O_5 sludge is ten times higher than that in the 40% P_2O_5 sludge.

The goal of the sludge pre-treatment research is to generate a solid material high in heavy REE content as a feed for downstream processing and recovery of the REEs. Since the phosphoric acid value in the sludge is many orders of magnitude higher than the REEs, maximizing recovery of the phosphoric acid value is critical to the overall economics. Settling test results indicate that the as-received sludge settles too slowly to use thickener or any other mechanical means for solid–liquid separation. When the sludge is diluted to 30% P_2O_5, it becomes readily treatable with mechanical solid–liquid separation techniques. Leaching characteristics of the sludge using dilute sulfuric acid solution are similar to those of phosphogypsum, but REE leaching is more challenging due to the presence of Fe and Al precipitates. However, 80% REE recovery is still possible by countercurrent leaching of alcohol-washed sludge.

The phosphoric acid sludge containing about 54% P_2O_5 is extremely viscous with a viscosity of 12 centistokes at room temperature. Traditional solid–liquid separation devices are not suitable for this material without some kind of pre-treatment. As shown in Figure 4.34, the as-received sludge settles extremely slowly.

Under the CMI project, numerous approaches were tested as pre-treatment to enhance solid–liquid separation, including flocculation, addition of sawdust, and acid dilution (Allaedini & Zhang, 2018, 2019a, 2019b). Figure 4.35 indicates that dilution of the sludge to 30% P_2O_5 increased the settling rate by nearly two orders of magnitude. Therefore, this method was selected as the main pre-treatment approach for phosphoric acid sludge. After dilution, solid–liquid separation was achieved using a filter press.

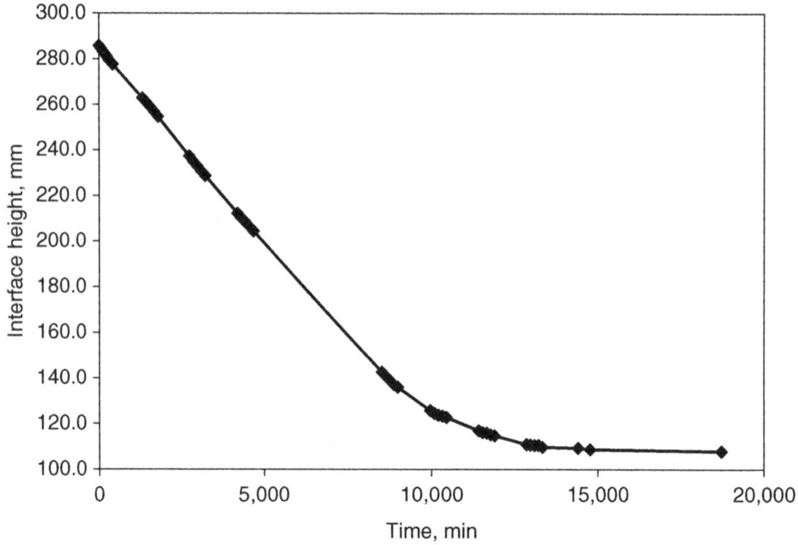

Figure 4.34 Settling rate of as-received phosphoric acid sludge in a graduated cylinder.

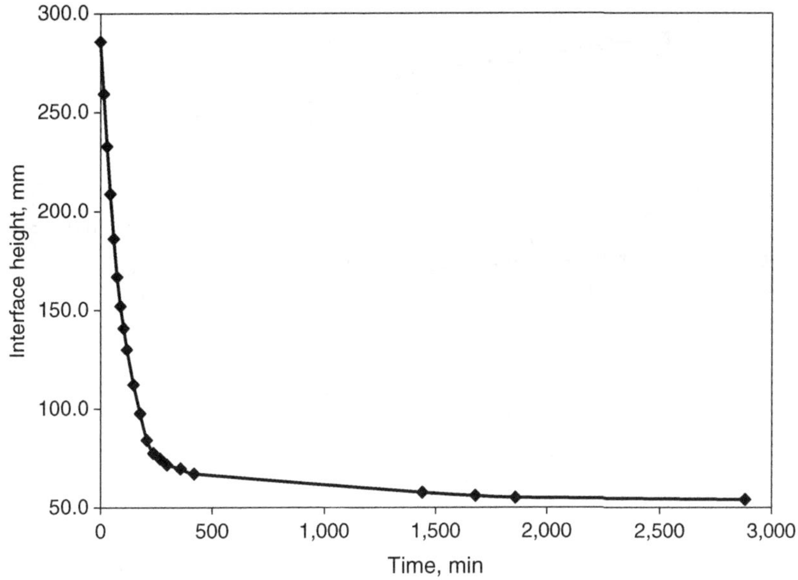

Figure 4.35 Settling rate of diluted phosphoric acid sludge (30% P_2O_5) in a graduated cylinder.

Standard settling testing was conducted on diluted sludge (to 30%, 35%, and 40% P_2O_5, respectively) in a graduated cylinder for 6 h. Parallel settling tests on the same samples were conducted to evaluate the effect of flocculation and sonication, respectively.

Based on the settling tests, a large volume of the treated sludge that renders the fastest settling rate will be prepared and tested using a laboratory filter press.

Leaching of Sludge Solids

Although a majority of the sludge is gypsum, the leaching characteristics of sludge are different from those of phosphogypsum. Therefore, some leaching tests on the sludge solids will be similar to leaching of phosphogypsum, and the others are quite different.

Again, three sets of leaching tests were conducted on the sludge. The first set of tests was designed to study the effect of acid concentration using sulfuric acid solutions of 0%, 1%, 2.5%, 5%, and 10% sulfuric acid by weight. The second set of tests was conducted mainly to study the effect of temperature and phosphorus content in the sludge. The third set of tests again examined countercurrent leaching, using the same procedure as for phosphogypsum.

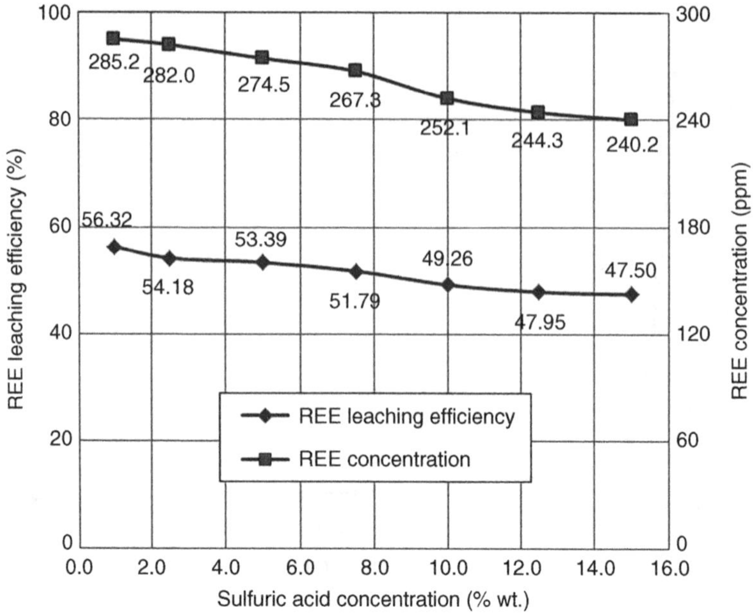

Figure 4.36 Effect of sulfuric acid concentration on leaching recovery.

Figure 4.36 shows the effect of sulfuric acid concentration on REE leaching recovery, while Figure 4.37 demonstrates the effect of leaching time. The best leaching results, again, were obtained by countercurrent leaching. Total REE concentration in the leachate reached about 290 ppm at a recovery of over 80%, under the following conditions: (1) sulfuric concentration, 2.5% (wt.); (2) leaching temperature, 25 °C; (3) liquid-to-solid ratio (wt./wt.), 3.5:1; and (4) leaching time, 2 h for each stage.

4.5.3. Recovery of REEs from Amine Flotation Tails

As is shown in Figure 4.2 in the earlier text, a majority of phosphate rock product in Florida is produced by a two-stage flotation process. The first stage is direct (rougher) flotation of phosphate using fatty acid, and the second stage is reverse flotation of the rougher concentrate using amine. Analysis of numerous samples showed that the fatty acid flotation tails contain very low content of REEs, but the amine flotation tails contain over 200 ppm REEs. Mineralogical analyses discussed in the earlier text indicated that REEs occur in amine tails mainly as monazite and xenotime. The specific gravity of these rare earth minerals is significantly

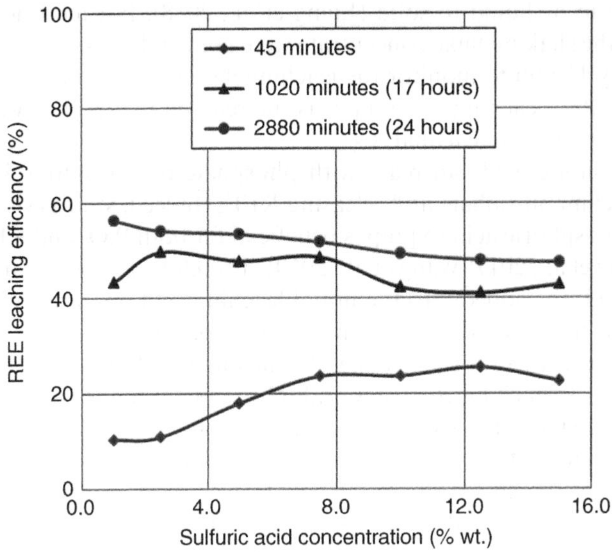

Figure 4.37 Effect of leaching time on leaching recovery.

Table 4.34 Results of one-stage shaking table separation of amine tails.

Product	Yield (wt%)	Total REE (ppm)	Total REE recovery (%)
Feed	100	201.90	100
Concentrate	12.51	657.39	40.73
Tailings	87.49	136.77	59.27

higher than that of silica (the main component of amine tails), making it possible to pre-concentrate the REE-containing minerals by gravity separation. In this case, a shaking table was used to separate rare earth and phosphate minerals from silica.

Table 4.34 shows what one stage of shaking table separation could achieve (Liang et al., 2018). The concentrate analysis was as follows: 187.38 ppm Y, 131.86 ppm La, 251.23 ppm Ce, 164.23 ppm Nd, and 29.96 ppm Pr.

The gravity separation concentrate was leached for REEs and P recovery based on the di-hydrate process of phosphoric acid production simulated in a 2.5-L water-bath batch reactor. The effects of four main factors on REEs and phosphorus leaching efficiencies were investigated, including phosphoric acid addition in the initial leaching pulp, leaching temperature, stoichiometric ratio of H_2SO_4 to calcium oxide (CaO), and weight ratio of liquid to solid (Liang et al., 2018). In each leaching test, 400 g of the shaking table concentrate were used and the solution was sampled every 15 min to monitor the leaching development through chemical analysis. After leaching, the pulp was filtered and the residue was washed three times for chemical analysis.

Phosphoric acid can react with phosphate rock resulting in soluble monocalcium phosphate and releasing REEs. Indeed, studies showed that adding phosphoric acid to pulp is beneficial for both P_2O_5 and REE leaching (Liang et al., 2017; Wang et al., 2010). Therefore, in this research, phosphoric acid was added into the initial leaching acid mixture with sulfuric acid, with the P_2O_5 concentration ranging from 0% to 20% in the initial pulps. Figure 4.38 indicates a significant impact of the phosphoric acid ratio on REE leaching efficiency and REE content in the solution. In this figure, L.E. stands for leaching efficiency (recovery).

The shaking table concentrate was upgraded further by flotation. In the flotation process, sodium carbonate was used as the pH adjustor, sodium silicate as the silica depressant, and fatty acid/fuel oil (at a ratio of 1:0.5) as the collector. Flotation results are shown in Table 4.35.

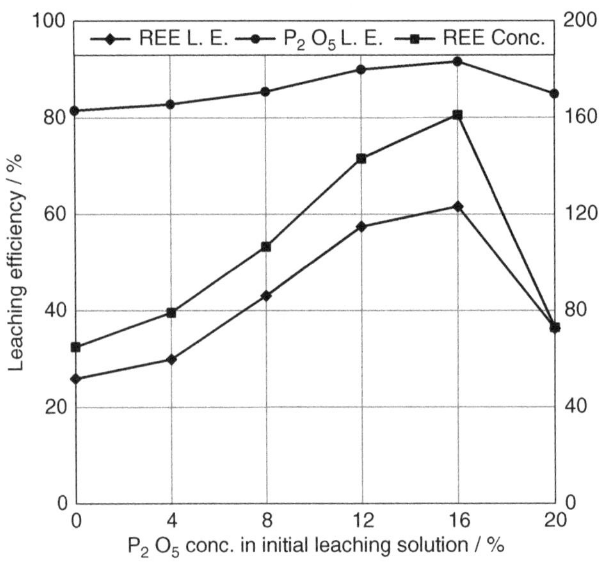

Figure 4.38 Effect of phosphoric acid concentration in initial pulp on leaching efficiency (leaching temperature = 75 °C; stoichiometric ratio of H_2SO_4 to CaO = 1.2; ratio of liquid to solid = 3.5) (Liang et al., 2018/MDPI/CC by 4.0).

Table 4.35 Results from flotation of shaking table separation concentrate.

Product	Yield (wt%)	Total REE (ppm)	Total REE recovery (%)	P_2O_5%	Percentage P_2O_5 recovery
Feed	100	621.75	100	8.12	100
Concentrate	25.28	1,809.50	73.18	26.37	82.07
tailings	74.72	219.88	26.82	1.95	17.93

4.5.4. Recovery of REEs from Phosphate Clay (Slime)

As mentioned earlier, phosphate beneficiation in Florida generates over 1 ton of phosphate clay (slime) per ton of phosphate rock produced. On that basis, the industry has been producing 20–40 million tons of waste clay annually depending on the rate of phosphate mining over the years. Since the beginning of large-scale washing and desliming practices for phosphate beneficiation, over 2 billion tons of such slime have been accumulated, which contain approximately 600 million tons of phosphate rock, 600,000 tons of REEs including 200,000 tons of yttrium, and 80 million kg of uranium.

The tremendous potential for recovering P, REEs, and U in phosphate clay has drawn numerous industrial partners to the CMI phosphate project led by the FIPR Institute.

Figure 4.39 summarizes the proposed processing steps by the FIPR Institute. Success of this ambitious research effort depends on three key elements: (1) successful removal of clay minerals, (2) upgrading of phosphate as well as REE-containing materials, and (3) removal of magnesium from the phosphate concentrate, thus making it suitable for producing high-value phosphoric acid. To accomplish these goals, we propose multi-stage cyclone separation to remove the clay minerals, innovative gravity separation and flotation to concentrate the phosphate, and smart chemistry to selectively leach magnesium from phosphate.

Selective leaching of Mg is usually conducted at pH 2.8–3.5, temperature 65–75 °C, and 30% solids. Several cyclone separation campaigns were conducted to separate clay minerals from non-clay minerals, both in the lab and on the plant site. Two hydrocyclone testing systems were used in those tests, with one using a 4-inch cyclone, and another with a 2-inch and a 1-inch cyclone.

One-Stage Separation of Using a 4-inch Cyclone and Downstream Processing

The 4-inch cyclone cuts relatively coarse and one-pass separation could recover roughly 20% of the mass with less than 20% of the REEs in the head sample. Table 4.36 shows results from one of the tests.

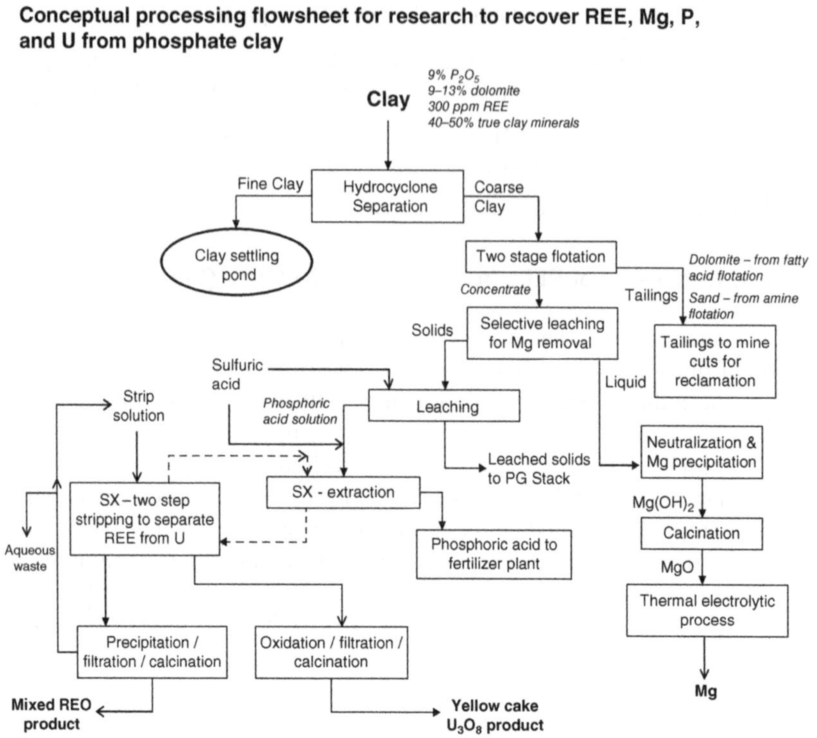

Figure 4.39 Proposed approach for REEs, U, and P recovery from phosphate clay.

Table 4.36 REE distribution in cyclone classification products using a 4-inch cyclone.

Product	Yield (wt%)	Total REEs (ppm)	Total REE recovery (%)
Feed	100	314.37	100
Underflow	20.68	266.46	17.53
Overflow	79.32	326.86	82.47

Fatty acid flotation of the underflow product produced a concentrate with 455 ppm REEs at a total REE recovery of 81.85%. Over 90% REE recovery was achieved by leaching the flotation concentrate at 230 °C with concentrated sulfuric acid at a stoichiometric ratio of H_2SO_4/CaO: 1.5–7.0 for 240 min.

Two-Stage Separation Using a 2-inch Cyclone

Since the product yield using a 4-inch cyclone was low (about 20%), a 2-inch cyclone was designed and tested for two-stage separation of the phosphate clay. In this scheme, the overflow from stage 1 was used as the feed for the second stage separation. Table 4.37 shows the first-stage separation results, while Table 4.38 shows the second-stage separation results.

The abovementioned two tables show that a much higher weight yield was produced from phosphate clay using the 2-inch cyclone. The first stage recovered about 62% of the total mass, and the second stage scavenged 14% from the stage-one tails, giving an overall weight yield of approximately 67%. Particle size distribution analysis using a laser analyzer shows that much of the fine clay particles were removed (Figures 4.40 and 4.41).

Mineralogical analysis using X-ray diffraction also indicates efficient removal of clay minerals using a cyclone. As shown in Table 4.39, after using a cyclone, the underflow product is free of three of the clay minerals, kaolinite, montmorillonite, and illite.

Table 4.37 First-stage separation of phosphate clay—two duplicate tests.

		REE concentration (ppm)				
Sample	Wt% yield	Y	Nd	La	Ce	Total
Under flow (UF) 1	61.5	49.3	29.5	32.1	54.5	205.6
OF 1	38.5	71.3	49.0	51.1	85.3	320.2
UF 2	62.2	51.7	30.9	33.6	57.1	215.9
OF 2	37.8	74.9	52.1	54.0	88.6	336.4

Table 4.38 Second-stage separation of phosphate clay—two duplicate tests.

		REE concentration (ppm)				
Sample	Wt% yield	Y	Nd	La	Ce	Total
UF 3	14.4	61.96	40.05	43.90	79.60	226.18
OF 3	85.6	63.02	40.72	44.98	80.25	227.80
UF 4	13.9	61.13	39.98	42.79	78.24	279.53
OF 4	86.1	65.32	42.67	45.67	83.02	298.76

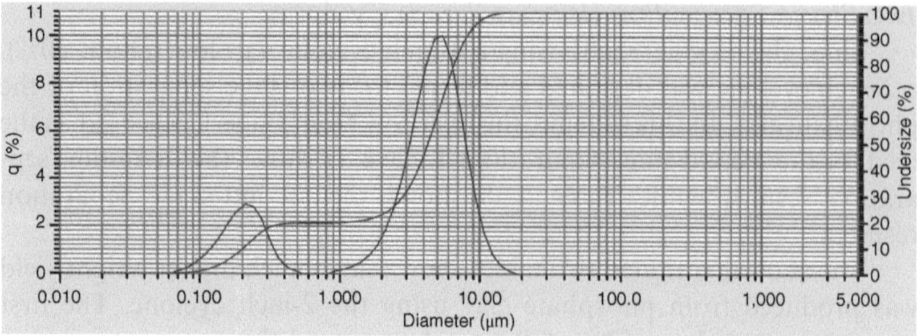

Figure 4.40 Size distribution of overflow from 2-inch cyclone testing—mean size 4.294 μm; cumulative 10%: 0.206 μm; cumulative 50%: 4.294 μm.

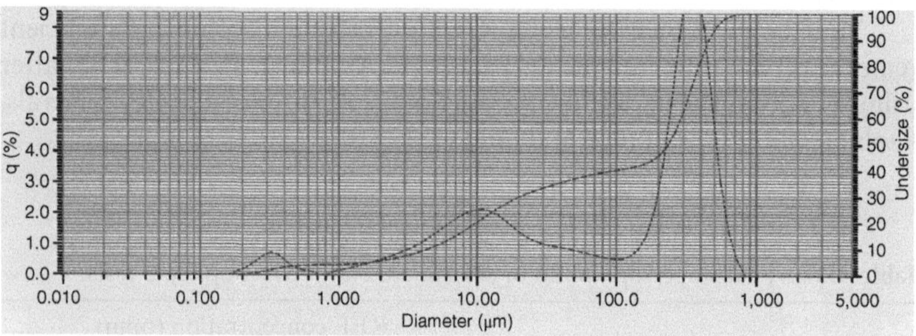

Figure 4.41 Size distribution of underflow from 2-inch cyclone testing—mean size 213.298 μm; cumulative 10%: 4.469 μm; cumulative 50%: 232.947.

Table 4.39 XRD analysis of cyclone products using the 2-inch cyclone.

		Weight %	
Phase	Chemical formula	Overflow	Underflow
Quartz	SiO_2	3.7	54.8
Apatite	$Ca_5P_3O_{12}(OH, F)$	15.0	13.7
Wavellite	$Al_3P_2O_{16}H_{13}$	0.6	0.0
Dolomite	$CaMgC_2O_6$	2.7	5.8
Kaolinite	$Si_2Al_2O_9H_4$	10.6	0.0
Montmorillonite	$Al_4Si_8CaO_{24}H_2$	2.7	0.0
Illite	$Al_4Si_2KO_{12}H_3$	0.7	0.0
Amorphous content		64.1	25.8

4.5.5. Recovery of REEs from Phosphate Rock

The wet process phosphoric acid manufacturing can be represented by the following simplified reaction (Zhang, 2014):

$$Ca_5(PO_4)_3F + 5H_2SO_4 + 5nH_2O = 5CaSO_4 \cdot nH_2O \downarrow + 3H_3PO_4 + HF - \text{exothermic} \quad (2)$$

Depending on the n value in Equation (2), the process is defined as the di-hydrate (n = 2) process, hemi-hydrate (n = 1/2) process, or anhydrate process. The di-hydrate process is most widely used worldwide because it is more tolerant to rock feed variations and easy to operate (Becker, 1983). The process uses phosphoric acid to react with ground phosphate rock containing 28–32% P_2O_5. The reaction results in a slurry of phosphoric acid and calcium sulfate di-hydrate (phosphogypsum). This slurry is filtered commonly using pan filters to produce a filter phosphoric acid containing about 28–30 P_2O_5 and the by-product PG. The filter cake (PG) undergoes a three-stage countercurrent washing prior to being pumped to the disposal stack. The PG washing solution containing around 17% P_2O_5 is sent back to the reactor. In some operations, the filter acid is concentrated to about 54% phosphoric acid by two-stage evaporation. The concentrated acid is processed using a centrifuge to recover the solid precipitates.

Several investigations including the CMI study found that in the phosphoric acid manufacturing process, about 30% of the REEs in phosphate rock report to phosphoric acid and 70% end up in phosphogypsum (Al-Thyabat & Zhang, 2015a, 2015b, 2016; Liang et al., 2017; Wang et al., 2010).

The most recent study by Liang is perhaps the most comprehensive one on this topic (Liang et al., 2017). In this research, the effect on REE leaching efficiency of five major parameters during the di-hydrate wet process was investigated, including leaching time, temperature, initial phosphoric acid concentration in the reactor, H_2SO_4-to-CaO ratio, and liquid-to-solid ratio. It was found that the initial phosphoric acid concentration in the pulp had the highest impact on REE leaching efficiency from phosphate rock, as is shown in Figure 4.42.

Liang's investigation indicated that up to 70% of the REE in phosphate rock could be dissolved into the acid solution by adjusting the processing parameters during the di-hydrate acidulation process.

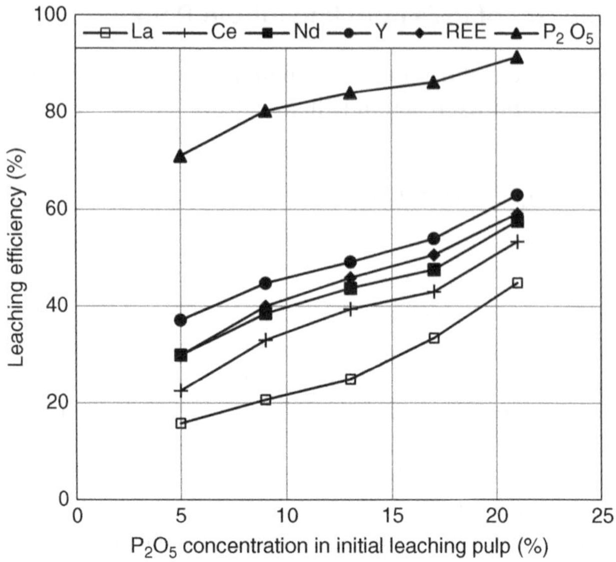

Figure 4.42 Effect of phosphoric acid concentration in initial pulp on leaching efficiency (leaching temperature 70 °C, stoichiometric ratio of H_2SO_4 to CaO 1:10, ratio of liquid to solid 3:5).

4.6. Summary

REEs in Florida phosphate have much higher proportions than traditional rare earth ores of heavy rare earths and some of the more critical elements such as yttrium and neodymium. REEs occur in phosphate ore in three major forms, xenotime, monazite, and calcium substitution in the phosphate crystals. These critical elements in phosphate ore got distributed into five major streams during mining and processing, including phosphate clay, flotation tails, phosphoric acid/phosphoric acid sludge, and PG. Depending on ore properties and mining operations, about 30%–40% of the REEs in phosphate ore report to phosphate clay, 10% to flotation tails (mainly the amine tails), 15%–20% to phosphoric acid/phosphoric acid sludge, and 35%–40% to PG.

Considering the easy integration of sulfuric acid system with the existing phosphoric acid manufacturing process, recovery of REEs from PG focused on leaching with sulfuric acid solutions. Experimental results indicated that REE leaching efficiency increased rapidly, reached a maximum, and then began to decrease with sulfuric acid concentrations ranging from 0% to 10% and at temperatures ranging from 20 °C to 70 °C. At a sulfuric acid concentration of 5% and leaching temperature of 50 °C, REE leaching efficiency obtained a maximum value of approximately 43%. Increasing the leaching time or liquid/solid ratio further increased the leaching efficiency. Leaching recovery of over 80% was achieved by countercurrent leaching of PG. Leaching study also showed that a 2.5% sulfuric acid solution was as effective as a 5% solution.

Phosphoric acid sludge offers the best opportunity for REE recovery from phosphate processing for three reasons: (1) it contains highest REE concentration (over 2,000 ppm) among the processing streams, (2) processing of this stream would not trigger current environmental regulations, and (3) processing of this stream would not interfere with the phosphoric acid manufacturing process, the main business of the phosphate industry. Pre-treatment and recovery of phosphoric acid are critical to economic recovery of REEs from the sludge.

No matter what further processing approach is selected, gravity separation ought to be the first step to pre-concentrate REE-containing minerals, mainly xenotime and monazite in amine tails, because these minerals are much heavier than silica that is the major gangue mineral in this stream. Flotation may well be the best option for further processing the gravity separation concentrate to upgrade both phosphate and REE minerals. One-stage shaking table separation could concentrate 41% of the REEs in a 12.5% mass, increasing REE concentration from 202 ppm to 657 ppm. The shaking table concentrate was further upgraded by flotation achieving 73% REE recovery and 82% phosphate recovery and enriching REE concentration to 1,809 ppm.

Due to its ultrafine particle size and complex mineral composition, the phosphate clay is extremely difficult to process. Based on this multi-year research funded by CMI, an ambitious processing flow sheet was devised (Figure 4.42) for comprehensive recovery of both REEs and other valuable elements from the clay. The flow sheet involves four major approaches: (1) removal of clay minerals using small hydrocyclones, (2) upgrading of phosphate as well as REE-containing materials from the cyclone underflow product by innovative gravity separation or flotation, (3) selective leaching of the concentrate from step two to remove and recover magnesium, and (4) stage wise leaching of the Mg-free product for REE recovery and phosphoric acid production.

Acknowledgments

Much of the technical data presented in this chapter was drawn from a major research project supported by the Critical Materials Innovation (CMI) Hub (formerly Critical Materials Institute, an Energy Innovation

Hub funded by the U.S. Department of Energy, Advanced Materials & Manufacturing Technologies Office. The guidance and leadership provided by Bruce Moyer, CMI area lead, and David DePaoli, CMI project lead, are highly appreciated. Substantial matching funds were provided by the FIPR Institute, Florida Polytechnic University. The Mosaic Co. is particularly acknowledged for their technical input, large in-kind support, and sample collection efforts. We want to express our gratitude to the following Mosaic employees and former employees for their help: Nicole Christiansen, Paul Kucera, Marcos Ortiz, Chris Dennis, Glen Oswald, Chaucer Hwang, Cameron Weed, Trevor Andrews, and Gary Whitt. The CMI project would not have progressed smoothly without strong support of Mosaic's senior managers, especially Karen Swager, at some critical junctures of the project. Professor Mei Yang of China University of Geosciences (Wuhan) made substantial contributions to the mineralogical analysis of amine flotation tails.

References

Al-Thyabat, S., & Zhang, P. (2016). Extraction of rare earth elements from upgraded phosphate flotation tailings. *Minerals & Metallurgical Processing*, *23*(1), 23–30.

Al-Thyabat, S., & Zhang, P. (2015a). REE extraction from phosphoric acid, phosphoric acid sludge, and phosphogypsum. *Mineral Processing and Extractive Metallurgy*, *124*(3), 143–150.

Al-Thyabat, S., & Zhang, P. (2015b). In-line extraction of REE from dehydrate (DH) and hemihydrate (HDH) wet processes. *Hydrometallurgy*, *153*, 30–37.

Allaedini, G., & Zhang, P. (2018). Treatment of phosphoric acid sludge for rare earths recovery II: effect of sonication and flocculant solution temperature on settling rate. *Separation Science and Technology*. doi: 10.1080/01496395.2018.1536715.

Allaedini, G., & Zhang, P. (2019a). Treatment of phosphoric acid sludge for rare earths recovery I: Effect of polymeric flocculant properties on filtration and recovery. *International Journal of Surface Engineering and Interdisciplinary Materials Science*, *7*(2), 1–18.

Allaedini, G., & Zhang, P. (2019b). Treatment of phosphoric acid sludge for rare earths recovery III: Comparison of sawdust filter aid as body feed with precoating. *Eurasian Mining*, *2*, 49–53.

Altschuler, Z. S., Berman, S., & Cuttitta, F. (1967a). Rare earths in phosphorites—Geochemistry and potential recovery, USGS Prof. Paper 575-B, 1–9.

Altschuler, Z. S., Berman, S., & Cuttitta, F. (1967b). Rare earths in phosphorites—Geochemistry and potential recovery, Intermountain Assoc. Petrol. Geol Annu. Conf. Guidebook No. 15, pp. 125–135.

Becker, P. (1983). *Phosphates and phosphoric acid: Raw materials, Technology, and Economics of the wet process*. New York: Marcel Dekker.

Bliskoyskii, V. Z., Mineev, D. A., & Kholodov, V. N. (1969). Composition of lanthanides in phosphorites. *Dokl. Akad. Nauk. SSSR, 186*(4), 932–935. *Chemical Abstracts* 71, 52347-d.

Hill, W. L., Marshall, H. L., & Jacob, K. D. (1932). Minor metallic constituents of phosphate rock, *Industrial Engineering Chemistry*, *24*, 1306–1312.

Jarosinski, A., Kowalczyk, J., & Mazanek, C. (1993). Development of the Polish wasteless technology of apatite phosphogypsum utilization with recovery of rare earths. *Journal of Alloys and Compounds*, *200*, 147–150.

Jorjani, E., Bagherieh, A. H., Mesroghli, S. H., & Chehreh, C. S. (2008). Prediction of yttrium, lanthanum, cerium, and neodymium leaching recovery from apatite concentrate using artificial neural networks. *Journal of University of Science and Technology Beijing*, *15*(4), 367–374.

Kremer, R. A., & Chokshi, J. C. (1989). Fate of rare earth elements in mining/beneficiation of Florida phosphate rock and conversion to DAP fertilizer. Research report, Mobil Mining and Minerals Company, Nichols, Florida.

Liang, H., Zhang, P., Jin, Z., & DePaoli, D. (2017). Rare-earth leaching from Florida phosphate rock in wet-process phosphoric acid production. *Mineral & Metallurgical Processing*, *34*(3), 146–153.

Liang, H., Zhang, P., Jin, Z., & DePaoli, D. W. (2018). Rare earth and phosphorus leaching from a flotation tailings of Florida phosphate rock. *Minerals*, *8*, 416. doi:10.3390/min8090416.

May, A., & Sweeney, J. W. (1983). Evaluation of radium and toxic element leaching characteristics of florida phosphogypsum stockpiles. USBM Report of Investigation #8776.

Preston, J. S., Cole, P. M., Craig, W. M., & Feather, A. M. (1996). The recovery of rare earth oxides from a phosphoric acid by-product. Part 1 Leaching of rare earth values and recovery of a mixed rare earth oxide by solvent extraction. *Hydrometallurgy*, *41*, 1–19.

Wang, L., Long, Z., Huang, X., Yu, Y., & Cui, D. (2010). Recovery of rare earths from wet-process phosphoric acid. *Hydrometallurgy*, *101*, 41–47.

Watts, M. (2010). Shock and ore: The scramble for rare earth self-sufficiency. *Industrial Minerals*, *6*, 35–43.

Xia, X. H. (2011). *The distribution characteristics and potential of endogenetic phosphate rock in North China*. Paper presented at Beneficiation of Phosphates VI conference, Kunming, China.

Zhang, P. (2014). Comprehensive recovery and sustainable development of phosphate resources. *Procedia Engineering*, *83*, 37–51.

Zhang, Z., Mei, Y., Zhang, W., Wu, Q., & He, D. (2012). Recovery of rare earths from phosphogypsum. In J. Zhang, J. D. Miller, & H. El-Shall, (Eds.), *Beneficiation of phosphates: New thought, new technology, new development* (pp. 149–152). Englewood, Colorado: Society for Mining, Metallurgy, and Exploration, Inc.

Zhang, P., Liang, H., Jin, Z., & DePaoli, D. (2017). The ultimate mineral processing challenge: recovery of rare earths, phosphorus and uranium from Florida phosphatic clay. *Minerals & Metallurgical Processing*, *34*(4), 183–188.

5

SOLVENT EXTRACTION OF RARE EARTH ELEMENTS FROM AQUEOUS SOLUTIONS

Rick Peterson[1], Darwin Argumedo[2], and Michael Heinrichs[3]

[1] Brightmark Energy, San Francisco, California, USA
[2] Brightmark Energy, Illinois Institute of Technology, Fort Wayne, Indiana, USA
[3] Battelle Memorial Institute, Columbus, Ohio, USA

Abstract

This chapter aims to provide a brief introduction to the principles of solvent extraction (SX) and its use for recovery and separation of rare earth elements (REEs). It begins with the basic operating principles of SX, including a comparison with other common rare earth separation processes and a discussion of common chemistries and equipment. It then addresses the challenges typical of alternative rare earth sources, and the basics of testing and designing a SX circuit. The fourth section covers a case study of REE recovery from coal fly ash to show the general development process. Finally, there is a discussion on the future of SX and emerging technologies that will advance and challenge it in the future.

Rare Earth Elements: Sustainable Recovery, Processing, and Purification, Special Publications 79, First Edition. Edited by Athanasios K. Karamalidis and Roderick Eggert.

DOI: 10.1002/9781119515005.ch5

5.1. Solvent Extraction Basics

5.1.1. Operating Principles

In concept, solvent extraction is a straightforward process, relying on two immiscible liquids to generate a separation between species, in the same way that oil and water will not mix. Generally, this involves an organic phase, like kerosene, and an aqueous solution, which will split with the organic phase typically floating on top of it. Special active extractant compounds are included in the organic phase. These compounds have a polar or ionic group that associates with targeted dissolved species in the aqueous phase but are primarily aliphatic carbon chains that ensure they are more soluble in the organic phase than the aqueous phase. When mixed together, the small droplets of organic or aqueous provide area for the active extractant ingredients to associate with target species in the aqueous phase, and when the two phases split, the extracted species will preferentially inhabit the organic phase (see Figures 5.1 and 5.2). They can then be

Figure 5.1 Mixing of organic and aqueous phases.

Figure 5.2 Separation of organic and aqueous phases.

separated from the aqueous phase by using overflow and underflow weirs in a gravity settler. This process of pulling target species into the organic phase is referred to as "extraction." The aqueous phase that is now depleted in the target compound is called the "raffinate," and in practice, any one cycle of mixing and then settling is called a "stage."

Once the target species are in the organic phase, they can be removed or "stripped" from the organic phase with a separate aqueous stream that has different characteristics than the first. Most often this is a change in the acidity, basicity, ionic strength, or concentration of other ions in the aqueous solution. This swing in solution characteristics will cause the extracted species to release from the organic extractant, and the exact conditions can be tailored to obtain a very selective recovery of extracted species in a particular stage. When all of the extracted species are thus removed from the organic phase, it is then recycled back to extract more species, generating very little organic waste. The preference for a compound to associate with the organic phase or aqueous phase is typically described by a partition coefficient, which is simply the ratio of the compound's concentration in the organic to its concentration in the aqueous phase. In most SX applications, there is a desire to separate one compound in solution from another, and so the separation factor is defined as the ratio of a desired compound's partition coefficient to an undesired compound's partition coefficient. Accordingly, separation factors that are farther away from unity are desirable to achieve good separation between components.

Figure 5.3 shows a simplified schematic of a SX system using only one stage of extraction and one stage of stripping. From a processing standpoint, it is easy to envision this process benefitting from multiple countercurrent stages to maximize the concentration differences that will drive target species into or out of the organic phase. Additionally, the organic and/or aqueous phases can be recycled within a stage or group of stages to concentrate them more fully. A full SX circuit will include many extraction and stripping stages with heavy recycling to achieve a high-purity and high-concentration product solution. When mixer/settlers are used for the process, it is common to use gravity to feed between the stages, sometimes with assistance from the mixers providing a small amount of head to help in liquid transfer in recycles or between stages.

The performance of any SX process will depend heavily on the chemistry in the system, and indeed the chemistry is where most of the complexity of a SX system lies. The organic phase needs to be designed so that it has very low solubility in water, but also accommodates the usually ionic extracted species. It should also have low viscosity with density very different than the aqueous phase to ease in mixing and separation, and have low volatility and toxicity so that it remains in the

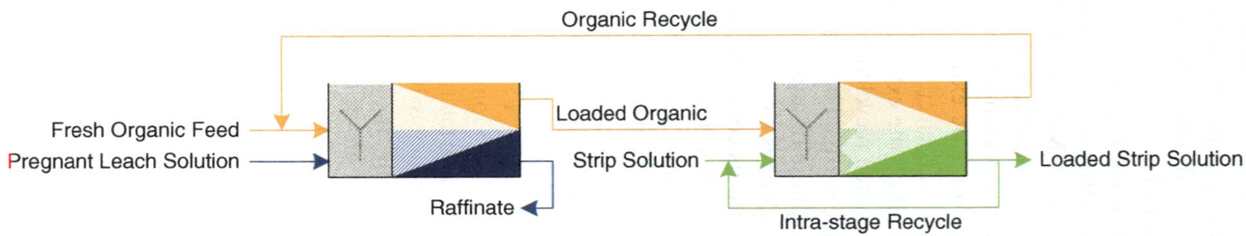

Figure 5.3 This is a simplified schematic of a SX system using only one stage of extraction and one stage of stripping. A small amount or organic is added to the recycled organic stream, and then mixed with pregnant leach solution. The mixed phases are allowed to separate in the settler, where the target compounds have been extracted into the organic phase, leaving a depleted aqueous phase, or raffinate. The organic is then mixed with strip solution in the stripping stage where the target compound is removed from the organic, allowing the organic to be recycled. The stripping stage in this case incorporates an intrastage recycle that allows for the stripping solution to be loaded to higher concentration with the target compound. In practice, the loaded strip solution usually contains the desired compound, and it is sent for further processing into a metal or oxide product.

plant and is not a hazard to plant operators. Generally, the organic phase consists of two or three components—a diluent, an extractant, and optionally a modifier. The diluent is typically a high-boiling-point distillate product that can be aliphatic or aromatic, and needs to be chemically stable to hold up to frequently harsh chemical conditions in a SX circuit. A diluent's role is to dissolve the extractant compound, lowering the viscosity and density in the process. The extractant is the active ingredient that will pull target species into the organic phase. A great deal of chemical design goes into the extractants, which are large hydrocarbon molecules with one or more functional groups that allow them to associate with target aqueous species. The different families of extractants will be discussed in a later section, but they often have highly branched carbon chains to improve their oil solubility and flow properties. The branched carbon chains can also be modified to sterically encourage selectivity for certain species. All extractants need to be chemically stable in the conditions of the SX circuit, and should have very low water solubility to avoid losses into the raffinate or stripping solutions. Modifiers are frequently a third addition to the extractant cocktail, used to improve phase disengagement and reduce aqueous carryover between stages. In many cases, the modifier compound will replace water in the extracted species, making the extracted complex more oil-soluble. Modifiers are typically polar organic compounds such as alcohols.

The chemistry of the aqueous phase also plays a significant role in operation of a SX circuit. Many common materials will create operational problems, often in the form of poor phase disengagement or third phase formation. One such contaminant is silica, which can cause formation of a viscous layer at the organic/aqueous interface, commonly referred to as "crud" or a "rag layer." Heavy transition metals, such as iron, can also cause formation of a rag layer (see Figure 5.4), or bind so strongly to the extractant that they cannot be easily stripped off, causing a decay in performance. The acidity or basicity of the aqueous phase has a large effect on the chemistry of the extracted species, and can change the stoichiometry of extraction, mechanism of extraction, and ultimately the selectivity of the extraction. The aqueous phase chemistry also has impacts on the process operability. It is common for low-ionic-strength solutions to take longer for phase disengagement, and often phase disengagement is more challenging when pH is nearer to neutral. Phase disengagement can also be impacted by the organic-to-aqueous ratio in the mixing compartment (which can be controlled by recycle of the organic or aqueous phases), with ratios nearer 1:1 typically causing more disengagement problems. Whether a mixture is aqueous or organic continuous also has a large impact in phase disengagement.

Figure 5.4 Picture of a rag layer formation due to the high concentration of iron in the aqueous phase.

5.1.2. Comparison with Solid Phase Ion Exchange and Selective Precipitation

Other common methods for separation/purification of aqueous solutions include solid phase extraction and selective precipitation. These will be discussed in relation to solvent extraction in the following sections.

In solid phase extraction, an active extractant group is fixed onto a solid support such as a polymer, silica, or carbon. As solutions flow through the solid bed, ions are captured by the functional groups, retaining them in the bed while the depleted solution exits the bed. The ions are then washed from the solid phase by flowing a solution with different chemical characteristics through the bed—often a solution with different acidity or basicity from the original solution. The process can be batch or continuous. In continuous chromatography, a derivative process, the solids are physically staged to allow for some portions to be loading with ions while others are stripped of ions with different solutions. Solid phase extraction allows for careful control of the eluting aqueous phase, allowing for very strong separations of ions, and because the active functional groups are fixed, there is very little physical loss of the active portions compared to soluble losses in solvent extraction. Solid phase processes can also be more tolerant of solids in the system, where they can be physically washed out of the bed, while solids in a SX circuit will frequently cause organic entrainment and organic losses with their removal.

Solid phase extraction does have significant drawbacks, however. The ease of handling two liquid phases in SX allows for significant solvent or aqueous phase recycling, allowing for higher concentration factors in the aqueous phases. Solid phase extractions are inherently limited in concentration factors by the volume of the solid phase bed. This leads to higher solution handling costs if they need to be thermally concentrated after stripping of the bed, or to higher waste disposal costs.

Selective precipitation is another common method for separation of ions in an aqueous solution, and it relies upon differential solubility of salts in solution across ranges of pH, counterion concentration, oxidation reduction potential, and so forth. A classic example of selective precipitation is the use of oxalate in the recovery of rare earth elements (REEs). Rare earth oxalates are very insoluble except at very low pH levels, while alkali and alkali earth and many transition metals such as iron and aluminum remain soluble as oxalates. This allows for rare earth oxalates to be precipitated from solution at relatively high purities, and in fact many heavy REEs are produced this way from ion-exchangeable deposits in South China. However, base metals in solution will "consume" oxalate ions by association without leading to rare earth precipitation, suggesting that oxalate precipitation is most economical from high-quality aqueous streams. Additionally, many heavy transition metals, notably including the radionuclides uranium and thorium, are also insoluble as oxalates, requiring additional purification before or after oxalate use. Application of oxalate for REEs is also best performed at low pH, but not high acidity, with precipitation generally more selective around pH 2–3. In cases where a strong acid is used to leach the REEs, the chemical cost of adjusting the pH higher can be significant.

Perhaps the biggest drawback for selective precipitation is the low separation factor between REEs in a selective precipitation scheme. For example, the solubility product for all the rare earth hydroxides is below 10^{-19} (Kremers, 1945), and the water solubility of all of the rare earth oxalates is below 5 mg/L (Stevenson & Nervik, 1961). This requires many stages of solid separation and redissolution to achieve separation of the rare earth series, making for a process that is chemical- and labor-intensive. Methyl oxalate precipitation, which can improve separation by varying dependence on methyl oxalate hydration, showed a separation factor of 1.3–1.5 between Nd and Pr (Weaver, 1953), which is similar to common SX reagents. However, it would still require many stages of precipitation and redissolution, which increases the chemical and equipment costs compared to solvent extraction.

5.1.3. Mechanisms and Extractants

This section will briefly explain the mechanism and some common examples of the main families of SX reagents. Readers are encouraged to refer to Gordon Ritcey's book "Solvent Extraction: Principles and Applications to Process Metallurgy" for more detail regarding the common

SX reagents (Ritcey, 2006). Two key properties for all reagents are high organic solubility and low aqueous solubility, which are typically achieved by inclusion of long organic chains with 8 or more carbons. Design of the compounds is a balance between high organic solubility and good operability, which are generally at opposite ends of the carbon chain length, with longer carbon chains leading to higher viscosity. Most compounds incorporate branched carbon chains to improve organic solubility with elevated melting point and reduced viscosity. Branching of the carbon chains can also introduce steric hindrances in the molecules that make active extractant ingredients more selective for the target species. Another consideration for all ingredients is chemical stability, as SX circuits often incorporate very acidic and or basic conditions.

Solvation

Extractants operating by solvation associate with neutral inorganic species, pulling them into the organic phase. They are used for the recovery of metals but can also be used for recovery of acids. Some of the most commonly encountered compounds are organic esters of phosphoric acid, such as tributyl phosphate (TBP) and trioctylphosphine oxide (TOPO), which have high polarity and can substitute for water in coordination complexes. Alcohols, ketones, amides, and sulfur containing compounds have also been used as extractants. Tributyl phosphate was one of the earliest extractants used for the extraction of REEs, having been utilized in the early days of nuclear research for recovery of uranium from monazite ores (Bostian, 1959; Peppard et al., 1953). In the case of REEs, TBP or TOPO can be used to extract rare earths from high-acidity nitrate solutions on the order of 6N, with stripping performed by contact with a low-acidity solution like water (Ritcey, 2006).

Chelation and Ion Exchange

Chelation and ion exchange extractants function by forming a compound with ionic species in the aqueous phase and pulling them into the organic phase. Naturally, the performance and behavior of the extractant differ with the functional groups present and the type of compounds formed. Chelating extractants will form a polydentate complex with metal ions, and commercially available compounds are often substituted derivatives of hydroxybenzophenone oximes or hydroxyquinolines. Ion-exchange extractants may be carboxylic acids such as neodecanoic acid, phosphonic/phosphinic acid such as di-(2-ethylhexyl)phosphoric acid (DEHPA), or tertiary and quaternary amines such as trioctylamine or methyltrioctylammonium chloride. Compound formation with each of these extractants is highly pK_a-dependent, and extraction/stripping is usually done with a pH swing.

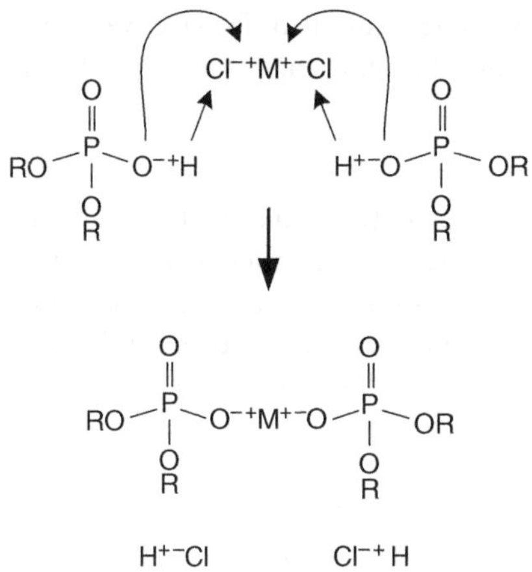

Figure 5.5 A simplified schematic of the ion-exchange mechanism showing in this case a divalent metal chloride in solution with an organophosphoric acid extractant. The hydrogen ions in the acid will associate with the chloride ions, while the metal cation will then associate with the anionic oxygen on the extractant. The resulting metal–organophosphate complex is oil-soluble and will partition to the organic phase.

In concept, the extraction mechanism is simple and straightforward, as illustrated in Figure 5.5; however, in practice, there is variation in the actual composition of the extracted species and associated stoichiometry, particularly for streams with a high concentration of contaminant compounds.

For REE extraction, it is common to use organic phosphonic and phosphinic acids, in particular ethylhexyl phosphonic acid, which has better separation factors than DEHPA. For these phosphate-derived extractants, there is typically higher affinity for the heavy REEs, decreasing with decreasing atomic number. Amines, such as methyltrioctylammonium chloride, or its tertiary analog trioctylamine, can also be used for rare earth extraction, and typically exhibit inverse affinity for the REEs with preference for the lighter elements.

Modifiers

Modifiers are organic compounds included in the organic cocktail to improve phase disengagement behavior. In some cases, the active ingredient may have limited solubility in the organic diluent without addition of a modifier, and in more common cases the extracted species has limited organic solubility before a modifier is added. Common modifiers include tributyl phosphate and isodecanol, which associate with the extracted species and increase its organic solubility. Modifiers can also displace water in the extracted species, reducing water carryover into the stages following extraction.

Diluents

Diluents usually form the bulk of the organic phase, acting as a continuous carrier for the active extraction ingredient and reducing viscosity of the organic phase. Some lighter extractant compounds, such as TBP or certain ketones, can be used neat without a diluent. The key properties for a diluent are low water solubility, low viscosity, low volatility, and good solubility for the active ingredient. Solvent extraction circuits are frequently operated with large swings in acidity, basicity, and oxidation/reduction potential, so the diluents also need to be chemically stable. Many circuits are run with simple kerosene as a diluent, but performance may be improved with incorporation of specialty solvents such as hydrotreated aliphatic distillates, aromatic solvents, or a mixture. In general, extractant solubility is better with aromatic solvents, but there are trade-offs in odor and toxicity when aromatic solvents are used.

5.1.4. Equipment

From the beginning, the workhorse of SX equipment has been the mixer settler. The mixer incorporates an impeller or propellers and baffles to ensure high contact efficiency between the organic and aqueous phases without introducing excessive shear rates that may lead to stable emulsions. Static mixers are also possible, and can reduce the footprint and inventory of solvent, with a trade-off in flexibility with mixing characteristics. For tank mixers, the mixture overflows by gravity into a settler. Settlers provide quiescent time for the two phases to separate, and current design incorporates tricks to ensure laminar flow and even distribution across the settler area. The size of the droplets affects the coalescing and settling rates, with many processes limited by the coalescing rate. It is possible therefore to include coalescing material, collecting the two phases into larger droplets so that the overall settling rate per area is improved. These coalescers come with a trade-off of maintenance to keep them clean, especially in cases where a third phase or "crud" can form in the circuit. Overflow and underflow weirs allow for continuous separation of the aqueous and organic phases, and adjustment in the heavy phase weir allows for control of the interface level in the settler (along with the associated residence time for each phase). Solvent extraction columns are also common, particularly in cases where many stages of extraction or stripping are required, and a column can have smaller footprint and be more cost-effective than multiple stages of mixer/settlers.

Centrifuges offer the benefit of low footprint, small solvent inventories, and short residence time. They especially find use in the nuclear industry where extended contact times with radionuclides will rapidly degrade the extractant, and in cases where an expensive extractant is required. There are many designs of centrifugal contactors available for solvent extraction. One design is the Podbielniak contactor, where the high-density phase is fed near the center of the rotor and the low-density phase near the outside of the rotor, allowing for phase contacting while the dense phase moves toward the outside of the rotor. Another centrifugal design incorporates mixing in the annular space before separation in the rotor. The trade-offs for centrifuges are higher power requirements and higher maintenance requirements.

5.2. Challenges of Alternative Rare Earth Sources

The key challenge for alternative rare earth sources is the inclusion of various contaminants that are not always encountered in high-quality ore sources. This section will describe contaminants that are notable for their ubiquity in or for their complicating effect on SX circuits.

5.2.1. Contaminants

For many alternative or marginal sources of REEs, the base minerals are best separated by physical methods before leaching and solvent extraction. This can typically exclude most of the aluminum, alkali and alkali earth metals, iron, and silica. In recycled sources, the REEs are typically already purified once, and bulk presence of these contaminants is limited. Some notable exceptions are iron, which is common in rare earth magnets, and aluminum and silica, which are common in catalyst supports. Raw mineral sources will often contain aluminum, silica, and the alkali and alkali earth metals due to their crustal abundance. Heavy transition metals can also occur in marginal and recycled REE sources, and many alternative sources can introduce organic materials into the process in the form of machine/lubricating oils, natural organic matter, or coal and hydrocarbons. The following paragraphs describe some of the impacts of these contaminants on a SX system.

The alkali and alkali earth elements may impact the extractability of REEs from a leach solution, but most extractants have good selectivity of REEs over alkali and alkali earth cations. At very high concentrations, one must be careful of solution entrainment in the organic phase causing contamination in the stripping stages. Transition metals, on the other hand, may extract in the same range as the REEs, and lead to contamination. In particular, zinc is known to behave very similarly to REEs in a SX circuit and should be excluded beforehand wherever possible. Aluminum can also extract and strip with the REEs. Heavier transition metals, with trivalent iron a notable example, are less likely to strip in the same range as the REEs and cause contamination, but are difficult to remove from the extractant. They may require a separate stripping step with aggressive acid concentrations, mixed acids, or a precipitative strip to maintain extractant health. Many extractants have a loading limit above which they have reduced solubility in the organic phase, leading to increased losses and operational challenges, and if heavy metals are not stripped, that loading limit may be exceeded upon recycling of the extractant. Most heavy transition metals can be precipitated selectively over the REEs with a careful pH swing. In the case of iron, it may also be dealt with by reducing ferric iron to ferrous iron, which extracts at much higher pH than the REEs and can be excluded in the extraction step.

Silica is common in mineral REE sources, and is known to cause rag layer formation in SX circuits at low concentrations. It is particularly problematic at neutral or high pH ranges, where its solubility increases, and in cases where the pH is adjusted down just prior to SX and not enough time is allowed for the silicic acid to precipitate. Silica is generally insoluble in acids, so it is less of a concern in leach circuits that are wholly acidic.

The impact of organic contaminants naturally depends on the specific organic impurity, but all will require increased maintenance of the solvent phase, most often in the form of solvent blowdown and makeup. Aliphatic and aromatic hydrocarbons or oils will tend to accumulate in the diluent of the solvent, reducing the concentration of active ingredients. Oxygenated and other polar hydrocarbons may interfere with phase disengagement between aqueous and organic phases and/or may interfere with the extraction of cations into the solvent. Charged organic compounds such as carboxylic acids or amines may bind with active extractant ingredients and reduce extractant efficiency. Some SX circuits have experienced microbial growth in the solvent phase, and in these cases, an antimicrobial additive must be selected that will not interfere significantly with extraction or phase disengagement.

5.2.2. Radionuclide Considerations

It is very common for mineral sources of REEs to be associated with the actinide series elements uranium and thorium, which are radioactive and must be dealt with in a safe manner. There is a market for uranium such that it may be further refined into a saleable byproduct from REE processing, but with the added expense of meeting regulatory requirements. The market for thorium, on the other hand, is very small and it is unlikely to be saleable in any meaningful quantities at this point in time. Both elements are heavy metals and can generally be precipitated out of a leach solution before the REEs with high selectivity, while keeping them "diluted" in other transition metals. The precipitated material will be technologically enhanced naturally occurring radioactive material and subject to regional regulations in its handling. Scandium is frequently cited as a valuable material in alternative mineral sources of REEs, but it tends to follow the radionuclides in the processing schema, which complicates its recovery to a saleable byproduct.

5.3. Testing and Design of Solvent Extraction Circuits

5.3.1. Extraction

Much of the preparation work for a SX circuit is done in the ore upgrading and leaching steps, which will dictate the purity, concentration, and other characteristics of the leach solution entering the SX circuit. Careful analysis of the leach solution composition can provide insight into selection of the extractant, as the contaminants present and the acidity of the leach solution will direct toward certain extractants. Many of the major extractant suppliers can provide recommendations for an extractant chemistry from leach solution analysis. It is good to remember that there are trade-offs for extractability and strippability. For example, DEHPA is commonly used as an extractant for metal species due to its low pK_a that allows it to quickly and strongly give up hydrogen ions and bind metal cations. However, because it binds metals very strongly in the extraction step, it also requires high acidities in the stripping step to release the metals and regenerate the protonated form of the extractant. Some heavy transition metals, such as ferric iron, can bind so strongly to DEHPA that it is cost-prohibitive to remove them from the extractant. Selection of an extractant composition ultimately requires a holistic approach considering extraction and stripping along with operational factors such as phase disengagement.

Screening of extractant chemistries can be quickly done in batch tests contacting the extractant composition with leach solution and analyzing the aqueous phase for target removal and selectivity. These screening tests can provide a quick understanding of which rough extractant chemistries will provide satisfactory extraction and selectivity performance. It is important to note that most extractant compounds are not sold in a completely pure form, and the impurities can impact extraction and phase disengagement

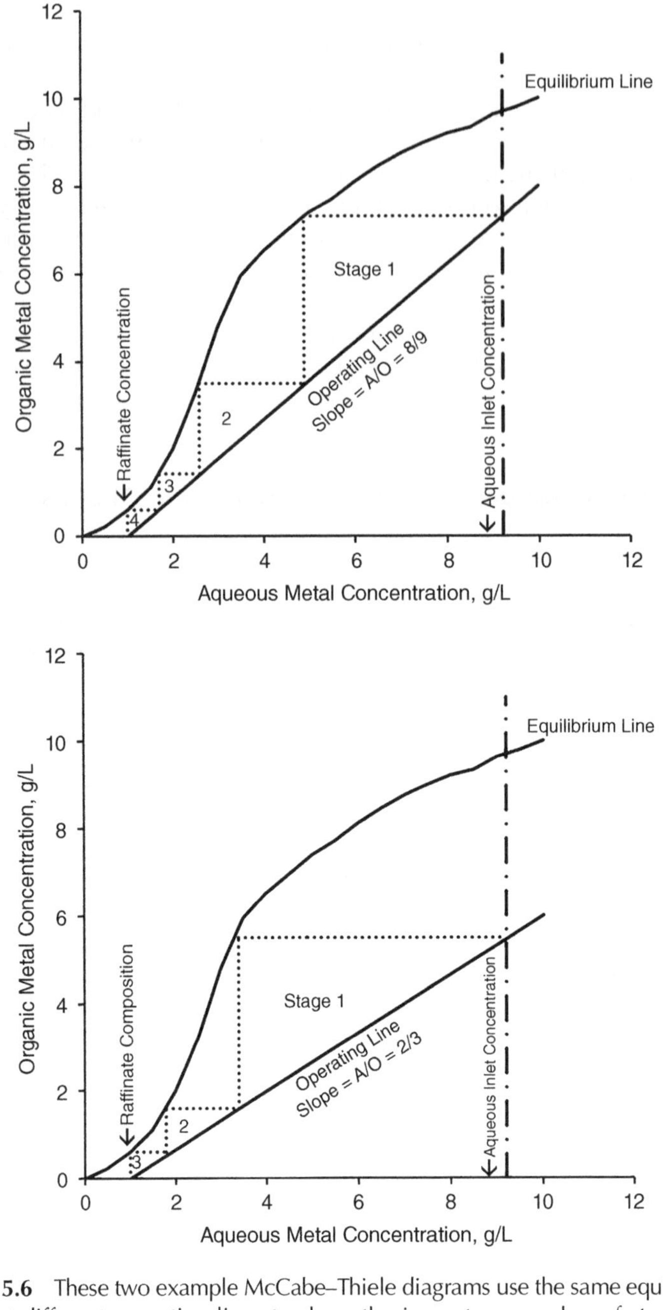

Figure 5.6 These two example McCabe–Thiele diagrams use the same equilibrium line but different operating lines to show the impact on number of stages. The operating line has a slope equal to the aqueous-to-organic-phase ratio, and crosses the *x*-axis at the raffinate outlet concentration. Stages are "stepped off" starting from the point of aqueous inlet concentration on the operating line. The stages are stepped first horizontally to the equilibrium line, and then vertically back down to the operating line. The diagram on the top has an A/O of 8/9,

Figure 5.6 (Continued) or less organic phase per volume of aqueous phase compared to the diagram on the bottom, which has an A/O of 2/3. As can be seen, with more organic phase in the system, approximately the same raffinate concentration can be achieved in fewer stages. However, those stages will be larger to accommodate the higher organic volume, and the concentration in the organic phase will be lower heading into the stripping section, making for larger stripping circuit stage volumes as well. The shape of this particular equilibrium line makes it difficult to approach zero concentration in the raffinate, requiring an aggressive A/O ratio and many stages.

performance. It is good practice to first wash the solvent composition to be tested with an aqueous solution similar to the leach solution to wash out impurities. Once the candidates have been narrowed down, design can be tailored to improve diluent selection, the impact of phase ratios (high organic volumes to low aqueous volumes and vice versa), and the impact of a modifier. Modifier addition is often done dropwise until a third phase is pulled into the organic phase, or phase disengagement performance is improved. An example of this procedure is provided in Section 5.4. Batch tests are also a good opportunity to check the aqueous phase for soluble organic losses, which can be quickly found by total organic carbon analysis. Batch testing is also a good time to understand extraction kinetics both with regard to overall efficiency and selectivity for target species. This phase of testing naturally lends itself well to a design of experiment to reduce the number of tests and analyses required for selection of the final solvent recipe.

Once the solvent cocktail has been selected, equilibrium data can be obtained for design of the extraction circuit. This is done by measuring the partition of metals between the organic and aqueous phases at varying overall loading rates. In practice, this can be done by contacting the leach solution with fresh organic phase at different organic to aqueous (O:A) ratios, or by serial contacts of organic with the leach solution until the loading limit is reached. Measurement of the species concentrations in the aqueous phase allows for generation of an equilibrium curve that plots concentration in the organic phase against concentration in the aqueous phase. This curve is the first step in construction of a McCabe–Thiele diagram, which is a graphical representation of the mass balance and allows for estimation of the number of stages by "stepping off" the stages, as shown in Figure 5.6. The equilibrium line is unique to the conditions of the solutions used to generate it, including the concentration of contaminants, temperature, pH, organic phase composition, and mixing conditions, to name a few. A McCabe–Thiele diagram is useful for visually explaining and understanding the selection of the number of stages, but in practice the calculations can be computed quickly without construction of the diagram.

5.3.2. Stripping

Stripping refers to removal of species extracted into the organic phase back into the aqueous phase, and is essentially the reverse of the extraction step. Often, stripping is accomplished by contacting with an acidic aqueous phase, where high acidity allows for hydrogen ions to replace cationic species on the organic extractant. However, there are other options such as reductive stripping, where cations are reduced and have less affinity for the extractant, causing them to partition to the aqueous phase. Precipitative stripping is also an option, where an anion is added to create an insoluble compound with the extracted metal. The surface charge of the precipitate needs to be controlled, typically by pH adjustment, to ensure that it will fully disengage from the organic phase.

Equilibrium data for stripping are generated in a similar manner to the extraction step, and can also be presented graphically in a McCabe–Thiele diagram but with the axes switched. Typically, the goal in a stripping operation is to achieve as high of a concentration in the stripping solution as possible to reduce chemical consumption and equipment sizes downstream. However, there is also consideration for fully regenerating the organic phase for recycle back to the extraction stages. In some cases, an extra step follows the stripping stage to prepare the extractant for recycle, such as neutralization of the extractant to a sodium salt to allow for pH control in the extraction step.

In typical solvent extraction separation of REEs, the stripping steps are complex with multiple stages of extraction and stripping with different active extractants. This is because the REEs exhibit low separation factors between adjacent elements. There are a few shortcuts that may be exploited; notably the ability to oxidize cerium and/or reduce samarium and europium, which greatly affects their extractability/strippability and eases their separation. Additionally, there is a natural gap between neodymium and samarium, since promethium exists in quantity only in nuclear reactors. Gupta and Krishnamurthy provide discussion of some separation and purification systems used in rare earth processing (Gupta & Krishnamurthy, 2004).

5.3.3. Considerations for Low Purity Streams

Most commercial SX circuits work with relatively high-purity pregnant leach solutions; a natural benefit of working with high-grade ores. When marginal resources are accessed, it leads to leach solutions with higher contaminant loadings that will enter the SX circuit. In the extraction step, careful pH and oxidation-reduction potential (ORP) control should be maintained to omit as much of the contaminant species as possible. In general, alkali and alkaline earth metals should be readily separable from rare earth metals by pH control, with transition metals extracting at much lower pH ranges. Transition metals with higher valence than the REEs, such as uranium, thorium, zirconium, etc., should be precipitated before SX where possible, as they have very high affinity for cationic exchange extractants, and can be difficult to strip.

Inevitably, some lighter metals and some heavier transition metals will end up in the loaded organic with the REEs. Accordingly, for low-purity rare earth sources, it is wise to incorporate scrubbing stages to improve the purity of the REE product. A scrub of the loaded extractant at a pH higher than the stripping pH of rare earths can remove extracted calcium, nickel, magnesium, and some of the transition metals with minimal loss of REEs. The REEs can then be stripped at a slightly lower pH, leaving behind the heavy transition metals. These heavy metals need to be removed from the organic phase to preserve the strength of the extractant as it is recycled to the extraction stage. A high-acidity scrubbing step should be able to remove these metals, and for recalcitrant metals a mixed stripping acid can be effective, or a swing to higher pH to precipitate the heavy metals. Heavy transition metals can also lead to rag layer formation in the extraction step, which can hinder phase disengagement and often contains active extractant compounds and target species. This rag layer may be avoidable with good removal of transition metals beforehand, or it can be handled in mixer settlers with some considerations to ensure all the rag layers proceed to the stripping section to avoid product and extractant losses.

5.4. Case Study—Rare Earth Recovery from Fly Ash

In this case study, experiments were conducted for the extraction and stripping of REEs from a pregnant leach solution derived from fly ash via a closed-loop acid digestion process (ADP). A pulverized coal combustion (PCC) plant fly ash was selected as the target feedstock for the process based on results from sampling and characterization work (Battelle, 2017). The plant selected for this study was operating on primarily Appalachian Basin coals and had a comparatively high total REE + Y + Sc concentration at 545 ppm ± 13 ppm. After processing the ash via ADP, roasting to remove the bulk of iron, aluminum, and some heavy metals, and leaching in water, the REE pregnant leach solution had a total REE + Y + Sc concentration of 5,000–7,000 ppm.

Table 5.1 Solvent extraction range-finding tests.

Description	Test 1	Test 2	Test 3
Reducing agent	Sodium metabisulfite	Sodium thiosulfate	Iron powder
Extraction	pH 3.35	pH 3.35	pH 3.35
REE stripping	pH 0.75	pH 0.75	pH 0.75
Extractant Regeneration	H_3PO_4:H_2SO_4 = 5%:5%	H_3PO_4:HNO_3 = 5%:5%	H_3PO_4:HCl = 5%:5%

Table 5.2 Percentage of extracted iron after addition of different reducing agents.

Reducing agent	Conditions	Percentage of iron extracted
Sodium metabisulfite	Starting pH 3.34 (equilibrium pH 2.38)	96.9%
Sodium thiosulfate	Starting pH 3.34 (equilibrium pH 2.39)	97.3%
Iron powder	Starting pH 3.35 (equilibrium pH 2.41)	99.5%

Range-finding tests were performed to inform the ranges used in an experimental design, and to understand the effect of adding reducing agents in the SX process (see Table 5.1). The reducing agent was to be added before the extraction step, intending to reduce the iron remaining in the solution from ferric to ferrous iron, rendering the reduced iron less extractable. In the extractant regeneration step (which takes place after REEs have been stripped from the extractant), the acid formulation is a key parameter to optimize for the stripping of residual heavy metals (Ti, Fe, As, Zr, U, and Th). This step is important because the extractant needs to be cleaned prior to being recycled to ensure good extraction performance and extractant life.

The results obtained showed that the addition of sodium metabisulfite, sodium thiosulfate, and iron powder did not impact the high percentage of iron extracted (see Table 5.2). However, it was also discovered that iron does not strip at the REE stripping step pH (pH 0.75) as originally expected. Therefore, REEs are able to be recovered in the stripped solution, and iron will remain in the extractant, which then will be stripped during extractant regeneration step (see Table 5.3).

During the extractant regeneration step, three different acid combinations were tested. The best combination based on the results obtained was

Table 5.3 Percentage stripped of REEs and iron during REE stripping step at starting pH 0.74 (equilibrium pH 0.84).

Species	Percentage stripped
Fe	0.61%
REE + Y + Sc out of total available	87.60%
REE + Y + Sc out of total measured	15.68%

Table 5.4 Percentage stripped of some heavy metals during the extractant regeneration step.

	Percentage stripped			
Conditions	Fe	As	Th	U
H_3PO_4:H_2SO_4 = 5%:5%	95.19%	100%	100%	19.85%
H_3PO_4:HNO_3 = 5%:5%	84.51%	67.14%	19.81%	8.31%
H_3PO_4:HCl = 5%:5%	74.77%	Detection limit	100%	18.51%

the combination of phosphoric acid and sulfuric acid (H_3PO_4:H_2SO_4 = 5%:5%). This combination yielded a high percentage stripped of heavy metals (see Table 5.4).

One complication encountered during the abovementioned range-finding tests was the slow disengagement of the aqueous and organic phases in the extraction step. The addition of a modifier to the extractant (15% Cyanex 572 in Solvent 467) was investigated to improve phase disengagement. The modifiers considered were isodecanol (Exxal 10) and TBP. Each modifier was added to the mixture in small doses, recording phase disengagement results between doses.

Figures 5.7–5.9 show disengagement between the organic and aqueous phases after extraction using unmodified extractant (15% Cyanex 572) and the two modifiers studied (Exxal 10 and TBP). For each case, extraction was performed at starting pH 2.5. Notice that the best phase disengagement between the organic and aqueous phases occurs after extraction with 13.2%:12% Cyanex 572:TBP extractant (see Figure 5.7). After this testing, two additional extraction tests were performed to investigate the difference between the performances of the unmodified and modified extractants (13.2%:12% Cyanex 572:TBP). The extraction tests were performed at starting pH 3.3. The results obtained are shown in Table 5.5. The results show that the modified extractant performance is superior to unmodified extractant. Tributyl phosphate concentration was incorporated into the extraction part of the design of experiments to further investigate the performance of the modified extractant.

Figure 5.7 After extraction with 13.2%:12% Cyanex 572:TBP extractant.

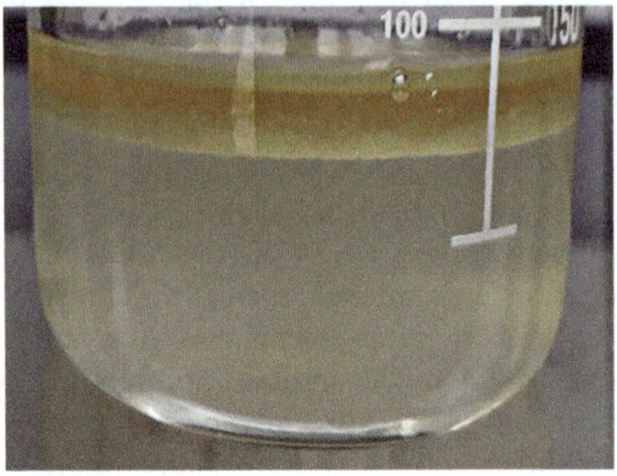

Figure 5.8 After extraction with 13.2%:12% Cyanex 572:Exxal 10 extractant.

Figure 5.9 After extraction with 15% Cyanex 572 extractant.

Table 5.5 Extraction results for 15% Cyanex 572 and 13.2%:12% Cyanex 572:TBP.

Elements	Percentage extracted	
	15% Cyanex 572 (equilibrium pH 2.43)	13.2%:12% Cyanex 572:TBP (equilibrium pH 2.29)
Sc	93.2%	93.2%
Y	84.2%	99.1%
La	10.2%	12.8%
Ce	24.1%	43.3%
Pr	24.8%	53.7%
Nd	24.6%	61.3%
Sm	50.5%	93.8%
Eu	58.8%	96.9%
Gd	54.3%	96.5%
Tb	76.7%	98.7%
Dy	85.6%	99.4%
Ho	86.6%	99.4%
Er	90.4%	99.2%
Tm	95.8%	99.4%
Yb	97.9%	99.4%
Lu	97.7%	99.3%
REE + Y + Sc out of total available	43.89%	63.9%
REE + Y + Sc out of total measured species	18.08%	20.28%

5.4.1. Rare Earth Extraction

Table 5.6 shows the factors and levels selected in the extraction testing. Mixing time, starting pH, and modifier concentration were selected for investigation, since they are expected to be the most impactful factors in the REE extraction operation. For simplicity, the starting pH was used as a variable; however, equilibrium pH is the key control variable for extraction, and was measured after each test. The experimental analysis can then be run using equilibrium rather than starting pH. To preserve some sensitive information, the REE recovery percentages and purity are presented as response variables; however, for SX design, the partition coefficients are the more useful variables.

A Box–Behnken design was selected, incorporating factors 1–3 pertaining to the extraction test factors. The REE stripping tests were treated independently, using the optimized factors from the REE extraction tests

Table 5.6 Factors and levels investigated in the REE extraction tests.

Extraction test levels			
Factor number | factor	Low level (−1)	Intermediate level (0)	High level (1)
1 **Mixing time**	1 min	10 min	20 min
2 **Starting pH**	2.5	3.3	4.0
3 **Modifier concentration (Cyanex 572:TBP)**	15%:0%	14.1%:6%	13.2%:12%

Table 5.7 REE extraction experimental design, based on a Box–Behnken design.

Test number	Mixing time	pH	Addition of TBP (Cyanex 572:TBP)
1	High	Mid	13.2%:12%
2	High	Low	14.1%:6%
3	Low	Low	14.1%:6%
4	Low	Mid	15%:0%
5	Low	High	14.1%:6%
6	Mid	High	13.2%:12%
7	Mid	Mid	14.1%:6%
8	Mid	Mid	14.1%:6%
9	Mid	Low	13.2%:12%
10	High	High	14.1%:6%
11	Mid	Mid	14.1%:6%
12	Mid	Low	15%:0%
13	Low	Mid	13.2%:12%
14	Mid	High	15%:0%
15	High	Mid	15%:0%

as the feedstock for the experiments. This experimental design was expected to collect the necessary data to develop surface response models featuring all linear, quadratic, and cross-terms for use in modeling and designing the REE extraction process (see Table 5.7).

Rare Earth Extraction Results

The design of experiment for REE extraction consisted of 15 tests. The results obtained are shown in Table 5.8 with best results highlighted in green. The experimental conditions for the best results were the following: equilibrium pH 2.65 and pH 2.64, mixing time of 10 min, and extractant composition of 14.1%:6% Cyanex 572:TBP. The data obtained from this experimental design were used to develop surface response models for

Table 5.8 Results obtained from the extraction design of experiments (single stage).

Test number	Starting pH	Equilibrium pH	Mixing time (min)	Addition of TBP (Cyanex 572:TBP)	Total REE recovery	REE purity
1	3.28	2.51	20	13.2%:12%	69.7%	12.1%
2	2.52	2.34	20	14.1%:6%	69.2%	19.5%
3	2.52	2.48	1	14.1%:6%	24.1%	26.6%
4	3.28	2.86	1	15%:0%	37.3%	7.5%
5	4.03	3.06	1	14.1%:6%	75.2%	9.4%
6	4.03	2.73	10	13.2%:12%	78.0%	14.3%
7	3.26	2.65	10	14.1%:6%	79.6%	17.7%
8	3.26	2.65	10	14.1%:6%	79.2%	22.6%
9	2.51	2.43	10	13.2%:12%	65.6%	17.4%
10	4.02	2.74	20	14.1%:6%	84.3%	13.4%
11	3.27	2.64	10	14.1%:6%	77.8%	24.8%
12	2.49	2.37	10	15%:0%	35.7%	12.7%
13	3.26	2.74	1	13.2%:12%	65.7%	13.0%
14	4	2.79	10	15%:0%	71.2%	7.4%
15	3.26	2.56	20	15%:0%	58.9%	15.3%

Table 5.9 Summary of the REE recovery response model (second-order model) for REE extraction showing the estimated coefficient for each variable of the model.

Variables	Estimates
Intercept	−6.565
pH	3.892
t	0.172
TBP	1.663
pH:t	−0.040
pH:TBP	−0.289
t:TBP	−0.019
pH^2	−0.519
t^2	−0.002
TBP^2	−0.524

REE recovery and REE purity (Tables 5.9 and 5.10 show a summary of the models for REE recovery and REE purity, respectively). Based on the two models, approximately 97% recovery of REEs can be achieved at equilibrium pH 3.06, mixing time 10.4 min, and Cyanex 572:TBP = 13.9%:7.6%, while a maximum REE purity of approximately 25% can be achieved at

Table 5.10 Summary of the REE purity response model (second-order model) for REE extraction showing the estimated coefficient for each variable of the model.

Variables	Estimates
Intercept	−2.842e+00
pH	2.386e+00
t	1.0099e-02
TBP	5.6446e-01
pH:t	−3.628e-03
pH:TBP	−5.600e-02
t:TBP	−3.900e-03
pH^2	−4.775e-01
t^2	−7.761e-05
TBP^2	−3.926e-01

Table 5.11 Optimized factors for one-stage REE extraction, equilibrium pH (pH), mixing time (t), and Cyanex 572:TBP concentration (TBP).

Results	Estimates
Maximum REE recovery	Approximately 97% at equilibrium pH 3.06, mixing time 10.4 min, and TBP of 0.55 (Cyanex 572: TBP = 13.9%:7.6%)
Maximum REE purity	Approximately 25% at equilibrium pH 2.46, mixing time 1 min, and TBP of 0.54 (Cyanex 572: TBP = 13.9%:7.5%)

equilibrium pH 2.46, mixing time of 1 min, and Cyanex 572:TBP = 13.9%:7.5% (see Table 5.11). It should be noted again that generally the more useful variable for SX design is partition coefficient, but recovery and purity are illustrative of performance.

For the validation of the response models for REE recovery and REE purity, fitted line plots of the models were graphed (Figures 5.10 and 5.11 are the fitted line plots for the REE recovery and REE purity model, respectively). It is notable that the response model for REE recovery is a good fit to the data gathered in the experimental design validating the second-order model obtained. The response model for REE purity shows potential non-random scattering of the empirical versus model data which could be improved with further testing. However, the second-order model for REE purity was also the model yielding the best linear fit.

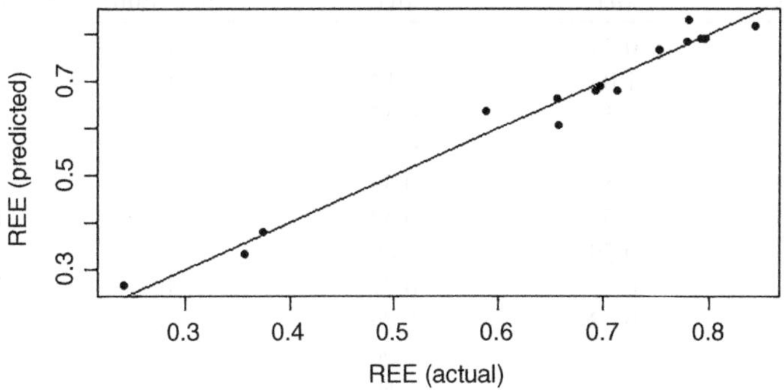

Figure 5.10 Fitted line plot of the REE recovery response model (second-order model) for REE extraction.

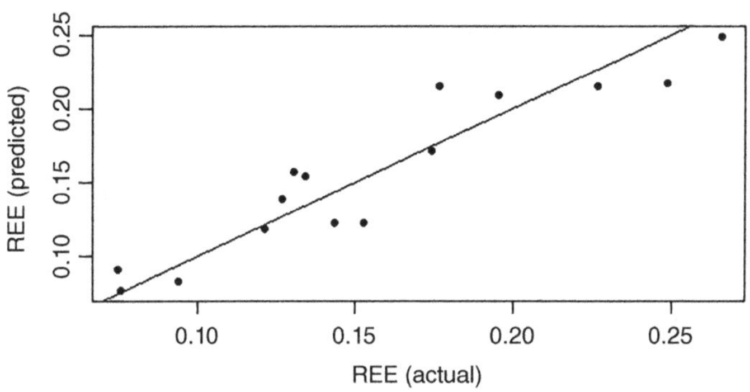

Figure 5.11 Fitted line plot of the REE purity (REEP) response model (second-order model) for REE extraction.

5.4.2. Rare Earth Stripping

Rare Earth Stripping Results

REE stripping is an important step in the SX process, since it performs much of the purification of the REEs. Therefore, two different experimental designs were performed. One experimental design investigated the stripping of REEs from the loaded extractant at low pH (0.5–1.0), and the other one at higher pH (1.4–2.4). The results obtained are shown in Tables 5.12 and 5.14 (low pH stripping and high pH stripping, respectively). The best result was obtained at high pH stripping with REE purity of approximately 55% at equilibrium pH 1.86 and mixing time of 10 min.

Table 5.12 Results obtained from the lower-pH REE stripping design of experiments (single stage).

Test number	Starting pH	Equilibrium pH	Mixing time (min)	REEs purity
1	0.74	0.76	10	30.5%
2	1.03	1.12	10	45.3%
3	0.52	0.57	20	19.0%
4	0.76	0.75	20	29.2%
5	0.52	0.59	10	22.4%
6	0.52	0.52	20	17.9%
7	0.50	0.60	1	35.4%
8	0.72	0.77	10	35.4%
9	0.72	0.76	10	35.5%
10	1.00	1.06	1	48.5%
11	0.52	0.59	20	19.6%
12	1.00	0.99	20	37.7%
13	0.74	0.78	1	41.0%

Table 5.13 Summary of the REE purity response model (second-order model) for low-pH REE stripping showing the estimated coefficient for each variable of the model.

Variables	Estimates
Intercept	−0.0109
pH	0.7713
t	−0.0155
pH:t	0.0055
pH^2	−0.2715
t^2	0.0002

Table 5.14 Results obtained from the higher-pH REE stripping design of experiments (single stage).

Test number	Starting pH	Equilibrium pH	Mixing time (min)	REE purity
1	1.80	1.86	1	53.7%
2	1.80	1.87	10	53.7%
3	1.80	1.86	10	55.1%
4	1.31	1.37	10	52.6%
5	1.79	1.84	10	53.0%
6	1.31	1.35	1	53.2%
7	1.29	1.28	20	34.6%
8	1.29	1.21	20	37.4%
9	2.32	2.35	20	21.8%
10	2.31	2.41	10	24.7%
11	1.78	1.70	20	46.2%
12	2.31	2.40	1	20.5%
13	1.28	1.25	20	38.9%

Table 5.15 Summary of the REE purity response model (second-order model) for high-pH REE stripping showing the estimated coefficient for each variable of the model.

Variables	Estimates
Intercept	−0.7738
pH	1.6824
t	−0.0008
pH:t	0.0046
pH^2	−0.5311
t^2	−0.0006

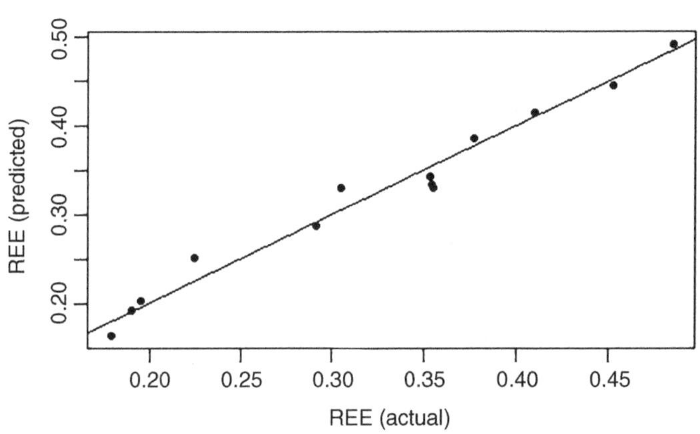

Figure 5.12 Fitted line plot of the REE purity (REEP) response model (second-order model) for low-pH REE stripping.

Using the data obtained from both experimental designs, surface response models were developed. Response models are summarized in Tables 5.13 and 5.15 for the low-pH REE stripping and high-pH REE stripping, respectively. Based on these response models, a purity of approximately 58% was predicted, within the conditions investigated, at equilibrium pH 1.61 and mixing time 6 min. Furthermore, validation of the second-order models obtained for low-pH and high-pH REE stripping is observed in Figures 5.12 and 5.13 showing strong fits to the data gathered in the experimental designs.

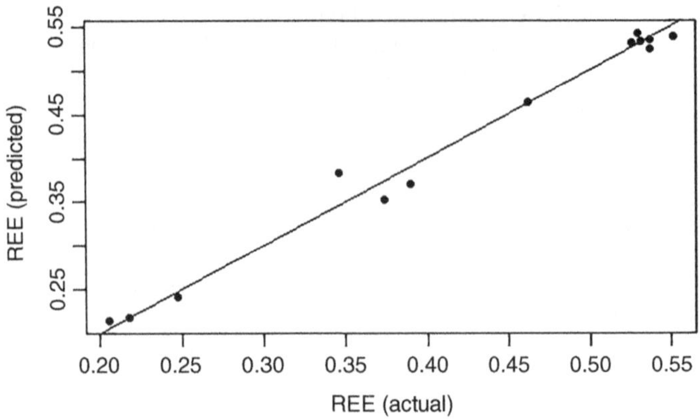

Figure 5.13 Fitted line plot of the REE purity (REEP) response model (second-order model) for high-pH REE stripping.

5.4.3. Case Study Discussion

This case study illustrated the approach for early-stage assessing of a SX process, especially for a complex leach solution. Design of experiment allows for optimization of the testing effort and can generate models that are useful for projecting performance and designing the eventual test circuit. The testing in this case revealed a key contaminant that limited the achievable purity in the stripping section, and the experiments did not suggest any conditions where good separation could be found between the contaminant and REEs. This ultimately led to other tests to remove the key contaminant by other means before the SX step.

Another important set of tests not shown in this case study is extractant loading and stripping solution loading tests. The partition coefficient for any given element will change as the concentration in the extractant or stripping solution changes. The loading level can be incorporated into the design of experiment as another factor. The best results are obtained if the extractant/stripping solution is loaded with a representative profile of elements, which can be done using larger volumes of pregnant leach solution and extractant.

5.5. Future of Solvent Extraction

The cost of a SX circuit naturally depends upon the number of stages required (directly impacted by the selectivity of the extractant) and the size of the stages (directly impacted by the loading capacity of the extractant and rate of extraction and phase disengagement). The number and size of stages also dictate the inventory of solvent and stripping solutions that must be accounted for in working capital. Chemical consumption is a trade-off with the number of stages, where more stages generally allow for more efficient utilization of the stripping solutions. The specific extractant design dictates how much acid or base is needed to effectively strip and regenerate the solvent for recycling in the process. Developments in SX will therefore likely focus on improved selectivity for REEs and in particular improved separation factors between the lanthanides. In the same vein, these extractants will be designed to release cations at lower acidities, reducing the required chemical consumption. This type of development work can largely be performed now using rational design with molecular modeling software, but with care to arrive at compounds that can be synthesized economically at scale.

There has been some work on the development of photoisomerizing extractants that change conformers upon exposure to certain wavelengths of light (Hamada et al., 2006). This mechanism could allow for cycling of extraction and stripping in SX using only solar inputs. Such a process could reduce the cost and logistical burden of delivering chemicals to remote mine or refining locations. This work is still at very early stage, but could be transformational in SX processes.

On the other hand, there have been developments in technologies that compete with SX to selectively separate and purify REEs. Solid phase extractants aim to solve the same problems of selectivity and chemical consumption as SX, but also need to gain ground in loading capacity to compete with solvent extraction. There are emerging technologies such as photophoresis, electrophoresis, capacitive separation, and other electrochemical processes. Photophoresis and electrophoresis rely on different rates of motion between molecules exposed to light or electric fields. Electrophoresis is common in laboratory separation of proteins and other macromolecules, and development is focused on extending the concept to the ion scale. Capacitive separation is similar to solid phase adsorption techniques, but the binding is created by electric charges on the sorbent rather than chemical sorption. It has the potential to reduce or eliminate the need for chemicals in the process, but developments are needed in selectivity of the capacitor electrodes. There are early stage companies developing and implementing electrochemical methods specifically to separate and purify REEs, and they will directly challenge SX over the next few years.

Acknowledgments

The authors would like to acknowledge the U.S. Department of Energy's Office of Fossil Energy National Energy Technology Laboratory for funding portions of this work in rare earth recovery under contract number FE-0027012. We also thank other contributors from Battelle, including Slawomir Winecki, Ann Lane, Kathryn Johnson, and Bruce Monzyk.

References

Battelle. (2017). *Recovery of rare earth elements from coal and coal byproducts via a closed loop leaching process: System design report*. U.S. DOE office of fossil energy. Contract number FE-0027012.

Bostian, H. E. (1959). *Mechanism of extraction of the rare earth nitrates by tributyl phosphate*. Ames: Iowa State University of Science and Technology.

Gupta, C. K., & Krishnamurthy, N. (2004). *Extractive metallurgy of rare earths*. Boca Raton: CRC Press.

Hamada, F., Masuda, T., & Kondo, Y. (2006). Photoregulated-metal binding with an Azobenzene-capped Calix[4]arene. *Supramolecular Chemistry*, *5*(2), 129–131.

Kremers, T. M. (1945). The basicity characteristics of scandium, yttrium, and the rare earth elements. *Chemical Reviews*, *37*(1), 97–159.

Peppard, D., Faris, J., Gray, P., & Mason, C. (1953). Studies of the solvent extraction behavior of transition elements. I. Order and degree of fractionation of the trivalent rare earths. *Journal of Physical Chemistry*, *57*, 294–301.

Ritcey, G. M. (2006). *Solvent extraction principles and applications to process metallurgy*. Ottawa: G.M. Ritcey and Associates.

Stevenson, P., & Nervik, W. E. (1961). *The radiochemistry of the rare earths, scandium, yttrium, and actinium*. Washington DC: U.S. Atomic Energy Commission.

Weaver, B. (1953). Fractional separation of rare earths by oxalate precipitation from homogeneous solution. *Analytical Chemistry*, *26*(3), 479–480.

6

SEPARATION OF RARE EARTH ELEMENTS BY CRYSTALLIZATION

Kerstin Forsberg and Michael Svärd

Department of Chemical Engineering, KTH Royal Institute of Technology, Stockholm, Sweden

Abstract

Precipitation and crystallization have long been common components of hydrometallurgical processes involving rare earth elements (REEs), for purposes of concentration, separation, and recovery. In this chapter, different possible crystallization-based techniques for separation of REEs as solid compounds from solution are reviewed. Some of the techniques are well-established, while others are promising alternatives of lower technological maturity. Fundamental theory common to all methods of crystallization of ionic compounds is covered, solubility data of different rare earth salts are presented, and important aspects of the underlying chemistry behind rare earth crystallization are highlighted, in order to rationalize the behavior of the different REEs and to give a deeper understanding of the phenomena at hand. The chapter supplies information helpful in the design of a suitable and efficient crystallization-based REE-separation process.

Rare Earth Elements: Sustainable Recovery, Processing, and Purification, Special Publications 79,
First Edition. Edited by Athanasios K. Karamalidis and Roderick Eggert.

DOI: 10.1002/9781119515005.ch6

6.1. Industrial Applications

Rare earth elements (REEs) can be recovered from ores, industrial process residues, and end-of-life products (Binnemans et al., 2015; Chen, 2011; Goodenough et al., 2018). The rare-earth-bearing minerals or materials are usually leached by an acid or a base. They are then separated from other compounds, and/or from one another, before being recovered in a suitable product form. To cite Callow (1967): "Discussing lanthanon separation is like discussing chess. There are a limited number of opening moves, which can be analyzed in detail. As the game develops, possibilities multiply enormously and can only be considered in the broadest outline." Before 1947, oxidation and reduction together with crystallization were the only techniques available for separating REEs. However, the procedures were relatively inefficient and laborious. Commercial rare earth separation starting in the 1950s also made use of ion exchange and solvent extraction, which provided a much higher purity of the end products obtained (Gupta & Krishnamurthy, 2005). Hydrometallurgical processes today generally consist of different steps of pre-treatment, in order to improve metal dissolution in the aqueous phase, followed by leaching, concentration, and/or purification, and finally recovery of different metals from the leach solution.

Precipitation can be used to separate the rare earths into a light and a heavy fraction of sparingly soluble salts, for example, rare earth oxalates or sodium rare earth double sulfates. The compounds are then converted into commercial rare earth salts by various additional process steps (Binnemans et al., 2015). Individual separation of the REEs is often performed by solvent extraction (Gupta & Krishnamurthy, 2005). In many commercial processes of today, the rare earths are stripped from loaded solvent extractants using aqueous solutions of inorganic acids. The dissolved rare earths are then precipitated as, for example, sparingly soluble oxalates and carbonates, from which oxides can be recovered by calcination (Gupta & Krishnamurthy, 2005). Commercial rare earth products include fluorides, nitrates, carbonates, oxides, alloys, and pure metals. Which product is produced is determined by the rare earth contents in the starting material and on the requirements of the target consumer (Chen, 2011). The value of the end product increases with the purity. The aim of crystallization processes is to achieve a high capacity, yield, and purity. Downstream processing, such as washing and filtration, has a large impact on the purity of the concentrate. These steps are directly dependent on the crystal size and morphology. In order to design an efficient crystallization process and to recover crystals of high quality in terms of purity, size distribution, morphology, etc., supersaturation control is of paramount importance.

6.2. Fundamentals of Crystallization

A crystallization process consists of several concurrent processes, most importantly nucleation and growth. The thermodynamic driving force behind these is the supersaturation. For an ionic compound forming the solid phase $A_nB_m(s)$ from n ions of A and m ions of B, the supersaturation S can be expressed as:

$$S = \left(\frac{(a_A)^n (a_B)^m}{(a_A^{eq})^n (a_B^{eq})^m} \right)^{1/(n+m)}, \quad (1)$$

where a_i denotes the activity of component i in solution and a_i^{eq} the corresponding activity at equilibrium with the solid (Mersmann, 2001). The activity is defined as the concentration modified by an activity coefficient describing the deviation from ideality. As activity coefficients are nontrivial to estimate, especially at supersaturation, the common assumption is to neglect them, replacing activities with concentrations. For a 1:1 compound, Equation (1) then reduces to:

$$S \approx \sqrt{\frac{[A][B]}{[A]^{eq}[B]^{eq}}} = \sqrt{\frac{[A][B]}{K_{sp}}}, \quad (2)$$

K_{sp} is the (activity-based) solubility product:

$$K_{sp} = (a_A^{eq})^n (a_B^{eq})^m, \quad (3)$$

which for a 1:1 compound, neglecting activity coefficients, is written:

$$K_{sp} \approx [A]^{eq}[B]^{eq}, \quad (4)$$

For any multicomponent solution, there will be a number of conceivable solid phases, each with their own solubility. From a given solution at a specific set of conditions, any solid phase for which the solution is supersaturated could theoretically emerge. Possibilities include hydrates as well as anhydrates, and different crystal structures of the same compound (polymorphs) as well as compounds with different stoichiometries.

Supersaturation is the most important factor governing the crystallization kinetics; both the growth rate and in particular the rate of primary nucleation exhibit nonlinear dependence on S. Controlling the supersaturation is crucial for controlling key properties of the solid product: particle size, morphology, purity and degree of crystallinity, and (not least) which crystal phase is obtained. In order to control the supersaturation and the product properties, knowledge of the solubility of the desired solid phase, as well as of any other potential solid phase that might form, is required.

Nucleation is the process whereby new crystals are generated from a supersaturated solution. Nucleation events occurring as a result of spontaneous aggregation of dissolved matter are termed primary, while nucleation occurring as a direct result of the presence of crystals, such as crystal breakage or shear-induced dislodgement of ordered aggregates from a growing crystal, is termed secondary. Primary nucleation can be homogeneous (nucleation occurring in the solution bulk) or heterogeneous (nucleation induced by the presence of a foreign surface). At low-to-moderate supersaturation, heterogeneous nucleation tends to dominate, while homogeneous nucleation becomes important at high supersaturation (Kashchiev, 2003).

There is an ongoing debate as to the mechanisms behind primary nucleation. According to the classical theory, originally based on the works of Gibbs (1876), nuclei arise as a result of the growth of clusters—small aggregates postulated to have a crystalline structure. As crystallization is a first-order phase transition, the formation of a crystal–solution interface is associated with an unfavorable energy, the interfacial energy. At supersaturated conditions, however, the formation of the crystal lattice itself is energetically favorable. This means that small clusters will tend to dissolve, and that there will be a threshold—a critical size–above which the total energy of the cluster will be favorable, and it will tend to grow into a crystal. The critical nucleus size is a function of the saturation ratio. As the supersaturation increases the critical nucleus size decreases, and thus nucleation at a high supersaturation is expected to yield a large number of small particles. The local concentration fluctuations of the continuous addition and removal of monomers (atoms, ions, or molecules) to a cluster are complex, resulting in the stochastic and unpredictable nature of primary nucleation.

More recent theories focus on the complex role and appearance of clusters (Gebauer et al., 2014). Recent advances in light-scattering-based detection techniques have allowed subcritical clusters to be studied, and gradually a more nuanced picture is emerging of the different types and structures of clusters possible. In the two-step theory of nucleation (Erdemir et al., 2009), fluctuations in concentration are separate from fluctuations in long-range order. Nucleation occurs in two steps, where the first involves the formation of dense, amorphous, liquid-like clusters, followed by crystal nucleation inside the clusters. Nucleation proceeding according to such a mechanism has been observed for various kinds of systems, ranging from proteins to salts. In some cases, classical-type nucleation seems to coexist with nonclassical nucleation. Nielsen et al. (2014) observed crystals of calcite and vaterite—two polymorphs of $CaCO_3$—apparently nucleating through different mechanisms.

Crystal growth, like nucleation, is a complex process where the underlying mechanisms are not well understood. For an ionic solid, it is likely to involve bulk as well as surface diffusion of growth units (which may be monomers or smaller aggregates), with associated counterdiffusion, desolvation, and finally integration into the lattice (Mullin, 2001). Depending on the relative rates of these processes, growth can be more or less controlled by mass transport or lattice integration.

The prerequisite for crystallization—supersaturation—can be generated either through actions on a solution of a product, such as cooling or heating (depending on the temperature dependence of the solubility), or evaporation of solvent, or alternatively by changing the system through addition of a chemical. This can be either another solvent or species chosen so as to reduce the solubility of the product, or one of the reactants in a reaction crystallization, that is, where a sparingly soluble, solid product is formed in solution through the combination of soluble reactants.

A term frequently encountered in conjunction with crystallization processes is 'precipitation.' The terminology is frequently unclear, and the terms 'precipitation' and 'crystallization' are often used loosely. 'Precipitation' generally refers to rapid crystallization processes, carried out under conditions of high supersaturation. Although in principle, this could apply to many types of processes, such as quench cooling, spray drying, or rapid antisolvent crystallizations, the term is commonly used synonymously with reaction crystallization, where, unless the products are not very soluble, the supersaturation with respect to the product tends to be high.

Traditionally, in precipitation processes, the focus has been on thermodynamics and solubility (Demopoulos, 2009), with the target being, for example, to maximize the yield of a particular product or the removal of a particular ionic species. However, while thermodynamics govern which phases can precipitate and to what extent, the nature and quality of the precipitating product also depend on kinetics. Because precipitation processes operate under high supersaturation, nucleation and growth rates will be high, which tends to reduce the product purity. The supersaturation in precipitation processes is generally difficult to control; local supersaturation gradients can be significant, and good mixing becomes crucial. This is of particular importance at the feed point of reactants or antisolvent (Torbacke & Rasmuson, 2001). In addition, the high supersaturation levels can also cause less stable phases to precipitate due to favorable kinetics. Such metastable phases could be other polymorphs of the same compound, or a compound (anhydrate or hydrate) with another stoichiometry. Amorphous (noncrystalline solid) phases are also possible. Phase transformations in solution to a more stable form may occur in minutes, or require years or more, depending on factors such as temperature and supersaturation (Spanos & Koutsoukos, 1998). Methods to promote the precipitation of the desired crystal phase, and promote crystal growth over primary nucleation to produce larger product crystals with increased purity, include seeding, ensuring adequate mixing to neutralize concentration gradients, and careful (slow) addition of reagent solution or antisolvent in a semi-batch process.

For precipitation processes, crystal growth through aggregation/agglomeration can be important. Aggregation is more important for small particles, that is, shortly after nucleation. If colloid stability is suppressed, small crystals (in the size range 0.01–0.1 μm) will tend to aggregate (Demopoulos, 2009). Aggregation is favored as the interaction potential energy of the two double layers decreases. This is achieved when the ionic strength of the solution increases.

Another process of importance is Ostwald ripening. With decreasing crystal size, the surface-to-bulk ratio increases, and the solubility becomes a function of the crystal size. This is described by the Gibbs–Thomson equation, and is particularly important for nanocrystalline precipitates:

$$C^{\mathrm{eq}} = C_{\infty} \exp\left(\frac{2\gamma V_{\mathrm{m}}}{rRT}\right), \tag{5}$$

where C^{eq} is the solubility of a crystalline particle with radius r (m), C_{∞} is the solubility of an infinitely sized crystal, γ is the specific interfacial energy (J m^{-2}), and V_{m} is the molar volume of the crystal (m^3 mol^{-1}). For example, it has been shown that the solubility of scorodite increases at least by a factor 2 as the crystallite size decreases from 200 to 10 nm (Demopoulos, 2009). The difference in solubility leads to a growth of larger crystals at the expense of smaller ones, also in saturated solutions.

6.3. Solubility Trends of Rare Earth Elements

All REEs most commonly occur as trivalent ions. The radii of the hydrated lanthanide trivalent ions decrease steadily from left to right in the periodic table. This is called the lanthanide contraction. The lanthanides differ with respect to the number of electrons in the 4f-shell, and the lanthanide contraction is caused by poor shielding of the nuclear charge by the electrons in this shell. This shell has little effect on bonding, however, and the significant difference between two lanthanides is often only their atomic or ionic radius. The lanthanide contraction results in a change of ionic radius of approximately 0.01 Å between each element (Diwu et al., 2012). The REEs can be divided into the light REE group (La to Eu plus Sc) and the heavy group (Gd to Lu plus Y) exhibiting slightly different properties introduced by the lanthanide contraction. The light trivalent lanthanides coordinate nine water molecules, while the slightly smaller heavier lanthanides and yttrium and scandium coordinate eight water molecules in aqueous solution (Helm & Merbach, 2005; Persson, 2010). Since the REE ions have very similar ionic radii, the lattice energy of their salts and the hydration energies of the ions are very similar, leading to similarities in solubility.

Discrepancies of more than an order of magnitude exist in reported solubility products for a given REE. Before the advances in solvent extraction and ion-exchange techniques in the 1950s, most rare earth samples were not very pure, which could explain discrepancies in older data sets. Some of the discrepancies arise because of differences in structure, degree of hydration, and crystallinity of the solids. In some cases, the nature of the solid has not been specified. In addition, the time needed for equilibration is often long and in some studies equilibrium might not have been reached at the time of sampling. The time it takes to reach solid–liquid equilibrium depends on a number of different factors such as temperature, pH, ionic strength, degree of crystallinity, surface area (particle size), solid concentration, and agitation. For example, Rard (1988) studied the aqueous solubility of Pr, Eu, and Lu sulfate octahydrates at 25 °C, and found that equilibration times up to 119 days were necessary. Furthermore, in determining solubility products, the speciation of the ions in solution must be known and these data are not always available or sufficiently accurate.

The solubility of rare earth salts often either increases or decreases along the series. The solubility trends along the lanthanide series for the rare earth carbonate (Firsching & Mohammadzadei, 1986), oxalate (Chung et al., 1998), hydroxide (Smith & Martell, 1976), phosphate (Liu & Byrne, 1997), and fluoride (Mioduski et al., 2014, 2015a, 2015b) salts are presented in Figure 6.1. As can be seen, all these salts are sparingly soluble in water.

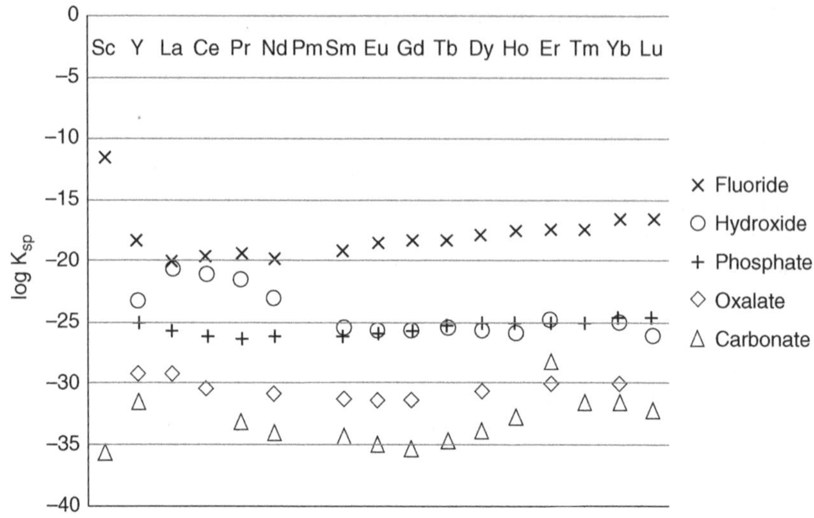

Figure 6.1 Solubility products, valid at 25 °C and zero ionic strength, for sparingly soluble REE salts (Chung et al., 1998; Firsching & Mohammadzadei, 1986; Liu & Byrne, 1997; Mioduski et al., 2014, 2015a, 2015b; Smith & Martell, 1976). K_{sp} is expressed in terms of $[REE]^n[L]^m$ for formation of the salt REE_nL_m.

The rare earths also form hydrated sulfate, chloride, and nitrate salts with much higher aqueous solubility (Mioduski, 1998; Mioduski et al., 2008, 2009a, 2009b; Siekierski et al., 1983). The solubility of these salts in pure water at 25 °C is presented in Figure 6.2. The figure shows the total concentration of the REEs in the aqueous phase at equilibrium with the solid phases when dissolved in pure water. The most stable hydrate form of the sulfate, chloride, and nitrate salts in pure water at 25 °C varies along the series. The lighter (larger) La, Ce, and Pr are most stable as chloride heptahydrates while the heavier (smaller) Nd, Lu, and Y form chloride hexahydrates at 25 °C (Das et al., 2017; Mioduski et al., 2008, 2009a, 2009b). The nitrate salts of La to Yb and Y are most stable as hexahydrates at 25 °C, while Lu forms a pentahydrate and Sc a tetrahydrate (Siekierski et al., 1983). The REEs can form a range of different sulfate hydrate salts (Cunningham et al., 2024; Das et al., 2019; Kazmierczak & Höppe 2011; Mioduski, 1998; Wickleder, 2002). Scandium is most stable as a pentahydrate and lanthanum and cerium are most stable as nonahydrates in water at 25 °C, while the heavier REEs form octahydrates. Polymorphs have been identified for some of the sulfate salts (Casari & Langer, 2007). Lower hydrates are obtained at higher temperature and when the respective acid concentration increases. The solubility of the REE nitrate and chloride

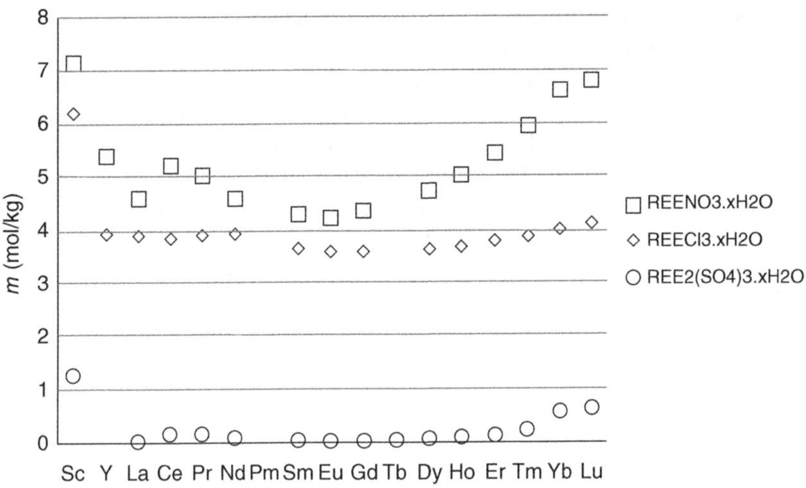

Figure 6.2 Solubility of the most stable hydrated sulfate, chloride, and nitrate rare earth salts in pure water at 25 °C (Mioduski, 1998; Mioduski et al., 2008, 2009a, 2009b; Siekierski et al., 1983).

hydrates increases as the temperature increases, while the solubility of the REE sulfate hydrates decreases as the temperature increases.

Although the most significant difference between two lanthanides is often their atomic or ionic radii, the solubility of a given rare earth salt cannot be directly related to the radii of the rare earth ions. The solubility of isostructural rare earth salts may increase with decreasing radius, decrease with decreasing radius, or change irregularly, see Figures 6.1 and 6.2. These observations have not yet been fully explained, and there is in fact no simple method of predicting solubility trends for the various rare earth salts (Chung et al., 1998).

Because of the electronic structure and large sizes of the trivalent lanthanide ions, the bonding is predominantly ionic. However, with decreasing ionic radius, lanthanide salts become somewhat less ionic. Thus, the covalent character of bonding increases along the series from La to Lu. Often differences in solubility of ionic solids can be explained in terms of attraction between the ions in solution and in the crystal lattice. Hydration of ions favors the dissolution of an ionic solid in water. The energy of hydration depends on the ionic radius. Smaller ions have a more concentrated charge and a stronger electric field that attract water molecules compared to larger ions. Therefore, the energy of hydration increases (which favors increased solubility) as the ionic radius of the REE decreases. It also increases as the charge of the ion increases.

Another important parameter that influences the solubility of ionic solids is the lattice energy. With increased lattice energy the solubility of an ionic solid decreases. The lattice energy is higher for more highly charged ions. It is also inversely proportional to the distance between the center of neighboring positive and negative ions. This distance depends on the radii of the ions in the solid. Thus, the lattice energy decreases (resulting in an increased solubility) as the radius of the ions increases. Ergo, the effect of lattice energy on solubility with changing ionic radii counteracts the effect of hydration on the solubility of a specific salt. The dehydration and lattice energies as well as the mean molal activity coefficients in concentrated solutions and the enthalpies of formation of the hydrates may all change smoothly with rare earth atomic number for isostructural rare earth salts. Therefore, solubility data for REE salts can be estimated based on data for isostructural REE salts, for which more accurate data exist (Guminski et al., 2011; Mioduski et al., 2013).

The basicity of the lanthanides is an important property that will affect how they will react with other elements. It can be considered a measure of the ease with which a species loses electrons, or equivalently the lack of attraction that a cation has for electrons or anions (Moeller & Kremers, 1945). Basicity decreases with decreasing atomic or ionic radius. Thus, lanthanum, having the largest ionic radius, is the most basic (among trivalent rare earths) and has the least attraction for anions. Tetravalent cerium and praseodymium are slightly more acidic in character compared to the trivalent rare earths due to their higher charge. Moeller and Kremers (1945) report a relatively small basicity difference between samarium and europium and between ytterbium and lutetium. This conclusion was based on the relatively small differences in the molecular volumes of the octahydrate sulfates of these rare earths, that is, based on the smaller size differences. As the basicity decreases along the series, the solubility is expected to decrease for the lanthanide hydroxides, carbonates, and oxalates. Basicity differences are also reflected in the hydrolysis of ions where less basic ions (heavier lanthanides) hydrolyze more readily. By decreasing the pH in the solution, hydrolysis can be prevented. The pH at which hydrates start to precipitate from an aqueous solution decreases gradually along the series moving from the lighter to the heavier lanthanides. Hydration can also play an important role for acid–base properties. The basicity can be reduced in aqueous solution due to hydration of the basic ion, which stabilizes the ion making it less basic. In some cases, four consecutive, smooth curves are obtained when plotting the logarithm of a certain, observable parameter (e.g., a distribution coefficient between two phases) versus the atomic number of the lanthanides (Andersson et al., 2006). This so-called tetrad effect can be explained by the stabilization energy related to the inter-electron repulsion energy for the 4f electrons. This effect can also explain some of the trends seen in the solubility along the lanthanide series (Firsching & Mohammadzadei, 1986).

6.4. Crystallization of Rare Earth Elements

Precipitation processes tend to be carried out at high supersaturation. Because of their chemical similarity, a high supersaturation of a compound of one REE generally means a high supersaturation with respect to similar compounds of the other REEs as well. For this reason, most precipitation processes are more suitable for separating a REE concentrate from other components in a solution than for separating the REEs into different fractions. Fractional crystallization refers to processes where individual components in a multicomponent mixture are selectively separated by crystallization. This requires knowledge of the phase diagram, and careful optimization of the process, which frequently has to be separated into several stages. Fractional crystallization can be accomplished by antisolvent addition, variation of the temperature, or evaporation.

The lanthanide contraction results in similar sizes of the lanthanides and the transition metals, which makes the separation of these elements more difficult. The separation is most often achieved by either adding a reagent (e.g., oxalic acid or a sodium salt) or by increasing the basicity of the solution (e.g., using magnesia or caustic soda). It is also possible to selectively crystallize the REEs by adding an antisolvent (Korkmaz et al., 2020). Selective precipitation can be used to achieve separation on the basis of light and heavy REEs. A redox reaction can also be utilized to selectively separate, for example, Ce (in tetravalent state) and Eu (in divalent state) by precipitation from the other REEs present in trivalent state.

The separation of REEs by precipitation is directly connected to the coordination chemistry of the REEs. The hard Lewis acid character of the trivalent rare earth ions (basicity), their large sizes, and their f electronic configurations all influence their coordination chemistry, making it rather different from that of the trivalent transition metal ions. In order to attract the large REE cations, the charge of the anions should be high, the size of the ions should preferably be small, or they should have chelating abilities. The large size of the trivalent rare earth ions leads to a high coordination number in the complexes formed. Usually, the coordination number is between 7 and 10 (Gupta & Krishnamurthy, 2005). Corresponding complexes of the trivalent transition metal ions form more stable complexes, that is, they are less dissociated (Huang, 2010).

6.5. Precipitation by Reagent Addition

Reactive precipitation of REEs with high yield by addition of a reagent is very common in hydrometallurgical processing. Sparingly soluble rare earth salts obtained by reagent addition include hydroxide, carbonate, oxalate, phosphate, fluoride, and sodium double-sulfate rare earth salts.

6.5.1. Hydroxides

Basicity precipitation (e.g., using magnesia or caustic soda) is widely used in industry. Ammonia has also been used in large scale to precipitate hydroxides from nitrate or chloride solutions. Although each metal has ideal precipitation conditions at different pH and temperature metals tend to coprecipitate. This can be due to locally higher pH in the feed point. Furthermore, by addition of caustic alkalis, there is a risk of contamination by occlusion of alkali and formation of basic salts. In addition, the precipitates are often gelatinous and difficult to process downstream. Magnesia (MgO) can be a better choice than caustic alkalis because it reacts more slowly (Callow, 1967).

6.5.2. Carbonates

Carbonates are almost insoluble in water, but readily soluble in dilute acids. Carbonates can be precipitated from aqueous solutions, for example, by addition of alkali carbonates, bicarbonates, or carbon dioxide gas. The concentrations of REEs at a given pH depend strongly on the carbon dioxide partial pressure and the temperature. Different carbonate hydrates and hydroxy carbonates are stable in aqueous solution, depending on the conditions (Meinrath & Takeishi, 1993). The precipitation rates of REE carbonates increase with increasing carbon dioxide pressure and with decreasing temperature (Konishi & Noda, 2001). Gentle heating will dehydrate the carbonate salts and stronger heating will decompose them to oxides.

In the presence of alkali metal ions (A^+), rare earth double carbonates, $AREE(CO_3)_2 \cdot xH_2O$, can precipitate (Kim et al., 2018). This is undesirable if pure lanthanide salts that are free of alkali ions are required. The double carbonate salts are thermodynamically stable in concentrated aqueous alkali carbonate solution, while at lower alkali or carbonate concentration, $REE_2(CO_3)_3 \cdot xH_2O$ and $REEOHCO_3$ compounds are thermodynamically stable for high and low CO_2 partial pressures, respectively (Philippini et al., 2008). These phases might also precipitate first in concentrated Na_2CO_3 solutions and then transform into the stable $AREE(CO_3)_2 \cdot xH_2O$ form.

6.5.3. Oxalates

The oxalates have a very low solubility in acidic solutions and the solubility, at least partly, follows the basicity series of the REEs, see Figure 6.1. The REEs can also be separated from other metal ions by oxalate precipitation, since many metals (e.g., Fe, Al, Ti, Zr, Nb, and Mo) remain in solution as soluble oxalate complexes (Gupta & Krishnamurthy, 2005). However, extensive formation of soluble oxalate metal complexes leads to a high consumption of oxalate in precipitation of the less-soluble REE oxalates. Oxalates are usually precipitated as hydrated salts ($REE_2(C_2O_4)_3 \cdot 10H_2O$) and the degree of hydration depends on the conditions used (Combes et al., 1997). The rare earths can also form ternary oxalate salts with alkaline metal ions or ammonium ion as the third component (Chapelet-Arab et al., 2006; Gammons & Wood, 2000). The solubility of REE oxalates is practically independent of temperature, while the solubility of the sodium REE oxalates decreases with increased temperature (Gammons & Wood, 2000).

6.5.4. Sodium Double Sulfates

The REEs also form a large number of sparingly soluble ternary rare earth sulfate salts with alkaline metal ions or ammonium ion as the third component (Lokshin et al., 2005; Wickleder, 2002). One of the most widely used precipitation methods to separate REEs from acidic solutions is by precipitation of sodium REE double-sulfate hydrates ($NaREE(SO_4)_2 \cdot H_2O$) through the addition of sodium salts such as NaOH, Na_2SO_4, or Na_2CO_3 (Gupta & Krishnamurthy, 2005). These salts are only slightly soluble in acidic solutions. The solubility decreases with increasing temperature. The solubility also increases with increased atomic number along the lanthanide series, and double-sulfate precipitation can be used to separate the light REE group from the heavy REE group. The different REEs can then be separated from each other in subsequent steps after first converting the double sulfates into a highly soluble compound, for example, REE hydroxides. Solubility data of $NaRE(SO_4)_2 \cdot H_2O$ can be found in the literature (Lokshin et al., 2005).

6.6. Redox Precipitation

Utilizing the redox properties for separation of the rare earths when applicable is common practice in industrial separation. Most rare earths are present as trivalent ions in aqueous solution. However, cerium, terbium, and praseodymium can also occur in tetravalent state and europium, samarium, and ytterbium can occur in divalent state. The +2 and +4 oxidation states are due to the presence of an either half-filled, completely filled, or empty 4f subshell. The reduction potentials of the Ln(IV)/Ln(III) couples for Tb and Pr are so high that they are reduced by water (Gupta & Krishnamurthy, 2005). Moreover, the divalent states of Yb and Sm are unstable to oxidation by water and air (Donohue, 1979). Complexing agents can stabilize the divalent or tetravalent lanthanides in solution; for example, Tb(IV) can be stabilized by a tetrametaphosphate ($P_4O_{12}^{4-}$) anion in alkaline aqueous solution (Li et al., 1991). Synthesis of nonclassical divalent lanthanides (i.e., not Eu, Sm, and Yb) was considered impossible until 1997 when the first stable complex of Th(II) was discovered. Since then, several more stable divalent lanthanide complexes have been detected. The divalent ions of alkaline earth metals and divalent lanthanide ions show a striking similarity in crystal structure of corresponding salts (e.g., fluorides, chlorides, and bromides). This is especially the case for Sr^{2+} and Ba^{2+}. Due to this fact, Sr^{2+} and Ba^{2+} can be used to carry divalent lanthanide ions for separation purposes.

Cerium is often separated from other REEs in its tetravalent state. In alkaline solutions, cerium can be oxidized from its trivalent state to its tetravalent state by bubbling oxygen through the solution (Kim et al., 2011). In acidic solutions, the oxidation can be performed by strong oxidants such as permanganate, persulfate, silver oxide, lead oxide, or bismuth or by electrochemical or photochemical processes (Donohue, 1979; Kim et al., 2011). Tetravalent cerium is smaller than trivalent REEs and can easily be precipitated as cerium hydroxide by increasing the basicity of the solution while the trivalent REEs remain in solution.

Europium can be reduced into divalent form chemically by zinc powder or zinc amalgam or electrochemically by graphite or titanium electrodes (Hung & Lee, 1992; Van de Voorde et al., 2018). Divalent Eu can then be precipitated as, for example, $EuSO_4$ or $EuCl_2{\cdot}2H_2O$ while leaving the trivalent REEs in solution (Van den Bogaert et al., 2015). However, chemical reduction by addition of zinc could contaminate the REE product. Electrochemical reduction in aqueous solutions is hampered by the evolution of hydrogen gas, which leads to a low current efficiency (Hung & Lee, 1992).

Redox reactions can also be initiated photochemically. Cerium, neodymium, samarium, and europium have been separated from other lanthanides in aqueous solution by photochemical oxidation to tetravalent (Ce) or reduction to divalent states (Eu, Sm, and Nd) followed by precipitation (Donohue, 1979; Tsushima et al., 1997; Van den Bogaert et al., 2015). Donohue (1979) used iodate as precipitation agent to separate Ce(IV) from an aqueous solution of other trivalent lanthanides after photooxidation, since Ln(IV) iodates are much less soluble than Ln(III) iodates. The solubility of $Ln(IO_3)_3$ is increased in strongly acidic solutions and in the experiments HCl or HNO_3 was added to increase the acidity.

6.7. Antisolvent Crystallization

In antisolvent crystallization, crystallization is effected through the addition of a second solvent to a solution. The principle is to make use of the fact that the activity coefficient, and thereby the solubility, of a compound depends on the solvent environment. An antisolvent is thus a solvent which when added to a solution reduces the—initially high—solubility. If the solubility is decreased more than the inevitable reduction in concentration caused by the increased amount of solvent, the solution can become supersaturated, leading to crystallization. After filtration to separate the solids, the solvents can be separated and recovered by *e.g.* distillation.

The antisolvent is often an organic solvent, which has to be miscible with the original solvent (typically water) under the crystallization conditions, such as alcohols. The dielectric constant of a solvent is a measure of its ability to solubilize ions. Alcohols, with lower dielectric constants than water, contribute to a decreased solvation, and thus reduce the solubility of ionic compounds. In addition, organic solvents can decrease the water activity and thus reduce the solubility of hydrate forms (Zhu et al., 1996). Peters et al. (2020) demonstrated a decreasing solubility of $(NH_4)_3ScF_6$ in aqueous ammonium fluoride with an increase in alcohol concentration for a range of studied alcohols. The decrease was shown to be correlated to a decrease in the effective dielectric constant of the solvent medium.

Antisolvent crystallization is widely used in *e.g.* the pharmaceutical industry to obtain different pure compounds in solid form. Among its advantages is the possibility to obtain controlled particle size distributions with good scalability. It has also been shown to be a potentially cost-effective alternative to evaporative crystallization (Zijlema et al., 2000). However, it is not a common process in the mineral and metallurgical industries. One reason for this is the large volume of antisolvent often needed to reach high yields (>90%), where a doubling of the total liquid volume is not uncommon (Moldoveanu & Demopoulos, 2014). In addition, there might be restrictions on introducing new organic solvents (e.g., alcohols) into established industrial processes.

Crystallization by antisolvent addition has the potential to be a viable option to obtain high-value elements, such as the REEs, in small scale. It can enable the precipitation of new REE salts, with high aqueous solubility, which are thus unsuitable for production using reaction crystallization. By adding a suitable antisolvent, these salts can be precipitated under acidic conditions at high yields, which is important in developing future processes for recovery of high-value end products from waste materials in sustainable processes with maximal resource recovery and minimal waste generation. Antisolvent precipitation could be an alternative to precipitation of REE oxalates or REE hydroxides, with the advantages that the solvents can be recovered and recycled and less acid is consumed. It can also be possible to achieve selective separation of REEs from elements present at higher concentrations by antisolvent crystallization. Recently, antisolvent crystallization has been applied to selectively separate REEs as a concentrate of sulfate hydrates from acidic media also containing high concentrations of, for example, iron, nickel and cobalt (Pawar et al., 2024; Korkmaz et al., 2020). The relatively high solubility of the REE sulfates in aqueous media allows for further processing of the concentrate and individual separation of the REEs by solvent extraction or elution chromatography. Antisolvent crystallization has also been applied to precipitate ammonium scandium fluoride, which is a good precursor for production of scandium metal, at high yield and purity (Peters et al., 2020).

Just as for reaction crystallization, however, addition of the antisolvent tends to create a high local supersaturation at the feed point. This can easily result in uncontrolled primary nucleation and—as a consequence—very small product crystals (<10 mm) (Myerson et al., 2019; Peters et al., 2019). This can in turn lead to difficulties in downstream dewatering and washing processes, resulting in possible impurity incorporation from entrained mother liquor. The mother liquor in hydrometallurgical precipitation of metal salts often contains ionic impurities. These problems can be mitigated through careful control of key process conditions, including location of antisolvent feed point, antisolvent addition rate, initial dilution of the antisolvent, and judicious seeding (Peters et al., 2022).

6.8. Crystallization by Temperature Modification or Evaporation

Crystallization can also be carried out through temperature modification or solvent evaporation. For many REE-containing compounds, however, the solubility is low, and because of the resulting low yield and productivity these methods are not suitable. For compounds where the solubility is high, and/or the temperature dependence of the solubility is sufficiently large, REEs can be selectively separated through fractional crystallization by either changing the temperature or evaporating the solvent. In a report from the U.S. Department of the Interior's Bureau of Mines from 1971 (Parker et al., 1971), this was considered the best method among the classical separation methods to obtain individual elements of high purity. The REEs with the highest potential to be separated by this technique are the ones at the beginning of the lanthanide series, where the differences in cationic radius are the greatest. The heavy lanthanides have very similar ionic radius, and as an example a bromate salt of a thulium fraction still contained erbium and ytterbium after being recrystallized 15,000 times (Gupta & Krishnamurthy, 2005).

Several salts have been used to separate REEs, most notably nitrates, bromates, and ethyl sulfates. The rare earth nitrates are highly soluble, including in many organic solvents. Individual trivalent REEs in high purity have been produced by separation as double nitrates. Especially, double ammonium nitrate salts ($REE(NO_3)_3{\cdot}2NH_4(NO_3)_2{\cdot}4H_2O$) and double magnesium nitrate salts ($2REE(NO_3)_3{\cdot}3Mg(NO_3)_2{\cdot}24H_2O$) have been widely used for fractional crystallization of light to medium REEs (Callow, 1967). Double manganese nitrates ($2REE(NO_3)_3{\cdot}3Mn(NO_3)_2{\cdot}24H_2O$), ethyl sulfates ($REE(C_2H_5SO_4)_3{\cdot}9H_2O$), and bromates ($REE(BrO_3)_3{\cdot}9H_2O$) have also been proven useful (Gupta & Krishnamurthy, 2005). The solubility of rare earth bromides, $REEBr_3{\cdot}xH_2O$, decreases strongly with decreasing temperature, making cooling crystallization suitable, but seeding is frequently necessary (Callow, 1967).

Complexing agents can also be used in fractional crystallization. The trivalent REE cations can form water-soluble complexes with amino polycarboxylic acids, for example, ethylenediaminetetraacetic acid (EDTA) and hydroxyethylenediaminetriacetic acid (HEDTA). The formation of these complexes is pH-dependent with a stability dependent on the ionic radius of the rare earth ions. This can be utilized for fractional crystallization; for example, a sodium REE EDTA salt has been used to separate Gd, Tb, Dy, and Y (Marsh, 1955).

6.9. Conclusions

Crystallization comprises a family of techniques that allow REEs to be separated from other elements and obtained as a concentrate, or as a solid end product in the form of a pure compound of a certain REE or a mixture of REEs. The similar properties of the REEs in most cases make crystallization unsuitable as a method to separate the REEs from each other. An exception is the separation of Ce(IV) and Eu(II) from the other trivalent REEs after oxidation or reduction, respectively, from the trivalent state.

The selection of crystallization technique must be chosen so as to take into consideration both upstream and downstream processing. By giving due consideration to the underlying REE chemistry, and the kinetics of the crystallization process, through the hydrodynamics and the rate of generation of supersaturation, a proper selection of crystallization technique and design of the process, ensuring the best possible product quality, can be made.

Currently, reaction crystallization is the dominating technique for separation of REEs by crystallization. Among the reviewed techniques, antisolvent crystallization has been identified as a future promising alternative for selective separation of REEs as a concentrate. By antisolvent crystallization it is possible to obtain REE salts with high solubility in aqueous solution, which could be advantageous in further processing, for example, for individual separation by other techniques. Further work is encouraged in order to investigate the potential for selective separation of REEs from common metals and how to design efficient hydrometallurgical unit operations and flow sheets based on antisolvent crystallization.

Acknowledgments

The authors would like to acknowledge funding from Formas (the Swedish research council for sustainable development), through grants numbers 2019-1150 and 2022-1452.

References

Andersson, S., Eberhardt, K., Ekberg, C., Liljenzin, J.-O., Nilsson, M., & Skarnemark, G. (2006). Determination of stability constants of lanthanide nitrate complex formation using a solvent extraction technique. *Radiochimica Acta*, *94*(8), 469–474.

Binnemans, K., Jones, P. T., Blanpain, B., Van Gerven, T., & Pontikes, Y. (2015). Towards zero-waste valorisation of rare-earth-containing industrial process residues: A critical review. *Journal of Cleaner Production*, *99*, 17–38.

Callow, R. J. (1967). *The industrial chemistry of lanthanum, yttrium, thorium and uranium*. New York, USA: Pergamon.

Casari, B. M., & Langer, V. (2007). New structure type among octahydrated rare-earth sulfates, β-$Ce_2(SO_4)_3{\cdot}8H_2O$, and a new $Ce_2(SO_4)_3{\cdot}4H_2O$ polymorph. *Zeitschrift für Anorganische und Allgemeine Chemie*, *633*, 1074–1081.

Chapelet-Arab, B., Duvieubourg, L., Nowogrocki, G., Abraham, F., & Grandjean, S. (2006). *Journal of Solid State Chemistry*, *179*, 4029–4036.

Chen, Z. (2011). Global rare earth resources and scenarios of future rare earth industry. *Journal of Rare Earths*, *29*(1), 1–6.

Chung, D-Y., Kim, E-H., Lee, E-H., & Yoo, J-H. (1998). Solubility of rare earth oxalate in oxalic and nitric acid media. *Journal of Industrial and Engineering Chemistry*, *4*(4), 277–284.

Combes, E., Sella, C., Bauer, D., & Sabot, J. L. (1997). Precipitation-stripping of yttrium oxalate powders from yttrium-loaded HDEHP organic solutions using an ultrasonic stirrer. *Hydrometallurgy*, *46*, 1–12.

Cunningham, S., Etherington-Rivas, M., & Azimi, G. (2024). Solubility of neodymium and dysprosium sulfates at different pH and temperature and the effect of yttrium sulfate, sodium sulfate, and ammonium sulfate mixtures: Strengthening the predictive capacities of the OLI software. *Hydrometallurgy*, *224*, 106253.

Das, G., Lencka, M. M., Eslamimanesh, A., Anderko, A., & Riman, R. E. (2017). Rare-earth elements in aqueous chloride systems: Thermodynamic modeling of binary and multicomponent systems in wide concentration ranges. *Fluid Phase Equilibria*, *452*, 16–57.

Das, G., Lencka, M. M., Eslamimanesh, A., Wang, P., Anderko, A., Riman, R. E., & Navrotsky A. (2019). Rare earth sulfates in aqueous systems: Thermodynamic modeling of binary and multicomponent systems over wide concentration and temperature ranges. *Journal of Chemical Thermodynamics*, *131*, 49–79.

Demopoulos, G. P. (2009). Aqueous precipitation and crystallization for the production of particulate solids with desired properties. *Hydrometallurgy*, *96*, 199–214.

Diwu, J., Grant, D. J., Wang, S., Gagliardi, L., & Albrecht-Schmitt, T. E. (2012). Periodic trends in lanthanide and actinide phosphonates: Discontinuity between plutonium and americium. *Inorganic Chemistry*, *51*(12), 6906–6915.

Donohue, T. (1979). Photochemical separation of cerium from rare earth mixtures in aqueous solution. *Chemical Physics Letters*, *61*(3), 601–604.

Erdemir, D., Lee, A. Y., & Myerson, A. (2009). Nucleation of crystals from solution: Classical and two-step models. *Accounts of Chemical Research*, *42*(2009), 621–629.

Firsching, F. H., & Mohammadzadei, J. (1986). Solubility products of the rare-earth carbonates. *Journal of Chemical and Engineering Data*, *31*, 40–42.

Gammons, C. H., & Wood, S. A. (2000). The aqueous geochemistry of REE. Part 8: Solubility of ytterbium oxalate and the stability of Yb III–oxalate complexes in water at 25 °C to 80 °C. *Chemical Geology*, *166*, 103–124.

Gebauer, D., Kellermeier, M., Gale, J. D., Bergström, L., & Cölfen, H. (2014). Pre-nucleation clusters as solute precursors in crystallisation. *Chemical Society Reviews*, *43*, 2348–2371.

Gibbs, J. W. (1876). On the equilibrium of heterogeneous substances. *Transactions of the Connecticut Academy of Arts and Sciences*, *3*, 108–248.

Goodenough, K. M., Wall, F., & Merriman, D. (2018). The rare earth elements: Demand, global resources, and challenges for resourcing future generations. *Natural Resources Research*, *27*(2), 201–216.

Gumiński, C., Voigt, H., & Zeng, D. (2011). Solubility of rare earth metal bromides and iodides in aqueous systems. *Monatshefte für Chemie*, *142*, 211–218.

Gupta, C. K., & Krishnamurthy, N. (2005). *Extractive metallurgy of rare earths*. Boca Raton, USA: CRC Press.

Helm, L., & Merbach, A. E. (2005). Inorganic and bioinorganic solvent exchange mechanisms. *Chemical Reviews*, *105*(6), 1923–1959.

Huang, C. (2010). *Rare earth coordination chemistry: Fundamentals and applications*. Singapore: John Wiley & Sons.

Hung, T. M., & Lee, C. J. (1992). Electrochemical reduction of Eu(III) using a flow-through porous graphite electrode. *Journal of Applied Electrochemistry*, *22*, 865–871.

Kashchiev, D. (2003). *Nucleation—Basic theory with applications*. Oxford, UK: Butterworth-Heinemann.

Kazmierczak, K., & Höppe, H. A. (2011). Syntheses, crystal structures and optical spectroscopy of Ln2(SO4)3.8H2O (Ln=Ho, Tm) and Pr2(SO4)3.4H2O. *Journal of Solid State Chemistry*, *184*, 1221–1226.

Kim, J.-Y., Kim, U.-S., Byeon, M.-S., Kang, W.-K., Hwang, K.-T., & Cho, W.-S. (2011). Recovery of cerium from glass polishing slurry. *Journal of Rare Earths*, *29*(11), 1075–1078.

Kim, P., Anderko, A., Navrotsky, A., & Riman, R. E. (2018). Trends in structure and thermodynamic properties of normal rare earth carbonates and rare earth hydroxycarbonates. *Minerals*, *8*, 106.

Konishi, Y., & Noda, Y. (2001). Precipitation stripping of rare-earth carbonate powders from rare-earth-loaded carboxylate solutions using carbon dioxide and water. *Industrial and Engineering Chemistry Research*, *40*, 1793–1797.

Korkmaz, K., Alemrajabi, M., Rasmuson, Å. C., & Forsberg, K. (2020). Separation of valuable elements from NiMH battery leach liquor via antisolvent precipitation. *Separation and Purification Technology*, *234*, 115812.

Li, X., Dong, W., Qi, Y., Wang, D., & Yang, R. (1991). Studies on the stabilization of terbium(IV) in aqueous tetrametaphosphate solution. *Polyhedron*, *10*, 1479–1483.

Liu, X., & Byrne, R. H. (1997). Rare earth and yttrium phosphate solubilities in aqueous solution. *Geochimica et Cosmochimica Acta*, *61*(8), 1625–1633.

Lokshin, E. P., Tareeva, O. A., Ivlev, K. G., & Kashulina, T. G. (2005). Solubility of double alkali metal (Na, K) rare-earth (La, Ce) sulfates in sulfuric—Phosphoric acid solutions at 20 °C. *Russian Journal of Applied Chemistry*, *78*(7), 1058–1063.

Marsh, J. K. (1955). The separation of lanthanons with the aid of ethylenediaminetetra-acetic acid ("enta acid"). Part V. The solubilities of some alkali lanthanon enta salts. *Journal of the Chemical Society*, *1955*, 451–452.

Meinrath, G., & Takeishi, H. (1993). Solid-liquid equilibria of Nd^{3+} in carbonate solutions. *Journal of Alloys and Compounds*, *194*, 93–99.

Mersmann, A. (2001). *Crystallization technology handbook* (2nd ed.). New York, USA: Marcel Dekker.

Mioduski, T. (1998). Identification of saturating solid phases in aqueous solutions of lanthanide(III) sulfates from solubility data. *Chemia Analityczna (Warsaw)*, *43*, 457–462.

Mioduski, T., Gumiński, C., & Zeng, D. (2008). IUPAC-NIST Solubility data series. 87. Rare earth metal chlorides in water and aqueous systems. Part 1. Scandium group (Sc, Y, La). *Journal of Physical and Chemical Reference Data*, *37*, 1765–1853.

Mioduski, T., Gumiński, C., & Zeng, D. (2009a). IUPAC-NIST Solubility data series. 87. Rare earth metal chlorides in water and aqueous systems. Part 2. Light lanthanides (Ce–Eu). *Journal of Physical and Chemical Reference Data*, *38*, 441–562.

Mioduski, T., Gumiński, C., & Zeng, D. (2009b). IUPAC-NIST Solubility data series. 87. Rare earth metal chlorides in water and aqueous systems. Part 3. Heavy lanthanides (Gd–Lu). *Journal of Physical and Chemical Reference Data*, *38*, 925–1010.

Mioduski, T., Gumiński, C., & Zeng, D. (2014). IUPAC-NIST solubility data series. 100. Rare earth metal fluorides in water and aqueous systems. Part 1. Scandium group (Sc, Y, La). *Journal of Physical and Chemical Reference Data*, *43*, 013105.

Mioduski, T., Gumiński, C., & Zeng, D. (2015a). IUPAC-NIST solubility data series. 100. Rare earth metal fluorides in water and aqueous systems. Part 2. Light lanthanides (Ce-Eu), *Journal of Physical and Chemical Reference Data*, *44*, 013102.

Mioduski, T., Gumiński, C., & Zeng, D. (2015b). IUPAC-NIST solubility data series. 100. Rare earth metal fluorides in water and aqueous systems. Part 3. Heavy lanthanides (Gd-Lu), *Journal of Physical and Chemical Reference Data*, *44*, 023102.

Mioduski, T., Gumiński, C., Zeng, D., & Voigt, H. (2013). IUPAC-NIST Solubility data series. 94. Rare earth metal iodides and bromides in water and aqueous systems. Part 2. *Bromides. Journal of Physical and Chemical Reference Data*, *42*, 013101.

Moeller, T., & Kremers, H. E. (1945). The basicity characteristics of scandium, yttrium, and the rare earth elements. *Chemical Reviews*, *37*(1), 97–159.

Moldoveanu, G. A., & Demopoulos, G. P. (2014). *Journal of Chemical Technology and Biotechnology*, *90*, 686–692.

Mullin, J. W. (2001). *Crystallization* (4th ed.). Oxford, UK: Butterworth-Heinemann.

Myerson, A. S., Erdemir, D., & Lee, A. Y. (Eds.). (2019). *Handbook of industrial crystallization* (3rd ed.). Cambridge, UK: Cambridge University Press.

Nielsen, M. H., Aloni, S., & DeYoreo J. J. (2014). In situ TEM imaging of $CaCO_3$ nucleation reveals coexistence of direct and indirect pathways. *Science*, *345*, 1158 1162.

Parker, J. G., Baroch, C. T., & Adams, J. W. (1971). *The rare earth elements, yttrium and thorium—A materials survey*. Bureau of Mines information circular 8476, U.S. Department of the Interior, Washington, D.C.

Pawar, N., Svärd, M., & Forsberg, K. (2024). *Recovery of rare earth sulfate hydrates using antisolvent crystallization*. In K. Forsberg, et al. (Eds.), *Rare metal technology 2024* (pp. 55–62). The Minerals, Metals & Materials Series. USA: Springer, Cham.

Persson, I. (2010). Hydrated metal ions in aqueous solution: How regular are their structures? *Pure and Applied Chemistry*, *82*(10), 1901–1917.

Peters, E. M., Kaya, Ş., Dittrich, C., & Forsberg K. (2019). Recovery of scandium by crystallization techniques. *Journal of Sustainable Metallurgy*, *5*, 48–56.

Peters, E. M., Svärd, M., & Forsberg, K. (2020). Phase equilibria of ammonium fluoride phases in aqueous alcohol mixtures for metal recovery by anti-solvent crystallization. *Separation and Purification Technology, 252, 117449.*

Peters, E. M., Svärd, M., & Forsberg, K. (2022). Impact of process parameters on product size and morphology in hydrometallurgical antisolvent crystallization. *CrystEngComm*, 2851–2866.

Philippini, V., Vercouter, T., Chaussé, A., & Vitorge, P. (2008). Precipitation of $ALn(CO_3)_2$,xH_2O and $Dy_2(CO_3)_3$,xH_2O compounds from aqueous solutions for A^+= Li^+, Na^+, K^+, Cs^+, NH_4^+ and Ln^{3+} = La^{3+}, Nd^{3+}, Eu^{3+} Dy^{3+}. *Journal of Solid State Chemistry*, *181*, 2143–2154.

Rard, J. A. (1988). Aqueous solubilities of praseodymium, europium, and lutetium sulfates. *Journal of Solution Chemistry*, *17*(6), 499–517.

Siekierski, S., Mioduski, T., & Salomon, M. (1983). *Scandium, yttrium, lanthanum and lanthanide nitrates*. Oxford, UK: Pergamon.

Smith, R. M., & Martell, A. E. (1976). *Critical stability constants, vol. 4: Inorganic complexes*. Boston, USA: Springer.

Spanos, N., & Koutsoukos, P. G. (1998). The transformation of vaterite to calcite: Effect of the conditions of the solutions in contact with the mineral phase. *Journal of Crystal Growth*, *191*, 783–790.

Torbacke, M., & Rasmuson, Å. C. (2001). Influence of direct scales of mixing in reaction crystallization. *Chemical Engineering Science*, *56*(7), 2459–2473.

Tsushima, S., Nagasaki, S., & Suzuki, A. (1997). Photochemical separation and co-precipitation of lanthanides in nitric acid solution. *Journal of Photochemistry and Photobiology A: Chemistry*, *106*, 57–60.

Van den Bogaert, B., Havaux, D., Binnemans, K., & Van Gerven, T. (2015). Photochemical recycling of europium from Eu/Y mixtures in red lamp phosphor waste streams. *Green Chemistry*, *17*, 2180–2187.

Van de Voorde, M., Van Hecke, K., Binnemans, K., & Cardinaels, T. (2018). Separation of samarium and europium by solvent extraction with an undiluted quaternary ammonium ionic liquid: Towards high-purity medical samarium-153. *RSC Advances*, *8*, 20077–20086.

Wickleder, M. S. (2002). Inorganic lanthanide compounds with complex anions. *Chemical Reviews*, *102*(6), 2011–2087.

Zhu, H., Yuen, C., & Grant, D. J. W. (1996). Influence of water activity in organic solvent + water mixtures on the nature of the crystallizing drug phase. 1. *Theophylline. International Journal of Pharmaceutics*, *135*, 151–160.

Zijlema, T. G., Geertman, R. M., Witkamp, G.-J., van Rosmalen, G. M., & de Graauw, J. (2000). Antisolvent crystallization as an alternative to evaporative crystallization for the production of sodium chloride. *Industrial & Engineering Chemistry Research*, *39*, 1330–1337.

7

AQUEOUS ELECTROCHEMICAL PROCESSING OF RARE EARTH ELEMENTS: A REVIEW

Eugene Engmann[1,2], Luis A. Diaz[1], Tedd E. Lister[1], and Haiyan Zhao[2,3]

[1] *Critical Materials Innovation Hub, Idaho National Laboratory, Idaho Falls, Idaho, USA*
[2] *Department of Nuclear Engineering and Industrial Management, University of Idaho, Idaho Falls, Idaho, USA*
[3] *Department of Chemical and Biological Engineering, University of Idaho, Moscow, Idaho, USA*

Abstract

This chapter is a comprehensive review of various technologies and methods related to the separation and extraction of rare earth elements (REEs) at low temperatures (<100 °C). Addressed are various topics such as redox-based processing, liquid-metal cathode (LMC) studies, and electrochemistry-driven REE magnet dissolution and recovery. The chapter begins with an introduction to REEs, their importance in technology, and the need for a more sustainable process for REE separation and capture. The second section addresses the redox-based processing techniques for Ce and Eu separation and details the Eu and Ce anomaly which allows them to be extracted from aqueous REEs

Rare Earth Elements: Sustainable Recovery, Processing, and Purification, Special Publications 79, First Edition. Edited by Athanasios K. Karamalidis and Roderick Eggert.

DOI: 10.1002/9781119515005.ch7

unlike the other REEs. The third section focuses on LMC studies using mercury (Hg) where REEs were extracted through electrolytic reduction or galvanic displacement. This section includes early work done prior to 1950, at Los Alamos National Laboratory, and in the former Soviet Union. The section also summarizes possible replacements for Hg. The thermodynamics for amalgam formation are also discussed. The fourth section covers electrochemistry-driven REE magnet dissolution and recovery. This section discusses dissolution of REE magnets to separate elements such as iron followed by REE precipitation. The final section provides perspective discussing the challenges and future directions in this field.

7.1. Introduction

Rare earth elements (REEs) are employed in metal alloys, rechargeable batteries, automotive catalysts, ceramics, cracking catalysts, polishing powders, permanent magnets, lighting displays, laser and wavelength amplifiers, military defense applications, and agricultural applications (Bünzli, 2013; Riba et al., 2016; Schüler et al., 2011; Song et al., 2013; Zhang et al., 2017)

As the world moves toward more environmentally benign chemical processing, factors such as waste prevention, reduction in the use of hazardous chemicals, energy efficiency, and the use of safer solvents are goals (Suresh & Sundaramoorthy, 2015). Electrochemical methods, which can leverage renewable power sourced from carbon-free energy sources, are gaining interest as paths to reducing environmental impacts of chemical processes. Metallurgical processes are one of those industrial sectors where electrochemical technologies can have significant impact (Su, 2020). For REEs, aqueous electrochemistry is challenged by the stability of the trivalent oxidation state. For example, the reduction potential of REEs to the metallic state is reported to be over −2.4 V versus the standard hydrogen electrode (SHE) potential which is more negative than that of water (Pourbaix, 1974). In an early work, Denis and Lemon investigated the direct electrolysis of REEs in low-concentration nitric acid solutions at a Pt electrode where the formation of REE hydroxides at the cathode surface was observed, a consequence of the hydrogen evolution reaction (HER) (Dennis & Lemon, 1915). This limitation has led most work in REE electrochemistry to be performed using either molten salt or ionic-liquid-based electrolytes due to a much wider window of electrochemical stability at negative (reducing) potentials (Arenas et al., 2016).

Despite the limitations of water, there is a long and rich history in electrochemically processing REEs from aqueous solutions. This work has taken advantage of metal alloying via liquid-metal cathodes (LMC) to shift REE reduction potentials into the stability window for water. Other methods have been developed to dissolve alloys from NdFeB magnets. Finally, some REEs such as Ce and Eu have redox potentials in the stability window of water, a phenomenon associated with the Ce–Eu anomaly which allows for facile separation (Cicconi et al., 2021).

7.2. Redox-Based Processing

The negative reduction potentials of REEs from the trivalent to the metallic state favor HER over metal reduction in aqueous solution (Abbott & McKenzie, 2006). However, REEs such as Ce and Eu have redox couples, Ce^{4+}/Ce^{3+} and Eu^{2+}/Eu^{3+}, that permit their separation from other aqueous REEs (Cicconi et al., 2021; Kronholm et al., 2013; Vickery, 1948, 1950). This is useful, as the separation of aqueous REEs in their trivalent oxidation state is challenging due to their similar chemical properties (Castor & Hedrick, 2006). As an example, Ce^{3+} is soluble, while Ce^{4+} precipitates to insoluble CeO_2 in aqueous solution at oxidative potentials (Arenas et al., 2016). Cicconi et al. provide an in-depth review on the redox behavior of Eu and Ce and the characteristics that lead to their atypical behavior (Cicconi et al., 2021).

7.2.1. Cerium (Ce) Processing from REEs

With a standard potential of 1.72 V versus reversible hydrogen electrode (RHE), Ce^{4+}/Ce^{3+} falls within the potential window of common aqueous electrolytes (Zhao et al., 2019). Furthermore, the strong relationship between the redox couple of Ce and ligand coordination, as well as composition of the electrolyte, permits the adjustment of anionic species or electrolyte composition to enhance electrochemical efficiency(Zhao et al., 2019). The popularity of Ce arises from its diverse applications, abundance, and low cost (Patnaik, 2003; Vasudevan et al., 2005). Equation (1) shows the electrochemical oxidation reaction of Ce^{3+}:

$$Ce^{3+} \rightleftharpoons Ce^{4+} + e^{-}. \tag{1}$$

Practical applications involving the separation of Ce from other REEs have been studied. As early as 1915, Dennis and Lemon described the electrolytic oxidation of Ce at Pt electrodes employing electrolyte solutions of cerium salts but provided no further details (Dennis & Lemon, 1915). Thereafter, Dennis and Meulen examined the decomposition potentials (potential difference between the cathode and anode needed for electrolysis to occur) of Ce among other REEs employing both neutral solutions of Ce chloride and Ce nitrate as electrolytes at a Pt cathode (Dennis & Meulen, 1915). The motivation of this study was to determine if differing applied potentials can be used to separate Ce from other REEs. Decomposition potentials of 2.18 and 2.36 V for the nitrate and chloride solutions, respectively, were observed.

The study of direct electrolytic oxidation of Ce appears to have gone dormant from the early 1900s to the early 2000s where Vasudevan et al. demonstrated Ce separation from other REEs in aqueous chloride solutions (Vasudevan et al., 2005). Electrolysis was performed at 500 A/m^2 for 6 h with up to 60% current efficiency, a REE purity of 95%, and an energy consumption of 3.5 kWh/kg of CeO_2. Arenas et al. published a comprehensive review and study on redox processes pertaining to soluble cerium species (Arenas et al., 2016). In their study, they discussed the effects of oxygen evolution, electrode kinetics, mass transport, and diffusion on the reduction and oxidation of Ce (Arenas et al., 2016).

7.2.2. Europium (Eu) Separation from REEs

With a standard potential of −0.429 V versus SHE, Eu^{2+}/Eu^{3+} is within the aqueous potential window for common electrolytes (Pourbaix, 1974). The electrochemical reduction of Eu^{3+} can be expressed as:

$$Eu^{3+} + e^- \rightleftharpoons Eu^{2+}. \tag{2}$$

The separation of Eu from other REEs has been reported using Zn amalgams, alkaline metal amalgams, and Zn powder (Atanasyants & Seryogin, 1995). However, low separation factors, low yields, difficulties in purification, and high costs are drawbacks to these reported processes. Furthermore, electrochemical studies of Eu reduction in aqueous media are limited.

Atanasyants and Seryogin investigated the reduction of Eu in hydrochloric acid using cyclic voltammetry. Both rotating disk electrode and temperature kinetics methods were employed (Atanasyants & Seryogin, 1995). During their study, they observed the following:

- The cathodic peak density is linearly related to the concentration of Eu.
- Linear dependence of cathodic peak density and scan rate show a purely electrochemical mechanism.
- Increase in scan rate shows a transition from a reversible to a quasi-reversible (kinetically limited) reaction.
- Current density of Eu reduction is dependent on the disk rotation speed, suggesting first-order reaction for Eu^{3+} to Eu^{2+}.

7.3. Liquid-Metal Cathode Studies

The use of LMCs as a means of processing lanthanides goes back many decades. Processing involves the formation of a metallic alloy, which effectively decreases the reduction potential of the metal to those accessible in aqueous solutions. LMCs are generally composed of elements with a high overpotential (additional potential needed due to slow kinetics) for HER. This offers a wider cathodic potential range to facilitate amalgam formation (Mikkelsen & Schrøder, 2003). For the purposes of this chapter, amalgams formed using mercury (Hg) were almost exclusively studied. However, electrochemical studies for Hg have largely ceased due to environmental and health concerns. LMCs offer ease of surface renewability and the ability to flow the electrode, which gives rise to various process options (Mikkelsen & Schrøder, 2003; Surmann & Zeyat, 2005).

From the early to mid-1900s, separation of REEs employing LMCs was achieved by either direct reduction or by galvanic displacement of less noble metals such as K, Li, or Na (Dennis & Lemon, 1915; Dennis & Meulen, 1915; Lavrukhina et al., 1960; Ryabchikov & Terent'eva, 1960). These studies reported promising results for REE separation before going dormant over half a century ago (Onstott, 1959, 1961; Shvedov & Antonov, 1964). The chlor-alkali process is the closest analog where reactive metals (such as sodium, (Na)) are reduced into an amalgam, as shown in Figure 7.1 (Bommaraju et al., 2007; Crook & Mousavi, 2016; Venkatesh & Tilak, 1983). This long-serving electrochemical process has been phased out, but produced caustic soda, chlorine gas, and bleach products.

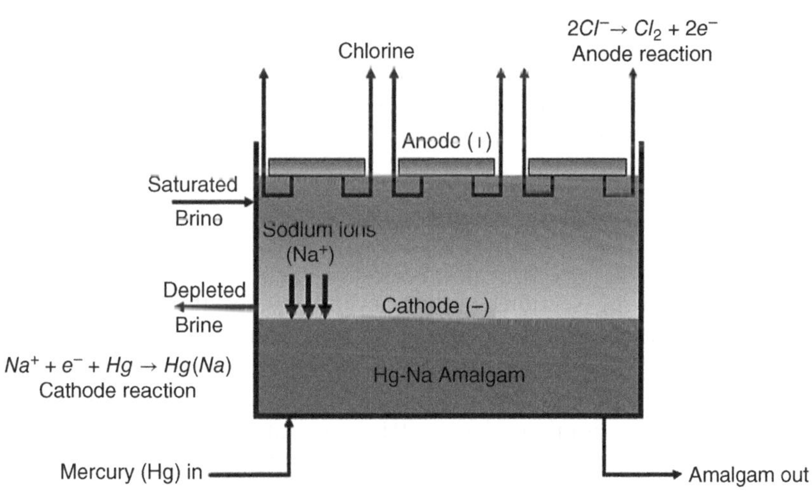

Figure 7.1 Chlor-alkali process for production of Na metal.

7.3.1. Application of a Liquid-Metal Cathode for REE Separations

The discovery of amalgam formation in a liquid-Hg cathode that permitted metal reduction created methods for reducing REEs in aqueous media. This gave rise to the reduction of REEs either by direct electrolysis (Equation (3a)) or by galvanic displacement (Equation (3b)), or by both approaches (McCoy, 1941a).

Direct electrolysis involves the application of a potential across an electrochemical cell to promote reduction (Zoski, 2006). This reaction can be expressed as

$$\mathrm{REE}^{3+} + 3\mathrm{e}^{-} + \mathrm{Hg} \rightarrow \mathrm{REE}^{0} - \mathrm{Hg}. \qquad (3a)$$

This reduction can be done with or without the presence of a complexing agent, such as citrate or tartrate ion in the electrolyte. One of the most vivid descriptions of the amalgamation process was given as part of a three-part investigation by McCoy where the formation of REE amalgams is described as solid crystals and not merely REE hydroxides (McCoy, 1941a). McCoy elaborated that the use of REE acidic aqueous solutions favors HER, while neutral solutions favor the deposition of hydroxides on the cathode. In both cases, little to no amalgam formation was observed. McCoy also stated that important contributors to achieving amalgam formation are the metals, such as K, citrate ion, and REE acetates. This combination gave the most amalgam at low voltage with encouraging current efficiencies.

Galvanic displacement involves the reduction of REEs by a redox reaction of a metal with a more negative reduction potential (Gough et al., 2016). For example, a reaction involving the use of lithium (Li) as an example can be expressed as

$$3Li - Hg + REE^{3+} \rightarrow REE - Hg + 3Li^{+}. \qquad (3b)$$

The following sections outline several studies that examined the amalgamation process as a means of separating REEs and were all reported prior to 1970. In 1968, Kolesov and Pankratova presented an in-depth review of Hg LMCs (Kolesov & Pankratova, 1968). This also includes work examining adjacent REE separation, with products such as praseodymium (Pr^{3+}) and neodymium (Nd^{3+}), which are difficult to separate due to their similar chemical and electrochemical properties (Onstott, 1961).

Early Work

For this review, we refer to early works reported before 1950, which generally provided fewer details in their approach and methods used. Further, work is segregated into the two general approaches: electrolytic and galvanic displacement.

Electrolytic Reduction

As early as 1906, electrolytic reduction of REEs from samarium (Sm) sulfates was reported by Marion (Kolesov & Pankratova, 1968). In 1911, Urbain and Burion reported isolation of a Eu compound from a Hg LMC (Kolesov & Pankratova, 1968). In 1915, Dennis and Lemon performed metal separation from various REE amalgams at a Hg cathode (Dennis & Lemon, 1915). Their work involved the separation of lanthanum (La) from La–Pr–Nd–Sm and La–Nd nitrate solutions, as well as the fractional separation of erbium (Er) from ytterbium (Yb). They observed that the precipitation of REEs depended on electrolysis time and the fractionation stage. Shortly thereafter, Dennis and Meulen investigated the fractional separation of various REEs at a Hg cathode (Dennis & Meulen, 1915). Some of the parameters studied were the decomposition potential and the effect of the number of fractionations on REE separation from neutral nitrate solutions.

In the late 1920s, Neckers and Kremers demonstrated the separation of La from La–Pr and La–Nd–Pr–yttrium (Y)–Sm chloride mixtures while employing a Hg cathode and a Pt anode (Neckers & Kremers, 1928). For the La–Pr solutions, electrolysis of >4 h and a potential >6 V yielded between 80% and 99% La. Similarly, with the La–Nd–Pr–Y–Sm solutions, greater La yields of up to 100% were seen at longer run times and potentials of >7 V. It was also observed that a 5% addition of Na to the La–Nd–Pr–Y–Sm solutions increased conductivity, which sped up the electrolysis without loss of separation efficiency. Attempts to separate Pr from Nd in Pr–Nd chloride solutions yielded no appreciable result.

In 1930, Yntema achieved the separation of Eu from a mixture of Eu, Sm, and gadolinium (Gd) chlorides after 6 h of electrolysis from a sulfate-based electrolyte (Yntema, 1930). Using a similar approach, Ball and Yntema successfully separated Yb to 98% purity from mixed REE chlorides containing Yb, Er, Th, and Y, in dilute sulfuric acid (H_2SO_4) (Ball & Yntema, 1930). Likewise, Audrieth et al. experimented with the electrolytic preparation of Nd and La amalgams at a Hg cathode. They observed that from aqueous solution, formation of an insoluble REE salt precipitate was favored over REE amalgams (Audrieth et al., 1931; Jukkola et al., 1934). However, when electrolyzed in ethanol solution, precipitate formation was avoided (Audrieth et al., 1931). This was successfully replicated by Hopkins and Audrieth to create amalgams of La, Cd, Nd, Sm, and Yb from ethanol solutions (Hopkins & Audrieth, 1934).

In the early 1940s, a three-part study conducted by McCoy investigated the separation of REEs at amalgam cathodes. The first study was reported in June 1941 where McCoy investigated the formation of Eu and Yb amalgams employing the use of REE acetates and potassium citrate as an electrolyte where simultaneous formation of K amalgams with Eu amalgams was observed (McCoy, 1941a). The presence of the K amalgam occurred either at very high current or during the continued application of a current after deposition of the REEs was complete. Following electrolysis of 6.25 h at a current density of 0.06 A cm^{-2}, 100% of the Eu was amalgamated as $Hg_{10}Eu$ and Hg_2Eu_3 crystals. Similarly, amalgams for Yb were also created at 8 V and 2 A.

A second paper reported the feasibility of separating Eu from Sm when present in equal concentrations in an electrolyte, which was a first of a kind study (McCoy, 1941b). Amalgams yielding 90% Eu were formed using REE acetates and potassium citrate complexing agents. The separation of Eu from other REEs was also investigated with reported yields between 60% and 94%. Furthermore, Sm acetate in potassium citrate also was investigated at an amalgam cathode and yields between 1% and 13% were observed. It should be noted that irrespective of the REE amalgam, K amalgam was always formed. It was also noted that complexing agents, such as citrates, likely prevented the REEs from forming hydroxide precipitates. The recovered REEs were found to be REE oxides (McCoy, 1941b).

Following the success of the first experiments with Eu and Yb, McCoy and Hammond reported a third study which also employed amalgam formation for the separation of Yb from other REEs. This experiment employed a greater amount of REE oxides (4 g) than the first experiment (McCoy & Hammond, 1942). This experiment's purpose was to demonstrate that Yb could be separated into amalgams in larger quantities. The electrolyte was an REE acetate with potassium citrate. Following electrolysis, the REEs in the amalgam were compared to the REEs in the electrolyte. REEs remaining in the electrolyte amounted to 0.1 g from the 4 g originally added. Without providing numeric quantification, it was reported that most of the REE yield in the amalgam was Yb with very little Eu (McCoy & Hammond, 1942).

In the late 1940s, Estee and Glockler reported the reduction of Nd at a dropping Hg electrode and observed only a single polarographic wave during reduction of aqueous Nd sulfates that they attributed to Nd^0 formation (Estee & Glockler, 1948). Other papers have reported polarography of REEs where reduction of Gd (Rabideau & Glockler, 1948), Pr (Swensen & Glockler, 1949), and Sm (Timnick & Glockler, 1948) was observed.

Galvanic Displacement (Reduction)

Employing galvanic displacement, West and Hopkins reacted Na amalgams with alcoholic REE chlorides to produce new amalgams with 1–3% REE content (West & Hopkins, 1935). Marsh also employed galvanic displacement using Na amalgams for the separation of Yb, Sm, and Eu in REE acetate and chloride solutions (Marsh, 1943a). In subsequent experiments, Marsh observed that >99.99% of Yb from solution was captured in the amalgam (Marsh, 1943a). This method provided a pathway to separate Eu, Sm, and Yb into amalgams by shaking in either chloride salts or acetic acid. One of the limitations to Marsh's experiments was the lack of thorough comparison between acetate and chloride salts, although they reported qualitatively that acetate performed better than chloride.

Other work reported by Marsh involved the separation of Nd from Sm and Gd (Marsh, 1943b), and the isolation of Eu (Marsh, 1943b) employing a Na amalgam. It was observed that for acetate, mixtures of Nd–Sm or Sm–Gd, Sm preferentially formed an amalgam. For Eu–Gd–Sm acetate mixtures, Eu and Sm can preferentially form amalgams when in contact with Na amalgam allowing the separation of Gd.

Following Marsh's work, Moeller and Kremers studied the formation of Yb amalgams in solutions containing different anions (Moeller & Kremers, 1945). They studied Yb with acetate, dichloroacetate, formate, chloride, and perchlorate anions. Dichloroacetate solution was found to yield no Yb amalgam; however, a >70% yield was observed for perchlorate, chloride, and formate anions following a single extraction time of approximately 15 min. Further, there was an increase in Yb amalgam stability from formate solution in comparison with the perchlorate solution. Following five extractions with acetate and formate solutions, each employing the use of fresh Na amalgams for each extraction, 97% of the original amount of Yb was extracted.

Electrolytic Studies at the Los Alamos National Laboratory

Onstott published a four-part study on REE separation at Hg and alkali metal amalgam cathodes. One reported the separation of Sm from Eu by electrolysis using REE acetates and lithium citrate (Li_3Citrate) solutions as the complexing agents at a Li amalgam cathode (Onstott, 1955). He observed that <10% of Sm from the electrolyte went toward amalgam formation, which is favored at low pH, high cathodic potentials, and low citrate concentrations. As an example, a 95% yield of Eu was obtained.

In a following work, Onstott studied the separation of Sm from Gd and Nd, and the purification of Eu from Sm at a Li amalgam cathode

(Onstott, 1956). The electrolytes were REE acetates with lithium citrate. For Sm–Gd electrolytes, separation factors as high as 6,200 and 2,300 were reported in the presence and absence of citrate, respectively. In this case, the separation factor was defined as the ratio of concentrations of one element to another in the amalgam phase with respect to the bulk electrolyte phase (Onstott, 1961). In the Sm–Nd system, Sm migrated to the amalgam, while Nd remained in the electrolyte. 99.98% purity of Sm was recovered from the amalgam. In a follow-up study to enrich and purify Eu from Sm_2O_3–Eu_2O_3, >99.9% purity of Eu_2O_3 was achieved after several cycles of electrolysis (Onstott, 1956).

Due to the success obtained from the separation of Eu from Sm, and Sm from Gd, Onstott extended the approach to the separation of the lanthanide series elements using a lithium amalgam (Onstott, 1959). A monazite mixture containing La, Ce, Pr, Nd, Sm, Eu, Gd, Tb, Dy, Ho, Er, Yb, and Th oxides was studied, as well as Y-earth (heavy REEs) and Tb–Dy mixtures. The electrolyte was either REE perchlorate/Li citrate or REE acetates/Li citrate. From the monazite mixture, 99.7% Eu was electrolyzed after the first fractionation independent of pH. Similarly, Yb was electrolyzed but at a much slower rate. At high pH, electrolysis of Sm and trivalent lanthanides is suppressed. For REEs in the Y-earth mixture, the rate of electrolysis for REEs heavier than Tb was slow and impractical. For the Tb–Dy mixture, Tb was electrolyzed over Dy for all scenarios tested. Some of the important conditions Onstott described that influence electrolysis of trivalent REEs are (Onstott, 1959):

- **Need for an amalgam cathode:** For the electrolysis of most trivalent REEs to take place, there needs to be a cation that can supply a free electron in the presence of Hg. Alkali metals such as K, Li, and Na are most suitable.
- **Need for a complexing agent:** Even in the presence of the alkali metals such as Li, appreciable electrolysis of inorganic salts of trivalent REEs was not achieved. However, when coupled with a carboxylic acid anion ligand, the rate of electrolysis was improved.
- **Dependence of electrolysis rates on pH:** The rate of electrolysis of trivalent REEs occurs best between pH 4 and 6 with exception of Eu, which is not impacted by pH.
- **Electrolysis being oxidation-dependent:** REE ions such as Eu, Sm, and Yb, with an oxidation number <3, are readily electrolyzed. For trivalent REEs, electrolysis is slower and is governed by the change in ionic radius, a characteristic known as lanthanide contraction. The smaller the ionic radius, the slower the rate of electrolysis.
- **Effect of solvent:** The use of solvents such as methanol, as compared to water, increases the rate of electrolysis while simultaneously increasing

the current density. This also improves efficiency because a majority of the REE species form an amalgam.

Finally, Onstott explored the separation of high-purity Yb from a mixture of heavy REEs and observed >99.9% removal of Yb from the REE mixture at a Hg cathode. The yield of Yb could be manipulated by changing either the electrolyte pH or the duration of the experiment (Onstott, 1960).

Later work examined the separation of Nd and Pr chlorides in lithium citrate electrolyte with respect to temperature, electrolyte composition, and current density (Onstott, 1961). This work was conducted due to the difficulty in separating trivalent REEs. As the concentration of Pr and Nd in the electrolyte increased, Pr preferentially formed an amalgam. When Pr concentration was lower, separation was improved. The highest separation factor achieved was 3, at current density of 50 mA cm^{-2}, with a $Li_3Citrate/(Pr + Nd)$ ratio of 2.5 and a Pr + Nd molarity of 0.4 (Onstott, 1961).

Electrolytic Studies in the Former Soviet Union

Antonov and Shvedov investigated the formation of Gd, Tb, Dy, and Ho amalgams (Shvedov & Antonov, 1964). They observed that the factors that drive the migration of REEs into the Hg phase depend on the ratio of the complexing agent to the concentration of REEs in the electrolyte; $\frac{C_{Li_3Citrate}}{C_{REE}}$. Using REE acetates and varying amounts of Li citrate as the complexing agents, they observed that at low (<0.15 M) $Li_3Citrate$ concentration, the precipitation of REE hydroxides limited amalgam formation. They further observed that optimal [Li]/[REE] for amalgam formation were 12–15, 13, 17, and 26 for Gd, Tb, Dy, and Ho, respectively. Furthermore, for every 100 mg of Hg employed for amalgam formation, 150 mg, 75 mg, 55 mg, and 18 mg of Gd, Tb, Dy, and Ho, respectively, were consumed in amalgam formation. Shvedov and Antonov also investigated the separation of Pr and Nd at a Hg cathode using Li tartrate as the complexing agent (Shvedov & Antonov, 1962). They observed yields of Pr and Nd reaching 100% in some cases and separation factors between 30 and 60. They conducted other studies to investigate the influence of factors such as the concentration of REEs, the current density, the time for electrolysis, and the $\frac{C_{Li_3Citrate}}{C_{REE}}$ ratio on amalgam formation (Shvedov & Antonov, 1967). They observed that a $Li_3Citrate$ to REE ratio of 9:1 was optimal for separation of Tb and Dy, and 4:1 for Tb and Ho. Separation factors as high as 7.8, 10, 66, 44.5, 75, and 60 were observed for Gd–Tb, Tb–Dy, Gd–Ho, Tb–Ho, Tb–Er, and Tb–Tm, respectively. Furthermore, when the concentration of $Li_3Citrate$ was increased, H_2 evolution was favored.

Summary of LMC Studies in Separating REEs

Distribution of REEs in the Hg phase is affected by the solvent properties of Hg toward metals such as K, Na, and Li to influence the formation of REE amalgams (through galvanic reaction) (Kolthoff & Lingane, 1952; Onstott, 1956). Li has proven to be the most effective in REE amalgam formation due to its high negative potential (Kolthoff & Lingane, 1952; Onstott, 1955). Researchers have employed various extractants and complexing agents such as acetates, citrates, and tartrates for preferential amalgamation (Kolesov & Pankratova, 1968; McCoy, 1941a; Mccoy et al., 1946). Table 7.1 is an example of various extraction percentages using Li, Na, and K as amalgams.

In evaluating the various studies, the work at Los Alamos and the former Soviet Union provides the most experimental details for evaluation. For adjacent REEs such as Nd and Pr, Onstott achieved a maximum separation factor of 3 with Li_3Citrate and REE chlorides (Onstott, 1961). However, Shvedov and Antonov reported separation factors from 30 to 60 for Pr and Nd employing REE acetates and Li tartrate (Shvedov & Antonov, 1962). These values seem unusually high given the similarity in reduction potential and lack of binding energy difference for the tartrate anion. Generally, separation factors and migration percentage of REEs into the amalgam for studies conducted by Shvedov and Antonov tended to be higher than those reported by Onstott. Onstott reported high separation factors for Eu from an Eu–Sm mixture at very high polarizations (Onstott, 1959), and for Sm from Gd–Sm–Dy mixtures (Onstott, 1956). It is noteworthy to mention that the

Table 7.1 Percentage extraction of rare earth elements by Li, Na, and K amalgams (C_{Ln} = 0.5 M) (Adapted from Kolesov & Pankratova, 1968).

	Amalgam*		
Rare earth	Li	Na	K
La	38	21	17
Ce	–	13	8
Pr	31	13	14
Nd	23	7	5
Sm	45	66	50
Eu	41	50	50
Yb	25	47	37

* 0.04 M amalgam was used.

work reported by Shvedov and Antonov was performed with pure Hg to form Hg–REE amalgams, while Onstott employed alkali amalgams (Li, K) (Onstott, 1960). The studies, however, showed that high separation factors can be achieved by methods such as amalgam formation at a Hg cathode.

7.3.2. Potential Substitutes to Hg LMC: Ga and Its Alloys

Recently, work was undertaken to investigate paths to replace the Hg LMC in REE separation chemistries. This led to a search for other liquid metals with similar characteristics to Hg but are nontoxic (Surmann & Zeyat, 2005). Liquid metals such as Ga, and alloys such as GaInSn, GaIn, and GaInSnZn have been investigated as potential substitutes for Hg (Khan et al., 2014; Morley et al., 2008; Surmann & Zeyat, 2005). Table 7.2 shows the physical properties for these alloys and Hg, as well as the reduction potential for Ga and Hg. Pure Ga has a melting point slightly above room temperature (29.8 °C) coupled with a very low vapor pressure of 1.3×10^{-6} atm at 965 °C. Hg possesses a relatively high vapor pressure which factors into the challenge of working safely with it (Loftness, 1951; Powell, 1968).

The ability of Ga to dissolve a host of other metals creates a pathway to produce amalgams of metal and metal oxides (Ghasemian et al., 2019; Han et al., 2021; Lertanantawong et al., 2018; Mayyas et al., 2020; Park et al., 2019; Wang et al., 2018; Wang et al., 2020). Furthermore, Ga is notorious for the spontaneous formation of a passivating oxide layer when exposed to oxygen-containing environments such as air or water, making it extremely difficult to use in aqueous electrolyte (Cademartiri et al., 2012; Doudrick et al., 2014; Khan et al., 2015; Regan et al., 1997; Scharmann et al., 2004). A recent study published by Engmann et al. harnessed the oxide capabilities of Ga, as a Hg LMC replacement, for the immobilization of Pr from aqueous solutions (Engmann et al., 2022). They observed that the mechanism involving REE immobilization was complex with a strong reduction feature in the presence of REEs that involved reaction of Ga oxides. The disproportionation reaction of Ga_2O with Pr formed a dark gray precipitate composed of Ga oxides and Pr as seen in Figure 7.2. The Pr entrained or immobilized was in the form of Pr_2O_3. Ga–REE alloys do not appear to have formed as observed by the Hg systems discussed earlier or were too reactive to observe. This work showed limitations for using pure Ga system as a replacement for Hg in separating REE mixtures as amalgams.

Table 7.2 Property comparison between Ga and Ga-alloys and Hg.

Property[a]	$Ga^{67}In^{20.5}Sn^{12.5}$	Ga	$Ga^{75.5}In^{24.5}$	$Ga^{61}In^{25}Sn^{13}Zn^{1}$	Hg
Melting point (°C)	10.5	29.8	15.5	7.6	−38.8
Boiling point (°C)	>1,300	2,204	2,000	>900	357
Density (kg/m^3)	6,360	6,080	6,280	6,500	1,353
Conductivity ($\Omega^{-1}\ m^{-1}$)	3.1×10^6	3.7×10^6	3.4×10^6	2.8×10^6	1.0×10^6
Viscosity (m^2/s)	2.98×10^{-7}[b]	3.24×10^{-7}	2.7×10^{-7}	7.11×10^{-8}	13.5×10^{-7}
Surface tension (N/m)	0.533[c]	0.7	0.624	0.5	0.5
Water compatibility	Insoluble	Insoluble	Insoluble	Insoluble	Soluble
Reduction potential (V vs SHE)[d]	–	−0.529	–	–	0.788

[a] Values taken from (Morley et al., 2008).
[b] University of California-Los Angeles (UCLA) measurements using a glass viscometer tube are typically closer to $4.0 \times 10^{-7}\ m^2/s$.
[c] Princeton Plasma Physics Laboratory (PPPL) measurements indicate that oxides can significantly reduce surface tension.
[d] Reduction potential data from Pourbaix (Pourbaix, 1974).

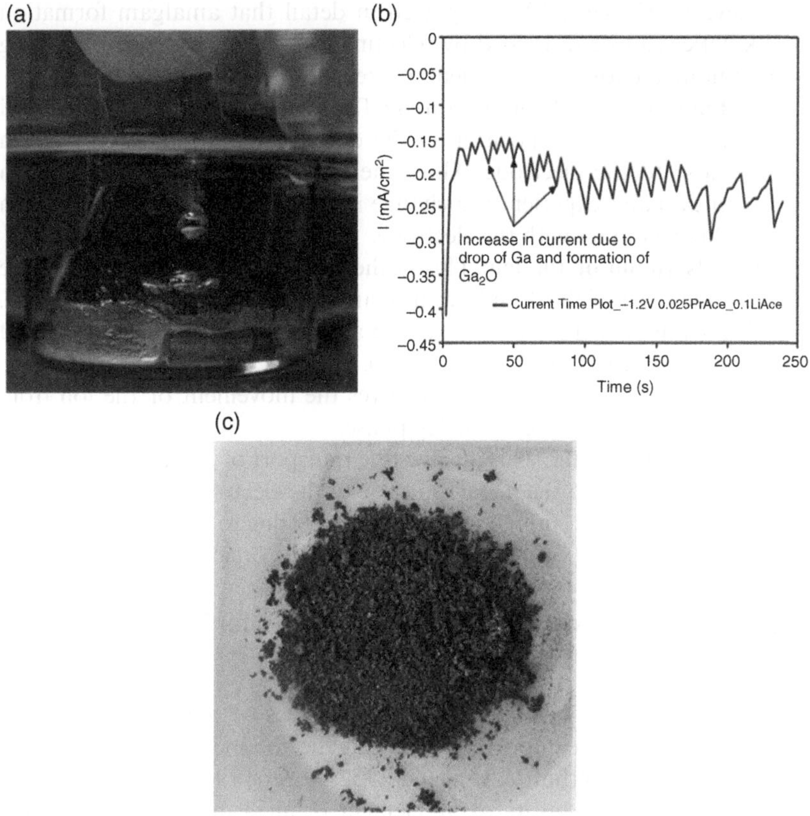

Figure 7.2 Recovery of REEs from dilute solutions. (a) Formation of dark gray precipitate while Ga was dropped into a Ga pool electrode in the presence of 0.05 M Pr acetate (OAc) and 0.1 LiOAc electrolyte at 30 °C. (b) Increase in current versus time as Ga drop hits the surface of liquid-Ga pool and (c) dark gray precipitate from electrolyte composed of Ga oxides and Pr after separation and drying (Engmann et al., 2022/IOP Publishing).

7.3.3. Mechanisms Involving Amalgam Formation

Electron and ion transfer have been speculated to govern the formation of amalgams and act as the rate-determining steps during the process (Fawcett, 1989). This process is governed by the equation, in the case of Hg, as:

$$M^{n+} + ne^{-} + Hg \rightarrow M(Hg). \tag{4}$$

Fawcett (Fawcett, 1989) reported in detail that amalgam formation and kinetics can be described by selecting the adequate rate-determining steps among the following elementary steps:

i. **Ion transfer without adsorption:** This occurs when an ion is moved from its current position in the double layer to another location close to the electrode. Thus, the ion loses some of its solvation sheath and experiences a change in potential but does not come in direct contact with the electrode.

ii. **Adsorption of the ion:** This is the movement of the ion from the inner part of the double layer to an adsorption site on the electrode. At this site, the ion is solvated partially by solvent molecules and Hg molecules (in the case of Hg).

iii. **Metal incorporation:** This involves the movement of the ion from the absorption site to the Hg phase.

iv. **Electron transfer:** This involves the transport of an electron from the metal phase to the metal ion close to the electrode. The metal ion is not directly in contact with the electrode due to solvation by solvent molecules and complexing of REEs by the organic solvent.

7.3.4. Thermodynamics and Theoretical Potential of Amalgam Formation

Earlier work involving the thermodynamics of amalgam formation suggests that it can be improved by varying the volume ratio of the electrolyte to electrode (Kolthoff & Lingane, 1952; Rogers, 1952). Rogers suggests that maximizing the purity of an element is much more important than complete recovery of the element from the electrolyte. This is particularly true when the half-wave potentials ($E_{1/2}$) of the two elements have similar values (<0.2 V) (Rogers, 1952). The formation of an amalgam with Hg is mathematically expressed as

$$E = E_{1/2} - \frac{0.060}{n} \log \frac{(1-x)/V_{ox}}{x/V_{red}}, \tag{5}$$

where E is the electrode potential, n is the number of electrons transferred by the atom, $E_{1/2}$ is the polarographic half-wave potential using Latimer (Latimer, 1938) sign convention, V_{ox} is the volume of the aqueous phase, V_{red} is the volume of the Hg phase, and x is the fraction of the element in the Hg phase. Assuming the volumes of the Hg and electrolyte are the same; $V_{ox} = V_{red}$, for $x = 50\%$ or 0.5, one would observe that:

$$E = E_{\frac{1}{2}}. \tag{6}$$

Furthermore, when $x = 0.5$ or 50%, the ratio of the volumes becomes the determining factor of $E_{50\%}$, expressed as:

$$E_{50\%} = E_{1/2} - \frac{0.060}{n} \log \frac{V_{ox}}{V_{red}}, \tag{7}$$

where $E_{50\%}$ is the potential at which half of the element of interest is deposited.

A closer look at Equation 7 suggests that for elements with different values of n, $E_{50\%}$ changes with the volume ratio, creating a condition for enhanced separation. This could be why certain elements such as Nd and Pr should be difficult to separate. Table 7.3 lists the half-wave potentials for the various REEs and alkali metal amalgams. The spread of the half-wave potentials for the REEs is not very large. For the specific separation of Pr and Nd, there is only a 0.01 V difference suggesting difficulty in separation (based solely on thermodynamics). However, for Pr and Eu, the half-wave potential difference is 0.32 V, indicating that effective separation between both elements is possible. In other words, the greater the difference between half-wave potentials, the greater the chances of separability.

Furthermore, tools such as the semi-log plot or the linear plot can be used to assess separability. The semi-log plot is a graph that provides information about the concentration of an element in the Hg and aqueous phase at a given potential. It is constructed by passing a solid line with a slope of $0.060/n$ through the polarographic half-potential of the element. An example of this plot has been reported (Rogers, 1952). Also, maximum enrichment can be calculated by taking the difference of the half-wave potentials for a more noble element (M_1) and less noble element (M_2) and dividing by $0.060/n$. For elements with the same half-wave potentials but different electron reactions, separability is obtained at potentials other than the half-wave potential. While semi-log plots allow for the calculation of potentials more than one unit away from $E_{50\%}$, their limitations lie in the fact that at values close to $E_{50\%}$, where the rate of change for x is linear, calculations using a semi-log plot can be awkward. In such instances, a linear plot is helpful (Rogers, 1952). The limitations of the method are that important parameters such as the activity coefficients and effect of overpotential (electrochemical kinetics) are not considered. Also, it is assumed that all reactive species are completely soluble in Hg.

Work reported by Bush and Densen suggests that for a given volume ratio, maximum separation is achieved when the fraction of M_1 in the Hg phase is equal to the fraction of M_2 in the aqueous phase (Bush &

Table 7.3 Half-wave potential of various rare earth elements, including Li, K, and Na (Kozlovskii, 1956; Kryukova et al., 1959).

Reaction/metal	$E_{½}, V$	Composition of electrolyte
$La^{3+} \rightarrow La^0$	−1.92	In the presence of 0.01 M LiCl
$Ce^{3+} \rightarrow Ce^0$	−1.97	In the presence of 0.01 M LiCl
$Pr^{3+} \rightarrow Pr^0$	−1.75	In the presence of 0.1 M LiCl, 0.1% gelatine, and 2.5×10^{-3} M Pr^{3+}
$Pr^{3+} \rightarrow Pr^0$	−1.82	In the presence of 0.1 M LiCl, 0.1% gelatine, and 8.9×10^{-3} M Pr^{3+}
$Nd^{3+} \rightarrow Nd^0$	−1.81	In the presence of 0.1 M LiCl, 0.1% gelatine, and 4.1×10^{-3} M Nd^{3+}
$Nd^{3+} \rightarrow Nd^0$	−1.77	In the presence of 0.1 M LiCl, 0.1% gelatine, and 1.6×10^{-3} M Nd^{3+}
$Sm^{3+} \rightarrow Sm^{2+}$	−1.77	In the presence of 0.1 M LiCl, 0.01% gelatine, and 5.0×10^{-3} M Sm^{3+}
$Sm^{3+} \rightarrow Sm^0$	−1.96	In the presence of 0.1 M LiCl, 0.01% gelatine, and 5.0×10^{-3} M Sm^{3+}
$Eu^{3+} \rightarrow Eu^{2+}$	−0.78	In the presence of 0.01 M LiCl
$Eu^{3+} \rightarrow Eu^0$	−2.14	In the presence of 0.01 M LiCl
$Gd^{3+} \rightarrow Gd^0$	−1.74	In the presence of 0.1 M LiCl and 8.0×10^{-4} M Gd^{3+}
$Tb^{3+} \rightarrow Tb^0$	−1.85	In the presence of 0.1 M LiCl and 0.005% gelatine
$Dy^{3+} \rightarrow Dy^0$	−1.85	–
$Ho^{3+} \rightarrow Ho^0$	−1.85	–
$Er^{3+} \rightarrow Er^0$	−1.85	–
$Tu^{3+} \rightarrow Tu^0$	−1.85	–
$Yb^{3+} \rightarrow Yb^{2+}$	−1.17	In the presence of 0.1 M LiCl, 0.005% gelatine, and 5.0×10^{-4} M Yb^{3+}
$Yb^{3+} \rightarrow Yb^0$	−2.05	In the presence of 0.1 M LiCl, 0.005% gelatine, and 5.0×10^{-4} M Yb^{3+}
$Lu^{3+} \rightarrow Lu^0$	−1.8	–
Li	−2.26	–
K	−2.12	–
Na	−2.04	–

Densen, 1948). The Nernst equation, while neglecting activity coefficients for a trivalent specie, then gives:

$$E = E^a - 0.02 \log C^a / C^e, \tag{8}$$

where E is the cathode potential, E^a is the standard amalgam potential, C^a is the concentration of lanthanides in the amalgam, and C^e is the concentration of the lanthanides in the electrolyte. For two lanthanides with a common electrolytic potential,

$$E_1^a - E_2^a = 0.02\log\frac{C_1^a C_2^e}{C_1^e C_2^a}. \tag{9}$$

One would observe that the arrangement of the concentration is similar to that used to describe the separation factor for ion exchange and solvent extraction (Onstott, 1956).

7.4. Electrochemistry-Driven REE Magnet Dissolution and Recovery

Another application of electrochemical processing of REEs involves the anodic dissolution of REE-based magnets. This section will examine work performed using chemical and electrochemical dissolution as a pathway for REE magnet recycling. Advantages to these methods are potential reductions in chemical use as either of those can be generated *in situ* or replaced by electrons. Studies involving electrochemical dissolution have taken place either in a single compartment or membrane-separated cells.

7.4.1. REE Recovery from Magnet Leachate

Venkatesan et al. explored the separation of Nd from neodymium–iron–boron (Nd-Fe-B) magnet leachate using three approaches with the aim of creating lower-impact recovery processes for REEs. The first approach explored the use of a single anode cell for the oxidation of Nd–Fe–B magnet leachate as seen in Figure 7.3. A nickel wire, titanium/platinum (Ti/Pt) mesh, and silver/silver chloride (Ag/AgCl) were the anode, cathode, and reference electrode, respectively. The process involved the application of current to oxidize Fe^{2+} to insoluble Fe^{3+} in the leachate followed by the precipitation of the REEs using oxalic acid (Venkatesan, Sun, et al., 2018). A current of 1.2 A was found to oxidize 98.9% of the Fe^{2+} within 4 h. The resulting leachate was filtered and the REEs were precipitated using oxalic acid yielding >97% recovery with a purity of >99%.

7.4.2. Nd-Fe-B Magnet as Anode

In 1994, Greenberg described an electrolytic method for Nd extraction from Nd–Fe–B metallic slag that involved using a modified plating tank filled with sulfamic acid (Greenberg, 1994). During electrolysis, Nd was

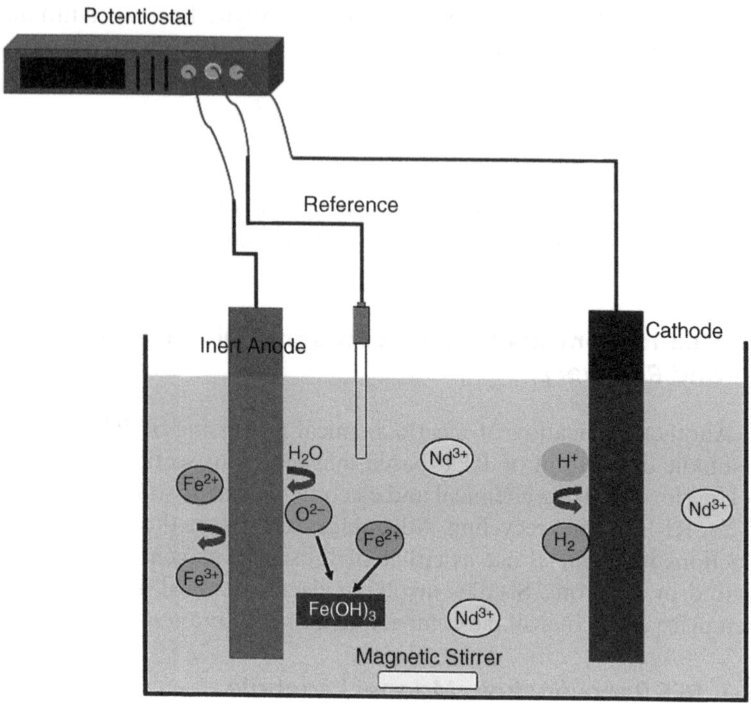

Figure 7.3 Single-anode electrolysis reactor for selective extraction of REEs from Nd–Fe–B magnets.

accumulated in the electrolyte and Fe was plated on the cathode. Hydrofluoric acid (HF) was then added to the electrolyte followed by filtration to produce NdF_3 which was then dried. This approach achieved >95% recovery for Nd with <3% moisture. More recently, the use of an Nd–Fe–B magnet as an anode was investigated.

The second approach employed by Venkatesan used a dual-anode system similar to Figure 7.4 to separate Nd from Nd–Fe–B magnets using NH_4Cl as electrolyte (Venkatesan, Vander Hoogerstraete, Binnemans, et al., 2018). The Nd–Fe–B magnet and a Ti/Pt dimensionally stable anode (DSA) electrode were employed as the active and inert anodes, respectively. During electrolysis, the Fe in the magnet was oxidized into the electrolyte, as Fe^{2+} then further oxidized at the Ti/Pt anode to Fe^{3+} (see Figure 7.4), where it was selectively precipitated at pH 3–3.5 (Lakshmanan et al., 2009; Önal et al., 2015; Önal et al., 2017; Yang et al., 2015). Following filtration of Fe^{3+}, REEs were selectively precipitated using oxalic acid. The purity of

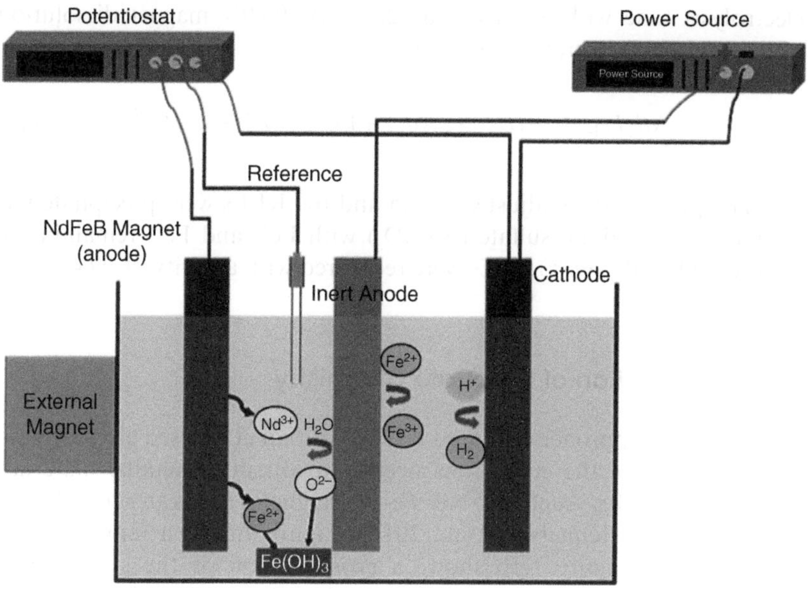

Figure 7.4 Dual-anode electrolysis reactor for selective extraction of REEs from Nd–Fe–B magnets.

the recovered REE oxalates was between 98.5% and 99% and energy consumption was 3 kWh/kg. It was observed that increasing the electrolyte concentration (NH_4Cl) increased magnet dissolution and decreased energy consumption.

Xu et al. demonstrated a two-step method for the simultaneous recovery of REEs and deposition of Fe from Nd–Fe–B magnets (Xu et al., 2020). The magnet was attached as the anode, Ag/AgCl electrode was the reference electrode, and the electrolyte was 0.6 mol L^{-1} $FeSO_4$, 0.4 mol L^{-1} $(NH_4)_2SO_4$, 0.175 mol L^{-1} trisodium citrate (Na_3Cit), and 0.4 mol L^{-1} boric acid (H_3BO_3). The role of the $(NH_4)_2SO_4$ was to complex REEs and maintain Fe^{2+} and Fe^{3+} as soluble species at negative potentials due to the effects of HER. The addition of Na_3Cit was to further prevent the formation of insoluble REE hydroxides. H_3BO_3 was added to prevent the formation of any metal hydroxides which interfere with the deposition of Fe at the cathode. For a 100% leaching efficiency, Xu observed that a current density <50 mA cm^{-2} was most favorable. At current densities >50 mA cm^{-2}, the oxidation of water is favored with decreased leaching efficiency. A current density of 12.5 mA/cm^2 was chosen for an in-depth analysis and after 8 h electrolysis time with a current of 200 mA, 99.9% magnet dissolution occurred with an overall reaction of:

$$Nd_2Fe_{14}B + 9H^+ \rightarrow 2Nd^{3+} + 14Fe + B^{3+} + 4.5H_2 \uparrow \quad (10)$$

The pH was then adjusted to 3.5 and the REEs were precipitated as sulfates using sodium sulfate (Na_2SO_4) with Fe^{2+} and Fe^{3+} remaining in solution. Overall, >93% REEs were recovered with a purity of 99.4%.

7.4.3. Generation of Leachate Chemistry

The third approach explored by Venkatesan et al. used electrochemistry to generate the conditions needed to dissolve metallic materials containing REEs, such as Nd–Fe–B magnets (Venkatesan, Vander Hoogerstraete, Hennebel, et al., 2018). Figure 7.5a is a representative system, while Figure 7.5b shows a cross section of the system with anodic reactions and ion flow depicted. The cell used an anion exchange membrane (AEM) that effectively allowed chloride and hydroxide to cross from the cathode side while preventing metal cations from reaching the cathode compartment. Prior to electrolysis, the magnet waste was partially leached by hydrochloric acid (HCl). The leachate and sodium chloride (NaCl) were the electrolytes in the anode and cathode compartments, respectively. Circulation of both the catholyte and anolyte was performed within the cathode and anode compartments, respectively (see Figure 7.5a,b).

Depending on the duration of electrolysis, chloride and hydroxide migrate from the catholyte to the anolyte. The chloride ions that migrate to the anolyte combine with protons produced in the anode compartment from HCl. This created unselective metal leaching and increased the amount of Fe co-extracted into the leachate (see Figure 7.5b). The role of the AEM during electrolysis was to prevent the migration of Fe^{3+} to the catholyte. Following leaching into the anolyte, the Fe^{3+} precipitates were removed through centrifugation. REEs were precipitated from the filtrate using oxalic acid. Inductively coupled plasma optical emission spectroscopy (ICP-OES) analysis showed that no metals presence in the catholyte suggesting the membrane was effective. At a current density of 50 A m^{-2}, REEs and cobalt extraction into solution was greater than 97%.

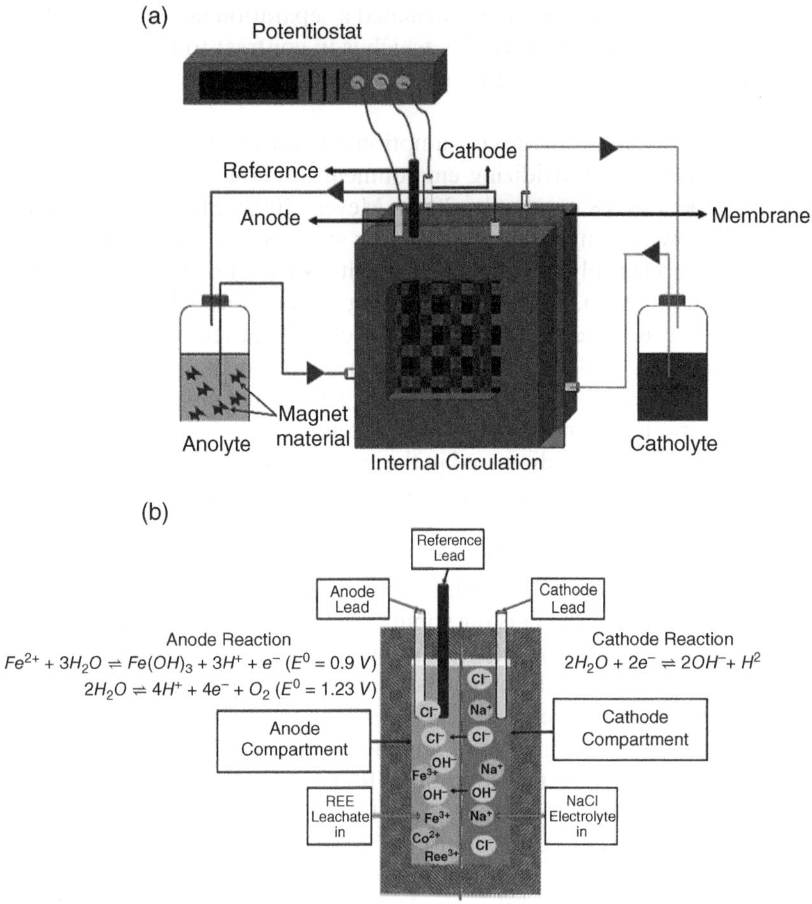

Figure 7.5 (a) Membrane cell for REEs separation and (b) cross section of membrane cell and reactions taking place in membrane cell.

7.5. Perspective and Conclusion

7.5.1. Discussion and Challenges

REEs play a crucial role in technology and will continue to be in huge demand as the green economy develops. Molten salt electrolysis (MSE), metallothermic reduction, and comproportionation (MRC) are established technologies for REE separation and extraction with purities as high as

99.8% (Gray, 1952). Uda et al. calculated a separation factor for a Nd–Sm mixture to be approximately 570, which is in contrast to the conventional solvent extraction method that gave 9.6 using comproportionation (Uda et al., 2000).

Scalability, high energy consumption, the use of very harsh chemicals, and the need for an oxidizing environment are concerns with MSE and MRC processes (Kingsnorth, 2015; Meyer, 2007; Sharma, 1987). For example, low-melting-temperature REEs such as La, Ce, Pr, and Nd are liquified via electrolysis with chloride salts between 650 °C and 1100 °C (Kobisk & Grisham, 1969). High-melting-temperature REEs such as Gd, Dy, and Y can be recovered using fluoride salts at temperatures ranging from 1300 °C to 1700 °C. However, due to environmental concerns, the use of REE chloride salts was halted due to hazardous off-gases, low yields, and excessive moisture adsorption (Liu et al., 2020).

From the research discussed in the earlier text, preferential amalgamation using Hg showed that effective separation of REEs was possible with high separation factors reported for some cases. However, due to its high toxicity, Hg has been banned in most places. There are other low-temperature alloys to be studied. Initial work using Ga was not successful. The biggest issue with Ga is the spontaneous formation of an oxide passivation layer which appears to be stable at very negative potentials (Engmann et al., 2022). If this can be controlled or overcome, this approach may have promise. There are few options available for liquid metals or alloys below the boiling point of aqueous solutions.

Furthermore, direct separation of REEs in aqueous media remains a challenge given the favorability of hydrogen evolution rather than REE reduction. However, the dissolution of REE magnets in acid followed by the electrochemical separation of Fe from the REEs is a path to reducing chemical waste, as most of the chemicals used in the process can be recycled back into the system. Attaching magnets as anodes to an electrochemical cell is viewed as challenging. It requires magnets to be isolated from e-waste streams. Hard disk drives, which are typically shredded into fragments, make isolation of the magnets a challenge (Diaz et al., 2016). Methods that generate chemicals through electrochemistry, such as through the membrane approach described earlier [90], are increasingly more attractive.

7.5.2. Future Directions

While REEs have limited redox activity in the electrochemical window of aqueous solutions, there has been some promising work. The options can be broken down into processes that directly act on REEs and those that act on other constituents to promote REE processing.

Direct processing: There could be opportunities to improve on existing methods that should be investigated, particularly for Ce which is the highest abundance REE. Another area that could be revisited is the use of alloying reactions to access zero-valent states within aqueous electrochemical windows. While briefly investigated (Engmann et al., 2022), there could be promise in pursuing Hg substitutes based on Ga or other alloys with melting points below the boiling point of aqueous electrolytes.

Indirect processing: Indirect methods that act on other constituents in an REE-containing solution can be almost unlimited. Indirect processing can involve manipulating bulk or localized chemical environments to remove unwanted species, generate reactive agents, or change the localized pH. A specific area that could be promising is the use of redox mediators that shuttle electrons from the anode or cathode to constituents in solution. This permits the electrodes to be located separately from the material being processed. An example of this would be the generation of oxidizing agents that can dissolve REE-based magnets. A redox dissolution scheme reported by Prodius et al. demonstrated galvanically driven dissolution of REE magnets using copper (Cu^{2+}) salts (Prodius et al., 2019). In the reaction, the magnet alloy was oxidized with Cu^{2+} to produce dissolved metals ions (including REEs) and Cu^0. It would be desirable to select a mediator that is ionic in both oxidation states, such as Fe^{2+}/Fe^{3+}, which can greatly reduce chemical consumption and limit introducing solid material to the waste.

Other options would be to manipulate local pH through HER to selectively precipitate dissolved species. This was recently demonstrated by O'Connor et al. where Eu and Nd were electrochemically precipitated at a conductive carbon nanotube filter cathode over a range of applied voltages and pHs (O'Connor et al., 2018). The electrolyte comprised 1 mM REE and 100 mM Na_2SO_4. Recoveries of Nd and Eu were 72% and 65%, respectively, at 3.0 V in the form of trivalent oxides. The reaction responsible for the trivalent oxides was the reduction of oxygen and water to hydroxides, followed by the formation of REE hydroxide species and then the subsequent dehydration to form metal oxides. Interestingly, REE oxides were also recovered as part of a process using a simpler three-electrode h-cell setup where REEs were immobilized by a simultaneous chemical and electrochemical process (Engmann et al., 2022). In scenarios where REEs are electrochemically precipitated to eventual oxides, the challenge arises with trivalent REEs where similarities in chemistry promote recovery as a mixed-oxide product.

Acknowledgments

This manuscript has been authored by Battelle Energy Alliance, LLC under Contract No. DE-AC07–05ID14517 with the U.S. Department of Energy. Writing of this chapter was supported by funding provided by the

Critical Materials Institute, an Energy Innovation Hub funded by the U.S. Department of Energy, Office of Energy Efficiency and Renewable Energy, Advanced Manufacturing Office. The authors wish to also express their sincere thanks to Dr. Frederick F. Stewart for offering his support and expertise during this project.

References

Abbott, A. P., & McKenzie, K. J. (2006). Application of ionic liquids to the electrodeposition of metals. *Physical Chemistry Chemical Physics*, *8*(37), 4265–4279.

Arenas, L. F., De León, C. P., & Walsh, F. C. (2016). Electrochemical redox processes involving soluble cerium species. *Electrochimica Acta*, *205*, 226–247.

Atanasyants, A. G., & Seryogin, A. N. (1995). The reaction of the electrochemical reduction Eu (III)+ e→ Eu (II) in hydrochloric solution. *Hydrometallurgy*, *37*(3), 367–374.

Audrieth, L. F., Jukkola, E. E., Meints, R. E., & Hopkins, B. S. (1931). Observations on the rare earths. XXXVII. Electrolytic preparation of rare earth amalgams. 1. Preparation of amalgams of lanthanum and neodymium. *Journal of the American Chemical Society*, *53*(5), 1805–1809.

Ball, R. W., & Yntema, L. F. (1930). The separation of ytterbium by electrolytic reduction. Observations on the rare earths. XXXVI. *Journal of the American Chemical Society*, *52*(11), 4264–4268.

Bommaraju, T. V, Orosz, P. J., & Sokol, E. A. (2007). Brine electrolysis. *Electrochemistry encyclopedia* (pp. 1–17). Cleveland, OH, USA: Ernest B. Yeager Center for Electrochemical Sciences.

Bünzli, J.-C. G. (2013). Lanthanides. *Kirk-Othmer encyclopedia of chemical technology* (pp. 1–43). Major Reference Works. doi:10.1002/0471238961.120114201901 0215.a01.pub3.

Bush, M. T., & Densen, P. M. (1948). Systematic multiple fractional extraction procedures. *Analytical Chemistry*, *20*(2), 121–129.

Cademartiri, L., Thuo, M. M., Nijhuis, C. A., Reus, W. F., Tricard, S., Barber, J. R., et al. (2012). Electrical resistance of AgTS–S (CH2) n−1CH3//Ga2O3/EGaIn tunneling junctions. *The Journal of Physical Chemistry C*, *116*(20), 10848–10860.

Castor, S. B., & Hedrick, J. B. (2006). Rare earth elements. *Industrial minerals and rocks* (pp. 769–792). Colorado, USA: Society for Mining, Metallurgy and Exploration Littleton.

Cicconi, M. R., Le Losq, C., Henderson, G. S., & Neuville, D. R. (2021). The redox behavior of rare earth elements. *Magma Redox Geochemistry*, 381–398.

Crook, J., & Mousavi, A. (2016). The chlor-alkali process: A review of history and pollution. *Environmental Forensics*, *17*(3), 211–217.

Dennis, L. M., & Lemon, B. J. (1915). The electrolysis of solutions of the rare earths. *Journal of the American Chemical Society*, *37*(1), 131–137.

Dennis, L. M., & Meulen, P. V. D. (1915). The electrolysis of solutions of the rare earths. *II. American Chemical Society*, *37*(9), 1963–1976.

Diaz, L. A., Lister, T. E., Parkman, J. A., & Clark, G. G. (2016). Comprehensive process for the recovery of value and critical materials from electronic waste. *Journal of Cleaner Production*, *125*, 236–244.

Doudrick, K., Liu, S., Mutunga, E. M., Klein, K. L., Damle, V., Varanasi, K. K., & Rykaczewski, K. (2014). Different shades of oxide: From nanoscale wetting mechanisms to contact printing of gallium-based liquid metals. *Langmuir*, *30*(23), 6867–6877.

Engmann, E., Diaz, L. A., Lister, T. E., Palasyuk, O., & Zhao, H. (2022). Electrochemistry of praseodymium in aqueous solution using a liquid gallium cathode. *Journal of The Electrochemical Society*, *169*(6), 063519.

Estee, C. R., & Glockler, G. (1948). Reduction of neodymium at the dropping mercury electrode. *Journal of the American Chemical Society*, *70*(4), 1344–1347.

Fawcett, W. R. (1989). Potential dependence of the elementary steps in the kinetics of electrode reactions involving amalgam formation. *Journal of Physical Chemistry (1952)*, *93*(6), 2675–2682.

Ghasemian, M. B., Mayyas, M., Idrus-Saidi, S. A., Jamal, M. A., Yang, J., Mofarah, S. S., et al. (2019). Self-limiting galvanic growth of MnO2 monolayers on a liquid metal—Applied to photocatalysis. *Advanced Functional Materials*, *29*(36), 1901649.

Gough, R. C., Dang, J. H., Moorefield, M. R., Zhang, G. B., Hihara, L. H., Shiroma, W. A., & Ohta, A. T. (2016). Self-actuation of liquid metal via redox reaction. *ACS Applied Materials & Interfaces*, *8*(1), 6–10.

Gray, P. M. J. (1952). The production of pure cerium metal by electrolytic and thermal reduction processes (vol. 61).

Greenberg, B. (1994). Neodymium recovery process. Google Patents.

Han, J., Mayyas, M., Tang, J., Mousavi, M., Idrus-Saidi, S. A., Cai, S., et al. (2021). Liquid metal enabled continuous flow reactor: A proof-of-concept. *Matter*, *4*(12), 4022–4041.

Hopkins, B. S., & Audrieth, L. F. (1934). The electrolysis of rare earth metal salts in non-aqueous solvents. *Transactions of The Electrochemical Society*, *66*(1), 135.

Jukkola, E. E., Audrieth, L. F., & Hopkins, B. S. (1934). Observations on the rare earths. XLI. Electrolytic preparation of rare earth amalgams. 3. Amalgams of lanthanum, neodymium, cerium, samarium and yttrium. metallic lanthanum, neodymium and cerium by thermal decomposition of their amalgams. *Journal of the American Chemical Society*, *56*(2), 303–304.

Khan, M. R., Trlica, C., & Dickey, M. D. (2015). Recapillarity: Electrochemically controlled capillary withdrawal of a liquid metal alloy from microchannels. *Advanced Functional Materials*, *25*(5), 671–678.

Khan, M. R., Trlica, C., So, J.-H., Valeri, M., & Dickey, M. D. (2014). Influence of water on the interfacial behavior of gallium liquid metal alloys. *ACS Applied Materials & Interfaces*, *6*(24), 22467–22473. doi: 10.1021/am506496u.

Kingsnorth, D. (2015). The global rare earth industry today—Plagued by illegal production in China. In *11th Rare Earth Conference, Metal Events Ltd.*, Singapore.

Kobisk, E. H., & Grisham, W. B. (1969). Application of reduction-distillation method for preparing high-purity rare-earth isotope metals. *Materials Research Bulletin*, *4*(9), 651–662.

Kolesov, G. M., & Pankratova, L. N. (1968). Separation of rare-earth elements by electrolysis at a mercury cathode. *Russian Chemical Reviews*, *37*(9), 701.

Kolthoff, I. M., & Lingane, J. J. (1952). *Polarography VI* (p. 243). New York–London: Interscience.

Kozlovskii, M. T. (1956). Mercury and amalgams in electrochemical methods of analysis [in Russian], Izd. Akad. Nauk KazSSR, Alma-Ata.

Kronholm, B., Anderson, C. G., & Taylor, P. R. (2013). A primer on hydrometallurgical rare earth separations. *Journal of Operations Management*, *65*(10), 1321–1326.

Kryukova, T. A., Sinyakova, S. I., and Aref'eva, T. V (1959). *Polyarograficheskii analiz. (Polarographic Analysis)* (pp. 412–504). Gos. Nauehno-Tekhn. Izd. Khim. Lit., Moscow.

Lakshmanan, D., Clifford, D. A., & Samanta, G. (2009). Ferrous and ferric ion generation during iron electrocoagulation. *Environmental Science & Technology*, *43*(10), 3853–3859.

Latimer, W. M. (1938). *Oxidation states of the elements and their potentials in aqueous solutions*. Prentice-Hall, Inc.

Lavrukhina, A. K., Kolesov, G. M., & Tang, H. (1960). Reduction of cerium group rare-earth elements on mercury cathode. In *Doklady Akademii Nauk SSSR*. Vernadskii Institute of Geochemistry and Analytical Chemistry USSR.

Lertanantawong, B., Lertsathitphong, P., & O'Mullane, A. P. (2018). Chemical reactivity of Ga-based liquid metals with redox active species and its influence on electrochemical processes. *Electrochemistry Communications*, *93*, 15–19.

Liu, H., Zhang, Y., Luan, Y., Yu, H., & Li, D. (2020). Research progress in preparation and purification of rare earth metals. *Metals*, *10*(10), 1376.

Loftness, R. L. (1951). *A vapor pressure chart for metals*. North American Aviation, Inc.

Marsh, J. K. (1943a). Rare-earth metal amalgams. Part III. The separation of ytterbium from its neighbours. *Journal of the Chemical Society (Resumed)*, 8–10.

Marsh, J. K. (1943b). Rare-earth metal amalgams. Part IV. The isolation of europium. *Journal of the Chemical Society (Resumed)*, 531–535.

Mayyas, M., Mousavi, M., Ghasemian, M. B., Abbasi, R., Li, H., Christoe, M. J., et al. (2020). Pulsing liquid alloys for nanomaterials synthesis. *ACS nano*, *14*(10), 14070–14079.

McCoy, H. N. (1941a). Europium and ytterbium amalgams. *Journal of the American Chemical Society*, *63*(6), 1622–1624.

McCoy, H. N. (1941b). The electrolysis of rare earth acetates and the separation of europium as amalgam from other rare earths. *Journal of the American Chemical Society*, *63*(12), 3432–3433. ACS Publications.

McCoy, H. N., & Hammond, R. P. (1942). The separation of ytterbium from accompanying rare earths by means of its amalgam. *Journal of the American Chemical Society*, *64*(4), 1009.

Mccoy, H. N., Morris, H., & Selwood, P. W. (1946). Europium amalgam: Separation of europium from rare earth mixtures. *Inorganic Syntheses*, 65–69. doi:10.1002/9780470132333.ch18.

Meyer, G. (2007). The reduction of rare-earth metal halides with unlike metals—Wöhler's metallothermic reduction. *Zeitschrift für anorganische und allgemeine Chemie*, *633*(15), 2537–2552. doi: 10.1002/zaac.200700386.

Mikkelsen, Ø., & Schrøder, K. H. (2003). Amalgam electrodes for electroanalysis. *Electroanalysis*, *15*(8), 679–687. doi: 10.1002/elan.200390085.

Moeller, T., & Kremers, H. E. (1945). Observations on rare earths extraction of ytterbium from rare earth mixtures with sodium amalgam. *Industrial & Engineering Chemistry Analytical Edition*, *17*(12), 798–800.

Morley, N. B., Burris, J., Cadwallader, L. C., & Nornberg, M. D. (2008). GaInSn usage in the research laboratory. *Review of Scientific Instruments*, *79*(5), 56107.

Neckers, J. W., & Kremers, H. C. (1928). Observations on the rare earths. XXVII. I. Fractional precipitation of the cerium group earths by electrolysis. II. Solubility of rare earth oxalates in nitric acid 1. *Journal of the American Chemical Society*, *50*(4), 950–954.

O'Connor, M. P., Coulthard, R. M., & Plata, D. L. (2018). Electrochemical deposition for the separation and recovery of metals using carbon nanotube-enabled filters. *Environmental Science: Water Research & Technology*, *4*(1), 58–66.

Önal, M. A. R., Borra, C. R., Guo, M., Blanpain, B., & Van Gerven, T. (2015). Recycling of NdFeB magnets using sulfation, selective roasting, and water leaching. *Journal of Sustainable Metallurgy*, *1*(3), 199–215.

Önal, M. A. R., Borra, C. R., Guo, M., Blanpain, B., & Van Gerven, T. (2017). Hydrometallurgical recycling of NdFeB magnets: Complete leaching, iron removal and electrolysis. *Journal of Rare Earths*, *35*(6), 574–584.

Onstott, E. I. (1955). The separation of europium from samarium by electrolysis. *Journal of the American Chemical Society*, *77*(8), 2129–2132.

Onstott, E. I. (1956). Separation of the lanthanons at amalgam cathodes. II. The separation of samarium from gadolinium and purification of europium at a lithium amalgam cathode. *Journal of the American Chemical Society*, *78*(10), 2070–2076.

Onstott, E. I. (1959). Separation of the lanthanons at amalgam cathodes III. Electrochemical fractionation of the lanthanons at a lithium amalgam cathode. *Journal of the American Chemical Society*, *81*(17), 4451–4458.

Onstott, E. I. (1960). Separation of the lanthanons at amalgam cathodes. IV. High purity ytterbium from mixtures of heavy lanthanons by aqueous electrolysis. *Journal of the American Chemical Society*, *82*(24), 6297–6299.

Onstott, E. I. (1961). Separation of the lanthanons at amalgam cathodes. Measurement of the separability of praseodymium and neodymium in aqueous lithium citrate electrolytes. *Analytical Chemistry*, *33*(11), 1470–1473.

Park, J.-E., Kang, H. S., Baek, J., Park, T. H., Oh, S., Lee, H., et al. (2019). Rewritable, printable conducting liquid metal hydrogel. *ACS nano*, *13*(8), 9122–9130.

Patnaik, P. (2003). *Handbook of inorganic chemicals* (Vol. *529*). New York: McGraw-Hill.

Pourbaix, M. (1974). Atlas of electrochemical equilibria in aqueous solution. *NACE, 307.*

Powell, C. J. (1968). Characteristic energy losses of 8-keV electrons in liquid Al, Bi, In, Ga, Hg, and Au. *Physical Review, 175*(3), 972.

Prodius, D., Gandha, K., Mudring, A.-V., & Nlebedim, I. C. (2019). Sustainable urban mining of critical elements from magnet and electronic wastes. *ACS Sustainable Chemistry & Engineering, 8*(3), 1455–1463.

Rabideau, S. W., & Glockler, G. (1948). The polarographic reduction of gadolinium. *Journal of the American Chemical Society, 70*(4), 1342–1344.

Regan, M. J., Tostmann, H., Pershan, P. S., Magnussen, O. M., DiMasi, E., Ocko, B. M., & Deutsch, M. (1997). X-ray study of the oxidation of liquid-gallium surfaces. *Physical Review B, 55*(16), 10786.

Riba, J.-R., López-Torres, C., Romeral, L., & Garcia, A. (2016). Rare-earth-free propulsion motors for electric vehicles: A technology review. *Renewable and Sustainable Energy Reviews, 57*, 367–379.

Rogers, L. B. (1952). Electrodeposition behavior of a simple ion: II. Precipitation into a liquid electrode (mercury). *Journal of The Electrochemical Society, 99*(6), 267.

Ryabchikov, D. I., & Terent'eva, E. A. (1960). Progress in methods for the separation of rare-earth elements. *Russian Chemical Reviews, 29*(10), 589.

Scharmann, F., Cherkashinin, G., Breternitz, V., Knedlik, C., Hartung, G., Weber, T., & Schaefer, J. A. (2004). Viscosity effect on GaInSn studied by XPS. *Surface and Interface Analysis: An International Journal devoted to the development and application of techniques for the analysis of surfaces, interfaces and thin films, 36*(8), 981–985.

Schüler, D., Buchert, M., Liu, R., Dittrich, S., & Merz, C. (2011). Study on rare earths and their recycling. *Öko-Institut eV Darmstadt, 49*, 30–40.

Sharma, R. A. (1987). Metallothermic reduction of rare earth chlorides. Google Patents.

Shvedov, V. P., & Antonov, P. G. (1962). Separation of praseodymium and neodymium by electrolysis of a mercury cathode. *Radiokhimiya*, *5*(3), 342–346. doi: 10.1017/CBO9781107415324.004.

Shvedov, V. P., & Antonov, P. G. (1964). *A study of the conditions for the formation of amalgams of gadolinium, terbium, dysprosium and holium during electrolysis on a mercury cathode*. Country unknown/Code not available. Available at: https://www.osti.gov/servlets/purl/4659589.

Shvedov, V. P., & Antonov, P. G. (1967). Separation of rare earth elements of the yttrium subgroup by electrolysis with mercury cathode. *Radiokhimiya, 9*, 478–3.

Song, X., Chang, M.-H., & Pecht, M. (2013). Rare-earth elements in lighting and optical applications and their recycling. *Journal of Operations Management, 65*(10), 1276–1282.

Su, X. (2020). Electrochemical separations for metal recycling. *The Electrochemical Society Interface, 29*(3), 55.

Suresh, S., & Sundaramoorthy, S. (2015). *Green chemical engineering*. Boca Raton: CRC Press.

Surmann, P., & Zeyat, H. (2005). Voltammetric analysis using a self-renewable non-mercury electrode. *Analytical and Bioanalytical Chemistry*, *383*(6), 1009–1013. doi: 10.1007/s00216-005-0069-7.

Swensen, A. W., & Glockler, G. (1949). The polarographic reduction of praseodymium. *Journal of the American Chemical Society*, *71*(5), 1641–1644.

Timnick, A., & Glockler, G. (1948). The half-wave potential of samarium. *Journal of the American Chemical Society*, *70*(4), 1347–1350.

Uda, T., Jacob, K. T., & Hirasawa, M. (2000). Technique for enhanced rare earth separation. *Science*, *289*(5488), 2326–2329. doi: 10.1126/science.289.5488.2326.

Vasudevan, S., Sozhan, G., Mohan, S., & Pushpavanam, S. (2005). An electrochemical process for the separation of cerium from rare earths. *Hydrometallurgy*, *76*(1–2), 115–121.

Venkatesan, P., Sun, Z. H. I., Sietsma, J., & Yang, Y. (2018). An environmentally friendly electro-oxidative approach to recover valuable elements from NdFeB magnet waste. *Separation and Purification Technology*, *191*, 384–391.

Venkatesan, P., Vander Hoogerstraete, T., Binnemans, K., Sun, Z., Sietsma, J., & Yang, Y. (2018). Selective extraction of rare-earth elements from NdFeB magnets by a room-temperature electrolysis pretreatment step. *ACS Sustainable Chemistry & Engineering*, *6*(7), 9375–9382.

Venkatesan, P., Vander Hoogerstraete, T., Hennebel, T., Binnemans, K., Sietsma, J., & Yang, Y. (2018). Selective electrochemical extraction of REEs from NdFeB magnet waste at room temperature. *Green Chemistry*, *20*(5), 1065–1073

Venkatesh, S., & Tilak, B. V (1983). *Chlor-alkali technology*. ACS Publications.

Vickery, R. C. (1948). The influence of basicity upon the efficiency of oxidation/hydrolysis procedures for the separation and purification of cerium. *Journal of the Society of Chemical Industry*, *67*(8), 333–336.

Vickery, R. C. (1950). Some reactions of cerium and other rare earths with chlorine and hypochlorite', *Journal of the Society of Chemical Industry*, *69*(4), 122–125.

Wang, D., Gao, C., Wang, W., Sun, M., Guo, B., Xie, H., & He, Q. (2018). Shape-transformable, fusible rodlike swimming liquid metal nanomachine. *ACS nano*, *12*(10), 10212–10220.

Wang, Y., Wang, S., Chang, H., & Rao, W. (2020). Galvanic replacement of liquid metal/reduced graphene oxide frameworks. *Advanced Materials Interfaces*, *7*(19), 2000626.

West, D. H., & Hopkins, B. S. (1935). Observations on the rare earths. XLV. Preparation of rare earth amalgams by displacement *Journal of the American Chemical Society*, *57*(11), 2185–2186.

Xu, X., Sturm, S., Samardzija, Z., Scancar, J., Markovic, K., & Rozman, K. Z. (2020). A facile method for the simultaneous recovery of rare-earth elements and transition metals from Nd–Fe–B magnets. *Green Chemistry*, *22*(4), 1105–1112.

Yang, Y., Wang, X., Wang, M., Wang, H., & Xian, P. (2015). Recovery of iron from red mud by selective leach with oxalic acid. *Hydrometallurgy*, *157*, 239–245.

Yntema, L. F. (1930). The separation of europium by electrolytic reduction. Observations on the rare earths. XXXV. *Journal of the American Chemical Society*, *52*(7), 2782–2784.

Zhang, K., Kleit, A. N., & Nieto, A. (2017). An economics strategy for criticality—Application to rare earth element Yttrium in new lighting technology and its sustainable availability. *Renewable and Sustainable Energy Reviews*, *77*, 899–915.

Zhao, H., Xia, J., Yin, D., Luo, M., Yan, C., & Du, Y. (2019). Rare earth incorporated electrode materials for advanced energy storage. *Coordination Chemistry Reviews*, *390*, 32–49.

Zoski, C. G. (2006). *Handbook of electrochemistry*. Elsevier.

8

BENEFICIATION OF RARE EARTH ELEMENTS: PROSPECTS FOR BIOTECHNOLOGY DEPLOYMENT

Yoshiko Fujita[1], Dan Park[2], Margaret Lencka[3], Andre Anderko[3], David Reed[1], Vicki Thompson[1], Gaurav Das[3], Ali Eslamimanesh[3], and Yongqin Jiao[2]

[1] *Critical Materials Innovation Hub, Idaho National Laboratory, Idaho Falls, Idaho, USA*
[2] *Critical Materials Innovation Hub, Lawrence Livermore National Laboratory, Livermore, California, USA*
[3] *Critical Materials Innovation Hub, OLI Systems Inc., Parsippany, New Jersey, USA*

Abstract

New methods for recovering rare earth elements (REE) from traditional and nontraditional sources are needed to meet growing global demand. Approaches based on microbiological processes or biologically produced reagents to recover REEs are among the most promising. Biological approaches can avoid many of the hazards and costs associated with corrosive or toxic chemicals used in hydrometallurgy, as well as the high energy, reagent, and capital costs and air emissions associated with pyrometallurgy. Microbial processes are considered environmentally friendly because they are "natural," although opportunities exist to improve on native capabilities

Rare Earth Elements: Sustainable Recovery, Processing, and Purification, Special Publications 79, First Edition. Edited by Athanasios K. Karamalidis and Roderick Eggert.

DOI: 10.1002/9781119515005.ch8

through application of synthetic biology. In this chapter, we focus on some important factors that have not been as widely discussed but which should be considered in development of biological approaches for recovery and purification of REEs. In particular, we focus on geochemical and biogeochemical constraints posed by REE-containing feedstocks, for bioleaching and biosorption, and emphasize the utility of aqueous equilibrium modeling for interpreting results and supporting design of recovery methods. We also discuss some important cost factors for REE recovery specific to biological processes.

8.1. Introduction

Multiple reviews have been published in recent years on the increasing global demand for rare earth elements (REEs) and prospects for REE recovery from both traditional and nontraditional sources (Binnemans, Jones, et al., 2013; Fathollahzadeh et al., 2019; Ganguli & Cook, 2018; Gaustad et al., 2021; Jyothi et al., 2020; Løvik et al., 2018; Nassar et al., 2015; Tunsu et al., 2015; Weber & Reisman, 2012; Zhang et al., 2020). Many have noted the need for ecologically sustainable alternatives to conventional pyrometallurgical and hydrometallurgical methods to recover REEs (Jha et al., 2016; Sovacool et al., 2020). Among the newer approaches that have garnered recent interest are those that rely on microbiological processes or microbiologically produced reagents to recover REEs (Barmettler et al., 2016; Hennebel et al., 2015; Zhuang et al., 2015). Biological approaches can often avoid many of the environmental and/or safety hazards associated with the corrosive (e.g., strong acids) or toxic chemicals (e.g., organic solvents) often used in hydrometallurgy, as well as costs related to the high energy, reagent and capital requirements, and potential air emissions associated with pyrometallurgy (Adesina et al., 2017; Cui & Zhang, 2008; Pollmann et al., 2016; Rasoulnia et al., 2020; Zhuang et al., 2015). Microbial processes are considered environmentally friendly because they are "natural," although opportunities also exist to improve on native capabilities by the application of synthetic biology (Adesina et al., 2017).

Microbial processes can be harnessed for both solubilization of metals from solid matrices ("bioleaching") (Bosecker, 1997) and concentration and purification of solubilized metals that occur in mixtures ("biosorption/desorption" or "bioseparation") (Moriwaki & Yamamoto, 2013). Figure 8.1 depicts the processes.

Bioleaching is a long accepted practice for commercial recovery of some metals, in particular copper (Bosecker, 1997), but recently a number of researchers have been evaluating its application for recovery of rare

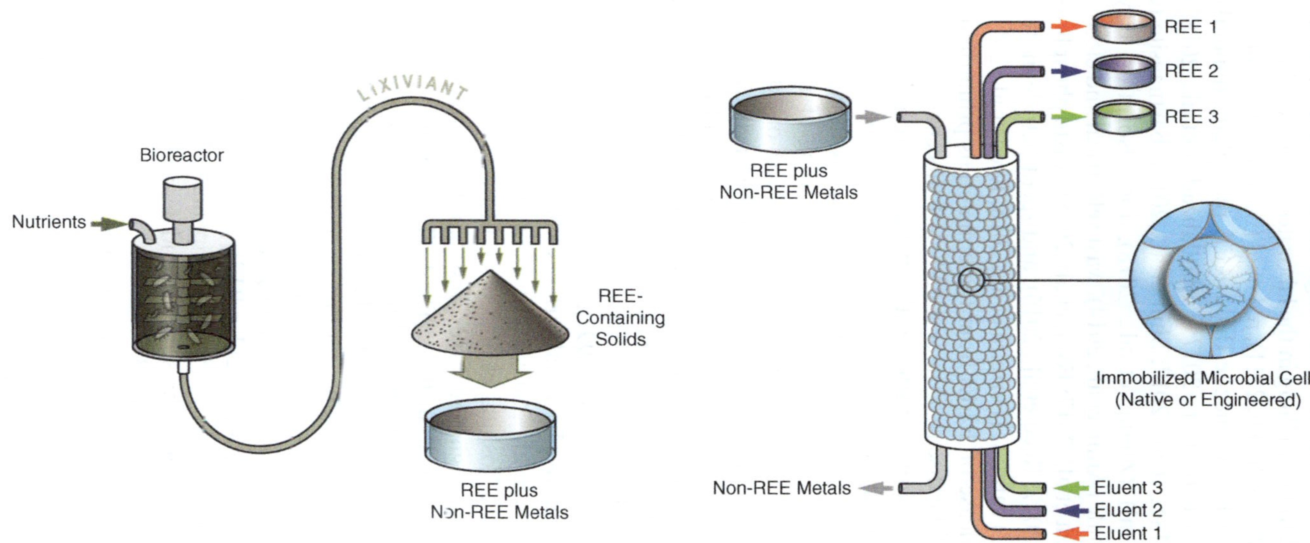

Figure 8.1 Schematic representation of bioleaching (left) and bioseparation (right). Graphic prepared by David Combs, INL.

earths, using both bacteria and fungi. REE-containing solids that have been subjected to bioleaching in the laboratory include gibbsite (aluminum ore) (Ibrahim & El-Sheikh, 2011), red mud (waste of aluminum production) (Qu & Lian, 2013), monazite (light-REE phosphate minerals) (Brisson et al., 2015; Corbett et al., 2017; Fathollahzadeh et al., 2018; Hassanien et al., 2014; Maes et al., 2017; Shin et al., 2015), bastnaesite (carbonate-fluoride mineral) (Zhang et al., 2018), ion-adsorption clay (Barnett et al., 2018), shale (Amin et al., 2014), waste electrical and electronic equipment (Di Piazza et al., 2017; Fonti et al., 2015), waste phosphors (lighting) (Hopfe et al., 2017; Reed et al., 2016), and spent petroleum fluid catalytic cracking (FCC) catalyst (Ferreira et al., 2018; Reed et al., 2016). Microbes mediate the release of REEs from these materials primarily by acid production, as will be discussed in Section 8.2 of this chapter.

While bioleaching aims to get the metals into solution, bioseparation and biosorption approaches are aimed primarily at concentrating and enriching (relative to other co-solubilized ions) the target metals once they are in solution. Biosorption relies on the binding of metals to selective ligands produced by and anchored on cell surfaces or extracellular biomaterials; after the metals are partitioned from the leachate solution onto the biological surfaces, they are desorbed (stripped) into another solution. Using a smaller volume of stripping solution results in concentration of the metals. Metal selectivity is imposed at the sorption step by the particular binding ligands, which typically incorporate carboxyl, phosphate, and hydroxyl groups (Fein et al., 1997). Numerous researchers have reported REE binding on native bacterial surfaces where such functional groups are common (Andrès et al., 2003; Moriwaki & Yamamoto, 2013; Takahashi et al., 2005, 2010; Yoshida et al., 2004). More recently, Park et al. have described bacteria engineered to express "lanthanide-binding tags" (LBTs) at high density on their surfaces. The LBTs have enhanced selectivity for binding of lanthanides compared to other non-REE metals (Figure 8.2), and they also exhibit preferential binding of heavy rare earth elements (HREEs) compared to light rare earth elements (LREEs) (Park et al., 2016, 2017). For REEs, because of their chemical similarity, separation of an individual REE is a particularly prized outcome. Selectivity (for REEs over non-REEs or for a particular REE) can also be imposed during the desorption step by the chemistry of the stripping solution. Factors important for optimal application of biosorption and desorption for REE separation will be discussed in Section 8.3.

In practice, leaching as well as sorption can result in some metal separation or enrichment. Unless the solid is completely dissolved, the relative abundances of metals in a leachate often differ from the original bulk solid phase (Corbett et al., 2017). For example, leaching by an organic acid that

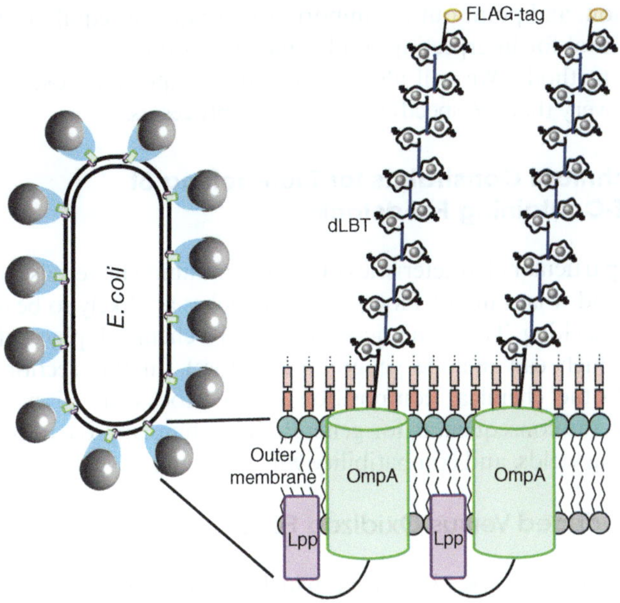

Figure 8.2 Surface display of LBT by *E. coli*. Cartoon depicting the Lpp–OmpA-mediated display dLBTx8. Blue lines represent the Muclb spacer. FlagTM Tag on the N-terminus. Reproduced from Park et al. (2017), / American Chemical Society.

has chelating properties (e.g., citric acid) can result in selective solubilization of metals that preferentially complex with the chelator. On the other hand, while solid dissolution by an agent such as hydrochloric acid might generally be considered nonselective leaching, the metal composition in the resulting leachate will still be a function of the stability of the metal-containing phases in the solid and the solubility of the individual metals in the leachate (and the kinetics of dissolution/precipitation). These concepts are expanded upon later in this chapter.

As noted in the earlier text, multiple reviews as well as many individual reports have been published recently regarding the potential for biological approaches to offer environmentally sustainable REE recovery. In this chapter, we will focus on some important factors that have not been widely discussed, but which should be considered in planning actual deployment of biological approaches for recovery and purification of rare earths, drawing on some of our own experience for examples. In particular, we will focus on geochemical and biogeochemical constraints posed by the feedstocks from which REEs may be extracted, for both bioleaching and biosorption, and point out the importance of aqueous equilibrium modeling as a tool for interpreting results and supporting design of biological recovery methods. We will also discuss some important cost factors for REE recovery that are specific to biological processes.

8.2. Technical Constraints for Bioleaching of REE-Containing Feedstocks

The particular characteristics of REE-containing materials will dictate whether and what kind of biological approaches are likely to be attractive for REE leaching. These characteristics can be chemical, physical, or economic (which can include geographic factors). In this section, factors related to the chemical composition of feedstocks are discussed, with respect to the consequences for general bioleaching approach, constraints on leaching yields, and compatibility with microorganisms.

8.2.1. Reduced Versus Oxidized Feedstocks

Biological processes have been widely accepted and practiced for the extraction of copper and gold from ores for decades, or, in the case of copper, perhaps centuries (Brierley, 2016). As much as 10%–15% of global copper production is currently estimated to result from bioleaching (Roberto, 2017). Other metals which have been commercially recovered by bioleaching include zinc, cobalt, uranium, and nickel (Brierley & Brierley, 2013; Kaksonen et al., 2018). These metals have in common that they frequently occur in sulfidic ores, providing conditions suitable for acidophilic iron- and sulfur-oxidizing microorganisms to produce ferric iron and sulfuric acid that promote solubilization of the target metals (Evangelou & Zhang, 1995; Kaksonen et al., 2018). For example, in copper bioleaching, iron-oxidizing organisms aid in the oxidation of ferrous to ferric iron (Equation (1)) (Rawlings, 2005). Elemental sulfur is oxidized to sulfate by acidophilic sulfur-oxidizing microbes and acidity is generated (Equation (2)). The ferric ion and protons can then chemically attack the solid copper sulfides to leach the metal into solution (Equation (3)). The ferric ion is also returned to ferrous iron by polysulfide oxidation (Equation (4)). The leaching microbes are involved in Equations (1) and (2):

$$2Fe^{2+} + 0.5O_2 + 2H^+ \rightarrow 2Fe^{3+} + H_2O. \tag{1}$$

$$0.125S_8 + 1.5O_2 + H_2O \rightarrow SO_4^{\ 2-} + 2H^+. \tag{2}$$

$$CuS + Fe^{3+} + H^+ \rightarrow Cu^{2+} + 0.5H_2S_n + Fe^{2+}, \text{where } n \geq 2. \tag{3}$$

$$0.5H_2S_n + Fe^{3+} \rightarrow 0.125S_8 + Fe^{2+} + H^+. \tag{4}$$

For gold, rather than directly leading to the dissolution of the target metal, microbial sulfur-oxidizing capabilities are utilized to dissolve the sulfidic mineral matrix that can otherwise shield the gold from chemical treatment. Subsequent gold leaching by nonbiological means, typically using cyanide solutions, follows. This gold ore pretreatment process is known as bio-oxidation, rather than bioleaching (Olson et al., 2003). Dissolution of minerals other than those actually containing the targeted metals is also a feature of approaches suggested by Johnson and colleagues, where reductive microbial metabolisms such as ferric iron or manganese oxide reduction that result in dissolution of oxidized minerals promote the release of incorporated or sorbed metals such as nickel or cobalt (Hallberg et al., 2011; Johnson, 2012).

Conventional rare earth ores (e.g., bastnaesite, monazite, xenotime, ion-adsorption clays) are typically oxidized and non-sulfidic (Van Gosen et al., 2017), and thus relying on acidophilic S and Fe oxidizers for *in situ* generation of acidic leaching agents for these ores would require external provision of suitable substrates (reduced sulfur compounds). Ibrahim and El-Sheikh added *Acidithiobacillus ferrooxidans*, ferric sulfate, and sulfuric acid to low-grade gibbsite ore to initiate lixiviant production, and they observed that when sulfur (0.5%) was added, the bioleaching efficiency increased 10% (Ibrahim & El-Sheikh, 2011). In our own work, we observed REE recovery using biolixiviant generated by mixed cultures of microorganisms enriched from Yellowstone National Park acidic geothermal features (Figure 8.3). The organisms were enriched using a medium that was first adjusted to pH 4 with sulfuric acid, and reduced sulfur was added as an electron donor for the microbes. With growth the medium pH decreased further to between 2 and 3. The growing enrichments were incubated with FCC catalyst for 6 weeks at 65 °C with shaking at 150 rpm. The lixiviants generated by enrichments from Frying Pan Spring and Nymph Lake resulted in 1.7-fold and 1.4-fold more leaching, respectively, than the abiotic control (uninoculated medium at pH 4) (Figure 8.3).

Unconventional sources for REEs might be more amenable to bioleaching based on such metabolism; for example, coals often contain significant levels of reduced sulfur (Calkins, 1994) and in fact REEs have been shown to be present in acid mine drainage from coal mines (Ayora et al., 2016; Stewart et al., 2017; Zhao et al., 2007). In those cases, bioleaching based upon sulfide oxidation would be attractive; the responsible organisms are likely to already be naturally present, and the primary requirements to promote leaching would be sufficient availability of oxygen and water.

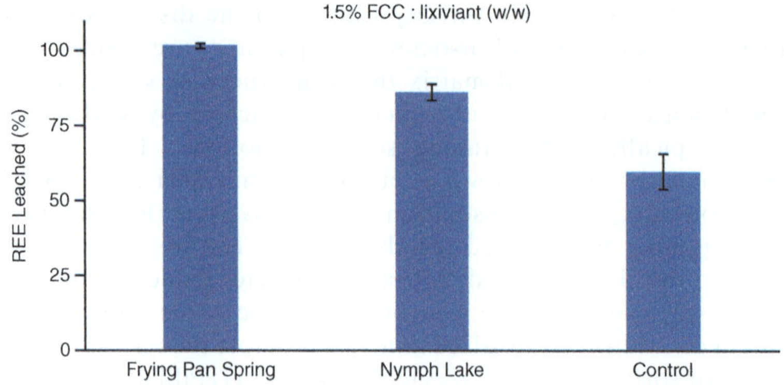

Figure 8.3 Total REEs leached from FCC catalyst (1.5%, w/w) after 6 weeks with microorganisms grown *in situ* with Sulfolobus medium (Leibniz Institute DSMZ–German Collection of Microorganisms and Cell Cultures), 25 mM $K_2S_4O_6$, and elemental sulfur (0.1% w/v). Error bars indicate standard deviations for three replicate experiments. Water samples for enrichment were collected from Frying Pan Springs (pH 2.0, 82 °C), and Nymph Lake (pH 2.8, 74 °C) in Yellowstone National Park, WY, in October 2013.

For non-sulfidic feedstocks, the most common biological approach proposed for solubilizing metals relies on microbes that partially oxidize added organic carbon compounds. These organic carbon-reliant organisms, known as heterotrophs, do not necessarily gain an advantage by growing in the presence of the ores or REE feedstocks, and indeed can be cultivated separately or in the presence of the solids containing the target metals. Each option has its advantages and disadvantages, as will be discussed in Section 8.2.3. Organic substrates are converted by the heterotrophs to compounds that promote mineral dissolution, typically organic acids and/or other chelators (the acids may themselves be chelators). Silica mineral dissolution can also be promoted by biological ammonia production from nitrogen-containing organic compounds (Jain & Sharma, 2004), but this mechanism is less likely to be useful for REEs given their low solubility at the higher pH engendered by ammonia production. In general, metal solubilization is generally slower with bioleaching compared to conventional hydrometallurgical approaches using more corrosive agents, but the potentially lower waste-disposal costs and environmental benefits, coupled with the fact that the carbon substrates could be derived from municipal, industrial, or agricultural wastes, could make heterotrophic bioleaching attractive for some REE-containing feedstocks (Jin et al., 2019; Thompson et al., 2018).

8.2.2. Complex Chemical Composition of Feedstocks Leads to Complex Leachates

Feedstocks with sufficient REE content to be potentially attractive for recovery include wastes from industrial mineral processing such as red mud (bauxite residue), phosphogypsum from phosphate production, kaolin production wastes, metallurgical slags and mine tailings, and wastes from energy production or waste management such as coal ash or incinerator ash (Abramov et al., 2018; Antonick et al., 2019; Binnemans, Pontikes, et al., 2013; Dehaine & Filippov, 2015; Emsbo et al., 2015; Funari et al., 2016; Klauber et al., 2011). End-of-life consumer products (e.g., fluorescent lamps, computer hard disk drives, mobile phones, various motors, etc.) and industrial processing materials (e.g., fluid catalytic cracking catalysts, glass polishing powders, etc.) have also been identified as possible targets for REE recycling (Binnemans, Jones, et al., 2013; Fonti et al., 2015; Tunsu et al., 2015). These solids are compositionally complex, and other matrix components besides REEs will interact with the biolixiviant and be extracted from the solids. The effectiveness of REE leaching will be a function of multiple factors, including pH, temperature, chelating agents, competing stable and metastable solids, and interactions with numerous species that result from the dissolution of the matrix in which the REEs were originally contained (Rasoulnia et al., 2020). Kinetics of course also plays a role. This complex situation explains why the relative abundances of metals in a solid feedstock are not always mirrored in the leachate composition. For example, Figure 8.4a shows the REE metal composition in retorted phosphor powder (RPP) solid compared to the metal composition in the leachate generated after 24 h of contact with a biologically produced lixiviant. In this case, yttrium accounts for 70% of the REEs in the original solid waste and almost 90% of the REEs in the leachate (Reed et al., 2016). The leachate is also enriched in europium relative to the original RPP, and cerium and lanthanum are much less abundant. Terbium is almost absent from the leachate. A reasonable explanation for these observations is that the Y and Eu in the RPP occur as oxides, which are likely more soluble than the La-, Ce-, and Tb-containing phosphates (Reed et al., 2016).

In another example, leachate from REE ore tailings exhibited relative enrichment in Ce and neodymium and a relative depletion of La compared to the original solids (Figure 8.4b). However, not all materials exhibit bias with respect to leaching; for example, the relative abundances of REEs in leachate of hard disk drive (HDD) magnets were similar to those in the starting material (Figures 8.4c).

The preceding examples illustrate the challenge of predicting leaching results for a given material using a given lixiviant. However, we do not have

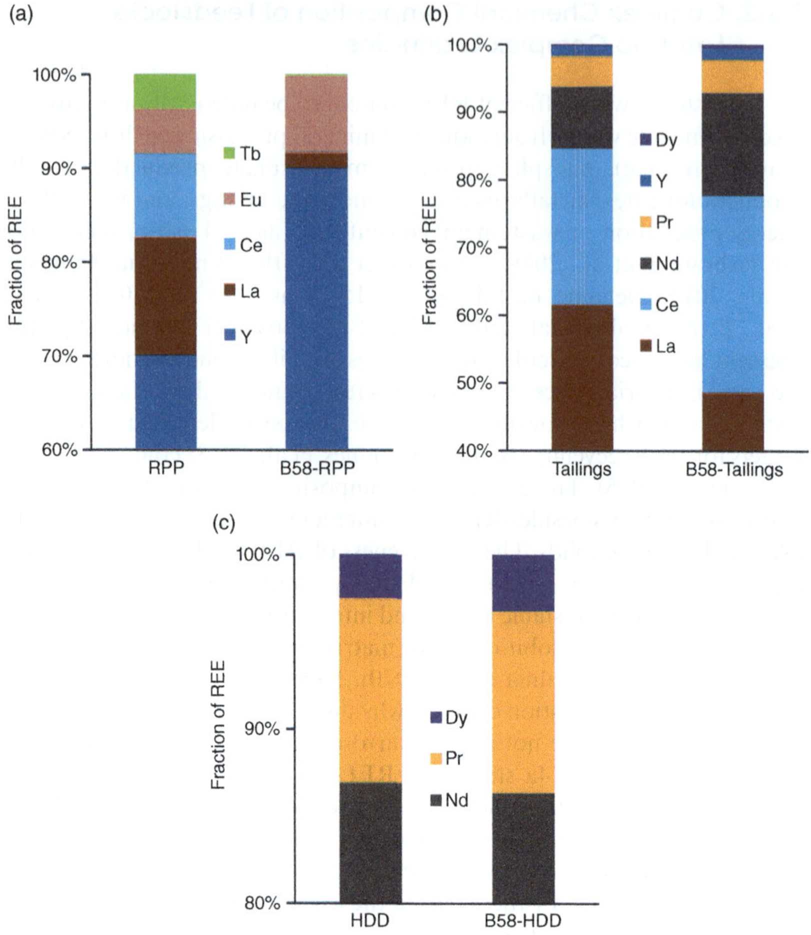

Figure 8.4 Relative distributions of REEs in solid materials (left bar) and bioleachates (lixiviant generated by *Gluconobacter oxydans* B58, denoted B58) after 1 day leaching (right bar) at 1.5% (w/v) pulp density of (a) retorted phosphor powder (RPP), (b) REE ore tailings, and (c) crushed hard disk drive (HDD) magnets. Y-axis shows the cumulative fraction of an individual REE.

to rely solely on empirical results. Geochemical modeling, based on thermodynamics, can provide critical aid in designing and optimizing conditions for leaching from complex materials. Equilibrium-based predictions facilitate the understanding of experimental results and can guide the selection of leaching conditions for empirical testing. Thermodynamic modeling of REE geochemistry is briefly reviewed next, followed by examples of applications to bioleaching.

Thermodynamic Modeling of REEs in Aqueous Solutions

A thermodynamic model is essential for capturing both chemical and phase equilibria in complex, multicomponent systems. Further, the equilibrium behavior can be used as a foundation for a kinetic analysis of leaching. To construct a credible thermodynamic model for aqueous systems containing REEs, it is necessary to parameterize and validate it using experimental data such as phase equilibria (i.e., solid–liquid and vapor–liquid equilibria) and solution speciation including acid–base equilibria and complexation equilibria. When necessary, the model should incorporate the effect of organic extractants that form liquid–liquid equilibria with aqueous solutions. An example of a model that satisfies these conditions and has been applied successfully to systems including REEs is the mixed-solvent electrolyte (MSE) model (Wang et al., 2002, 2004, 2006). The MSE model combines an equation of state for standard-state properties of individual species, an excess Gibbs energy model, and an algorithm for solving phase and chemical equilibria in multicomponent and potentially multiphase systems. Recently, the MSE model has been applied to simulate phase equilibria, caloric properties, and solution speciation in binary and multicomponent aqueous systems containing REE chlorides, sulfates, hydroxides, and various complexing agents (Das et al., 2017, 2019, 2023; Fujita et al., 2015; Park et al., 2016;). These systems have included microbiological media (Fujita et al., 2015, 2020).

The most practical way to establish a predictive thermodynamic model is to construct it using a bottom-up approach, that is, by developing model parameters for the constituent simple subsystems (primarily pure species and binary or ternary mixtures), followed by applying the model to more complex multicomponent systems. To obtain a good set of model parameters that accurately represent complex systems, reliable experimental data for multiple thermodynamic properties are necessary. In general, relatively plentiful experimental data are available for REEs in the presence of common inorganic ions. In particular, extensive and critically evaluated foundational data for inorganic REE compounds have been reported in a series of studies by Migdisov and co-workers and Mioduski and co-workers. Migdisov et al. have extensively investigated and established recommended values of standard-state properties, solubilities, and complexation constants at room and elevated temperatures (up to 300 °C) for various aqueous rare earth systems, including chlorides (Migdisov et al., 2008, 2009; Migdisov & Williams-Jones, 2002, 2006), fluorides (Loges et al., 2013; Migdisov & Williams-Jones, 2007; Migdisov et al., 2009), and sulfates (Migdisov et al., 2006; Migdisov & Williams-Jones, 2008). Mioduski and co-workers have compiled and evaluated experimental solubility data for aqueous binary and ternary rare earth chlorides (Mioduski et al., 2008, 2009a, 2009b), fluorides (Mioduski et al., 2014, 2015a, 2015b), iodides (Mioduski et al., 2012), bromides (Mioduski et al., 2013) and selected sulfates (Mioduski, 1999) based on multiple literature sources. These studies are of great importance, as they provide a detailed view of thermodynamic properties, phase behavior (i.e., solid–liquid and vapor–liquid equilibria), and complexation behavior within various families of inorganic REE salts. Based on such data sources, detailed thermodynamic models have been established for REE chlorides (Das et al., 2017), sulfates (Das et al., 2019), hydroxides (Das et al., 2023), carbonates (Kim et al., 2020), and phosphates (Fujita et al., 2020) using the MSE framework.

Thermodynamic behavior of rare earths in the presence of organic ligands is of particular importance, as it informs the selection of the best lixiviants and, hence, the microbe/metabolic pathways. Primary experimental data for REE–organic ligand systems are available to a lesser extent and cannot be, at present, as satisfactorily systematized as those for inorganic systems as discussed in the earlier text. Nevertheless, the MSE model has been applied to elucidate the leaching of REEs from phosphogypsum using gluconic acid (Antonick et al., 2019), the effects of citrate and acetate complexation on the recovery of REEs from bacterial surfaces (Park et al., 2016), and the effect of diethylenetriaminepentaacetic acid (DTPA) chelation on REEs in wastewater environments (Fujita et al., 2020). Here, we analyze representative examples of thermodynamic modeling of organic REE systems and compare the simulations with experimental data. The key requirement for the thermodynamic modeling of such systems is the simultaneous representation of speciation data (typically available from titration curves and/or spectroscopic measurements) and solubility measurements (typically as a function of both pH and ligand concentration). For this purpose, Figure 8.5 shows the behavior of rare earth systems in the presence of organic (citric, tartaric, and acetic) acids and illustrates the accuracy of the MSE model for capturing the complexation effects. Figure 8.5a presents the solubility of neodymium citrate in aqueous solutions containing 0.1 M $NaClO_4$ and sodium citrate as a function of pH, adjusted using $HClO_4$ and NaOH, at 25 °C, and compares the calculations with experimental data (Skorik et al., 1965; Skorik & Serebrennikov, 1965). The solubility diagram shows a strong dependence of the solubility of the neodymium citrate phase, $NdC_6H_5O_7{\cdot}3.5H_2O$, on pH, which is a consequence of the stability of citrate complexes. Remarkably, the solubility increases at pH values above neutral, which would not have been the case in the absence of complexation because the

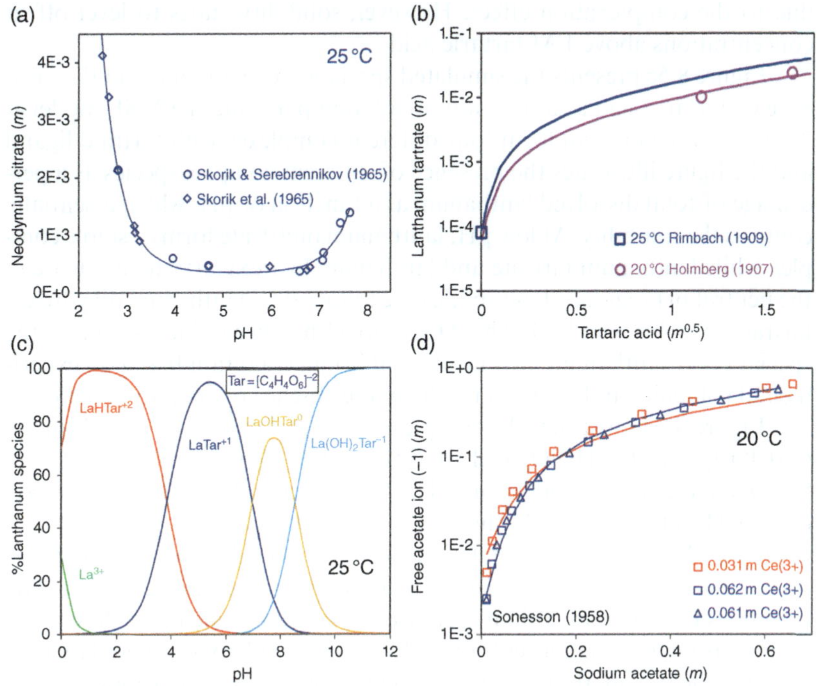

Figure 8.5 Calculated and experimental phase behavior and thermodynamic properties of rare-earth-containing systems in the presence of complex-forming organic ligands (i.e., citrate, tartrate, and acetate): (a) Solubility of Nd citrate ($NdC_6H_5O_7 \cdot 3.5H_2O$) in aqueous solutions as a function of pH at 25 °C; pH was varied using $HClO_4$ and NaOH, while $NaClO_4$ (0.1 M) was used as a background electrolyte; (b) solubility of lanthanum tartrate in tartaric acid solution around room temperature; (c) speciation of lanthanum (introduced as lanthanum chloride) in the presence of 1.0 *m* tartaric acid as a function of pH at 25°C; (d) free acetate ion concentration as a function of sodium acetate concentration in the $Ce(ClO_4)_3$–$NaClO_4$–$HClO_4$–CH_3COOH–$Na(CH_3COO)$ solution at 20 °C. The symbols represent experimental data (Holmberg, 1907; Rimbach & Schubert, 1909; Skorik et al., 1965; Skorik & Serebrennikov, 1965; Sonesson, 1958; Svoronos et al., 1981), whereas the curves represent calculated results.

precipitation of neodymium hydroxide would have interfered. Figure 8.5b shows the changes in lanthanum tartrate solubility with increasing tartaric acid concentration, which is in good agreement with experimental results (Holmberg, 1907; Rimbach & Schubert, 1909). Initially, the solubility increases significantly with the addition of a trace amount of tartaric acid due to the complexation effect. However, solubility starts to level off at concentrations above 1 M tartaric acid.

Figure 8.5c presents the simulated speciation of lanthanum in the presence of tartaric acid as a function of solution pH using the MSE model at 25 °C. Lanthanum can form four different complexes with tartrate ligand and the figure illustrates the presence of different complex species as a percentage of total dissolved lanthanum as a function of pH, which essentially controls the solubility. At low pH, lanthanum bitartrate forms a stable complex while lanthanum tartrate and lanthanum hydroxytartrate dominate in the neutral pH region. These species are followed by lanthanum dihydroxytartrate at pH beyond 10. The MSE model has been able to predict the speciation of lanthanum tartrate at equilibrium condition based on experimental solubility and titration results in the absence of speciation data.

Figure 8.5d illustrates the complexation effect of acetate ligands in the $Ce(ClO_4)_3$–$NaClO_4$–$HClO_4$–CH_3COOH–$Na(CH_3COO)$ system. $NaClO_4$–$HClO_4$ was used to set the ionic strength of the solution to 2.0 M; the concentration of $Na(CH_3COO)$ was varied, while the CH_3COOH concentration was kept constant at 2.5 M. In this case, acetate ions form considerably weaker complexes than citrates or tartrate ions. The concentration of free acetate ions directly quantifies the extent of cerium acetate complexation, which goes down with an increase in cerium concentration. Eventually, with excess $Na(CH_3COO)$ added to the system, the predicted concentration of free acetate ions plateaus in agreement with experimental data (Sonesson, 1958).

Calibration of the model against experimental solubility, titration, and speciation data makes it possible to make predictions of the effect of complexation on solubility in cases when direct solubility data in ligand-bearing solutions are not available. This is illustrated in Figure 8.6 for Nd in the absence and presence of complexing ligands. First, the solubility of Nd (as a representative REE) is calculated as a function of pH and compared with experimental data (Rao et al., 1996; Silva, 1982) at 25 °C in the presence of 0.1 M NaCl. The results indicate a strong pH dependence of the solubility, which is generally controlled by the precipitation of crystalline Nd hydroxide. This pH dependence is similar across the lanthanide series. Then, Figure 8.6 illustrates the effects of two organic chelating agents, that is, acetic and gluconic acids, on the availability of Nd in solution. To isolate the effect of complexation from the effect of pH, the concentrations of acetic and gluconic acids were fixed at 0.1 M and the pH was varied by adjusting the concentrations of HCl and NaOH. Thus, the increase in solubility at a fixed pH is due to the complexation of Nd with the organic ligands. Gluconic acid is a stronger complexing agent than acetic acid due to the presence of hydroxyl groups in addition to the carboxylic group. Figure 8.7 shows the speciation of Nd in the presence of gluconic

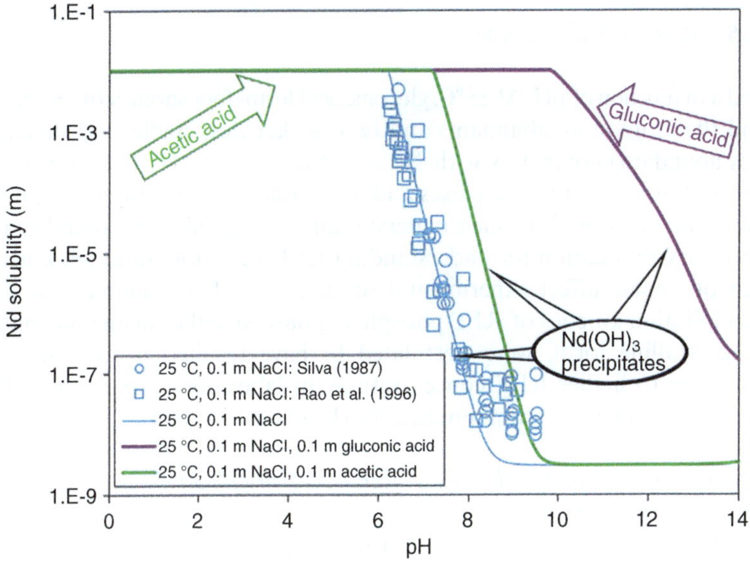

Figure 8.6 Solubility of neodymium hydroxide as a function of pH at 25 °C in a 0.1 M NaCl aqueous solution (teal) and in the presence of 0.1 M acetic acid (green) or 0.1 M gluconic acid (purple). The pH was adjusted by varying HCl and NaOH concentrations. The symbols represent experimental solubility data of $Nd(OH)_3$ and the curves represent the calculated results.

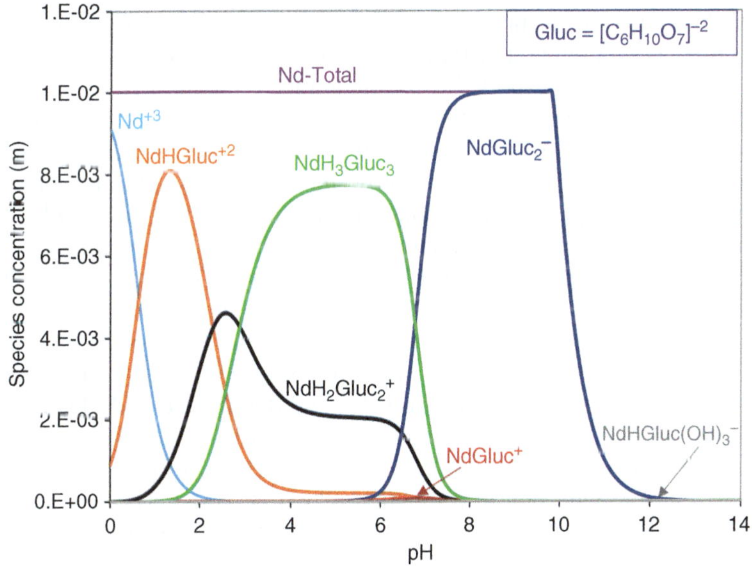

Figure 8.7 Calculated speciation for the system studied in Figure 8.1 in the presence of 0.1 M gluconic acid. The curves show the concentrations of various complexes between neodymium and gluconic acid as a function of pH.

acid as a function of pH. At 25 °C, gluconic acid forms four stronger (more stable, and therefore more abundant) and two weaker (less stable, and therefore less abundant) complexes with Nd as a function of pH. The sum of the concentrations of the complexes and uncomplexed neodymium is equal to the total Nd solubility. Such understanding of equilibrium solubility is a necessary foundation for understanding REE behavior, although kinetics can of course affect experimental observations. For example, the slow precipitation kinetics of REE phosphate phase(s), either in an amorphous or crystalline form, were postulated to have resulted in exposures of microbes to higher REE concentrations in laboratory experiments than expected from equilibrium predictions (Fujita et al., 2015).

Thermodynamic Modeling of REE Leaching from Industrial Feedstocks

Based on the parameterization for simpler, well-defined subsystems, the model can be applied to study the recovery of REEs from compositionally complex industrial wastes. Here, we present the example of a coal fly ash derived from an Appalachian Basin coal. In the fly ash, REEs occur in a matrix containing elements such as Si, Al, Fe, Ca, Mg, Na, K, Ti, Ba, Sr, and P. The behavior of the matrix elements is practically important and can also be simulated using the MSE model. In the simulation, the mixture of amorphous oxides of the matrix elements has been allowed to equilibrate with hydrochloric, acetic, or gluconic acid at varying concentrations. Nd was used as a representative REE; the dissolution and complexation behavior of other REE is expected to be qualitatively similar. Figure 8.8 shows the predicted total concentration of Nd dissolved from the fly ash sample as a function of either pH ranging from 0 to 6 (Figure 8.8a) or the concentration of the added acid (Figure 8.8b). For each acid system, the pH was adjusted by changing the acid concentration. At low pH values, all Nd is predicted to dissolve (cf. the horizonal line in Figure 8.8a), whereas the solubility of Nd at higher pH values is controlled by the formation of $NdPO_4$. $NdPO_4$ starts to form at a pH of about 1.6 for HCl and about 3.6 for both acetic and gluconic acids. Gluconic and acetic acids are more effective than HCl for the dissolution of Nd at pH values between 1.6 and 3.8 because of the formation of complexes between Nd and the organic acids. Due to the differences in complexation strength discussed earlier, gluconic acid is more effective than acetic acid in solubilizing Nd at pH above approximately 3.8. While the solubility plot as a function of pH (Figure 8.8a) illustrates the effect of complexation at a fixed pH, it is also of interest to simulate the concentration of the acid that is necessary to dissolve Nd. The latter simulation reflects the interplay of the effect of acid strength (which increases in the order acetic acid < gluconic acid < HCl) with the effect of complexation,

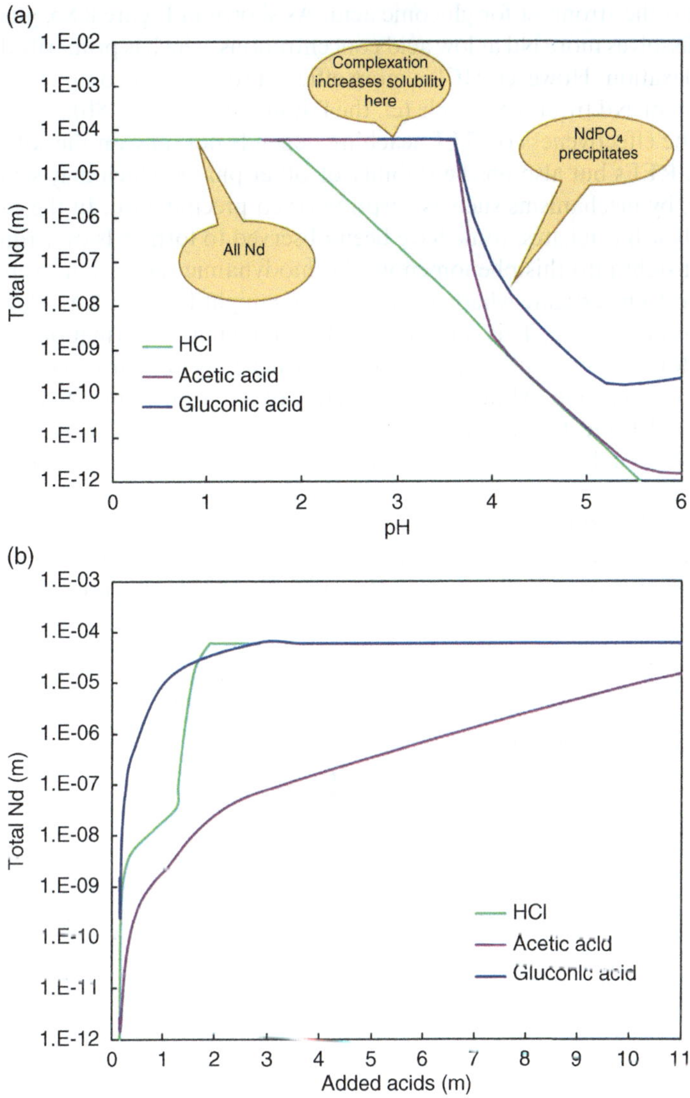

Figure 8.8 Total dissolved molality of Nd after equilibrating a coal fly ash sample with hydrochloric, acetic, or gluconic acid as a function of (a) pH or (b) molality of added acid.

which is the strongest for gluconic acid. As shown in Figure 8.8b, gluconic acid dissolves more Nd at low acid concentrations, which is primarily due to complexation. However, HCl is more efficient to obtain a complete solubilization of Nd from the sample (cf. the flat lines in Figure 8.8b).

The effectiveness of REE leaching depends not only on the solubility of the REEs but also on the stability of other phases which may scavenge REEs, by mechanisms such as sorption or co-precipitation. In the case of coal fly ash leachates, some have been observed to form gels over time. To gain insight into this phenomenon, thermodynamic simulations were performed to investigate whether certain solids might form with a delay after the initial leaching. This was simulated by allowing the formation of various additional crystalline silicate phases, which may form after a prolonged equilibration time. Although the kinetics of such phenomena cannot be predicted using thermodynamic analysis alone, the simulation makes it possible to indicate under what conditions gelation phenomena are possible. The results of the two simulations—without and with certain kinetically controlled crystalline silicate phases—are illustrated in Figure 8.9 for HCl leachates of the coal fly ash. The figure shows the molality of dissolved Si as a function of pH, which results from varying HCl concentration. These

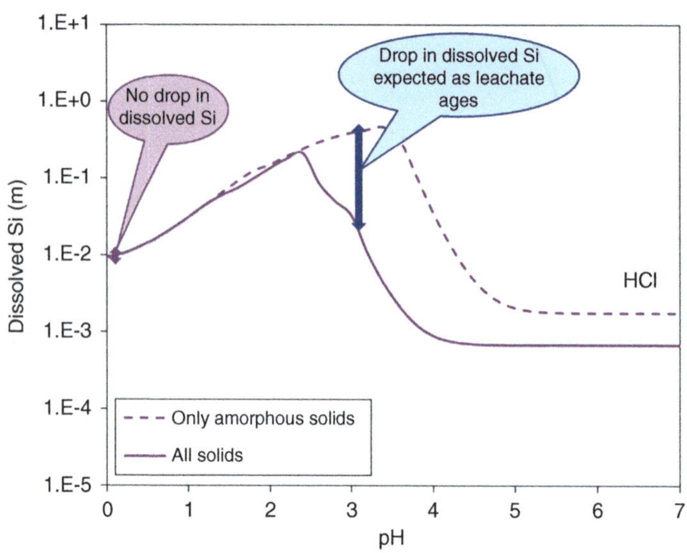

Figure 8.9 Total dissolved Si concentration predicted by excluding (dashed lines) or including (solid lines) the possible formation of crystalline silicate phases whose formation is kinetically controlled. The difference between the two lines at a given pH indicates the possibility of formation of solids as the leachate sample ages.

kinetically controlled solid phases start forming at pH above approximately 2, which leads to a drop in dissolved Si concentration due to the precipitation of additional crystalline silicates (i.e., $Al_2Si_2O_5(OH)_4$, $Al_2Si_4O_{10}(OH)_2$, $KAlSi_3O_8$, $Mg_3Si_4O_{10}(OH)_2$, and $Ca_2Mg_5Si_8O_{22}(OH)_2$) in addition to amorphous SiO_2. The formation of such phases may impair the separation of REEs due to adsorption phenomena.

As illustrated by the abovementioned examples, geochemical equilibrium modeling provides a powerful tool for understanding the complexities of the solution chemistry of leaching processes. However, care must be taken in the parameterization of the models to ensure that the predictions are accurate. This usually requires a critical analysis of the underlying phase equilibrium and speciation data, followed by a careful analysis of trends within the REE series. Furthermore, application of the models requires an understanding of the behavior of solid phases, some of which may form over the time scale of laboratory experiments, while others may appear long afterward.

8.2.3. Feedstock Compatibility with Bioleaching Microbes

In addition to limiting REE recovery through equilibrium or kinetic effects, a feedstock or leachate's chemical composition may limit REE recovery through inhibition of microbial activity. REEs themselves have generally been considered to have a low environmental impact in natural environments due in part to their typically very low concentrations and the perception that they are less likely to pose a problem than more common heavy metals (Gonzalez et al., 2014; Haque et al., 2014; Kurvet et al., 2017). However, in an industrial leachate, higher REE loads will be present, and some studies have suggested that high REE concentrations can inhibit microbial growth. Some early studies reported inhibition of bacteria and fungi provided with REEs at concentrations >10 mM (Talburt & Johnson, 1967; Wurm, 1951). More recently, other researchers have reported that high concentrations of REEs can disrupt membrane integrity of *Escherichia coli*, as observed by electron microscopy for Ce^{3+} at concentrations ≥0.25 mM (Chen et al., 2010; Chen et al., 2012) or by atomic force microscopy for Pr^{3+} at 1 mM (Peng et al., 2007) and La^{3+} at approximately 3 mM (Peng et al., 2004). High concentrations of Gd (≥0.3 mM) have also been reported to be toxic to the flagellate algae *Euglena gracilis*, protozoan *Tetrahymena thermophila*, and the bacterium *E. coli* in microcosms (Fuma et al., 2001). Additionally, Eu was shown to be inhibitory at ≥10 μM to Clostridium, but a cytoplasmic reduction mechanism for the REEs may have helped increase tolerance (Maleke et al., 2019).

In our own work, we observed inhibition of the ammonia-oxidizing bacterium *Nitrosomonas europaea* at nominal Y concentrations ≥0.56 mM and Eu concentrations ≥0.33 mM, and of the nitrifying bacterium *Nitrobacter winogradskyi* at ≥0.11 mM Y and ≥0.07 mM Eu (Fujita et al., 2015). Note that these concentrations are nominal, given that we observed significant precipitation of the REEs following addition to the media. Others have also pointed out that actual aqueous concentrations encountered by organisms are typically lower than the nominal concentrations provided under the testing conditions due to the limited solubility of lanthanides in many aqueous media (Gonzalez et al., 2014; González et al., 2015; Herrmann et al., 2016; Sneller et al., 2000). Gonzalez et al. conducted a comprehensive study of the ecotoxicity of lanthanides to a number of aquatic organisms, and for the bacterium *Aliivibrio fischeri* reported nominal (based on metal added to test cultures) EC_{50} values of 18 μM for Lu, >46 μM for Ce, and >41 μM for Gd (González et al., 2015). However, for the same *A. fischeri* organism, Weltje reported an EC_{50} value of 1.57 μM for the free Lu^{3+} ion (Weltje et al., 2004). Because of the sensitivity of REE speciation to geochemical conditions, as was discussed in the Section 8.2.2 on modeling, and the apparent differing sensitivity of particular microbes to different REEs (Fujita et al., 2015), it is difficult to determine general threshold concentrations of concern for REEs to bioleaching organisms. However, in a leachate, the REEs will be soluble, by definition, and therefore it is safe to assume that microbes present will be exposed to elevated soluble concentrations of REEs. Whether they are bioavailable however is a different question. Depending on the leachate composition, REEs may be present as complexes; for example, Figure 8.6 in Section 8.2.2 shows the predicted speciation of Nd in a system with gluconic acid, an organic acid identified as important in a number of REE bioleaching studies (Corbett et al., 2017; Glombitza & Reichel, 2014; Hopfe et al., 2017; Reed et al., 2016; Thompson et al., 2018). In this particular modeled system, Nd solubility is controlled largely by complexation with gluconate. According to the generally accepted model that biological responses to trace metals are best predicted by the free metal ion activity (Campbell, 1995; Morel & Hering, 1993), the vast majority of the Nd in the system would not be bioavailable at typical environmental pH. However, in the aforementioned studies with *N. europaea* in the absence of a known complexant, inhibition of ammonia oxidation was observed even when measured and modeled soluble concentrations of Eu were ≤2 ppb, representing <1% of the originally added Eu (nominally up to 660 μM, or 100 ppm) (Fujita et al., 2015). In contrast, more recent studies with *N. europaea* and Gd conducted under similar conditions measured that average soluble Gd concentrations as high as 90 ppb were not associated with observable inhibition, although high concentrations of Gd (up to 500 μM; equivalent to 79,000 ppb) complexed with DTPA seemed to cause at least temporary inhibition (Fujita et al., 2020). In the latter case, we hypothesized that DTPA complexation of other essential metals may have played a role. These observations illustrate that much remains to be elucidated with respect to interactions of REEs and microbes, and therefore the potential for inhibition of bioleaching organisms, whether directly by the free ion or by other REE species or associated ligands, should not be discounted.

REE-containing materials also often contain heavy metals and or radionuclides that can be solubilized along with the REEs (Cánovas et al., 2017; Nassar et al., 2015), and these may also pose a concern for bioleaching organisms (Qu & Lian, 2013; Xu et al., 2014). In end-of-life consumer products, plastic components or composites and organic glues may be present, and leached organic compounds could also affect the microorganisms, and or bioleaching efficiency. Potential toxicity of leachates to the organisms confers an advantage for bioleaching strategies that separate biological lixiviant production from the actual leaching of the metal-containing feedstock. Such strategies are referred to as "two-step bioleaching," as opposed to "one-step bioleaching" where the cells are grown in contact with the solids to be leached (Bosshard et al., 1996; Qu & Lian, 2013). In two-step bioleaching, organisms can be cultivated under optimal conditions with respect to lixiviant production, and then the culture medium would be applied to the REE-containing feedstocks. In an industrial-scale deployment, cells would generally not be removed from the medium, since that would require a costly processing step (e.g., filtration or centrifugation). As the cells would presumably have already produced an optimal amount of lixiviant, cell inhibition or toxicity is no longer a concern. Two-step bioleaching would be a good option for microorganisms that grow rapidly on inexpensive carbon sources where the cost of an extra unit operation to grow cells can be offset. For microorganisms that are difficult to cultivate, require expensive nutrients, or grow slowly, an alternative strategy would be cultivation of immobilized cells (note that possible approaches for immobilization are discussed in Section 8.3) in a reactor separate from the REE solids, where the cell-free lixiviant fluid can be drawn off to a separate unit for leaching. This allows cell cultivation to be maintained throughout the process and avoids the need for new batch cultivation for each leach cycle. This format would also be amenable to development of a continuous lixiviant production/bioleaching process.

In "one-step bioleaching," microbes are cultivated and produce the leaching agents in the presence of the solid feedstock. Separate unit operations for lixiviant production and bioleaching are not required, simplifying the system and reducing costs. In the case of bioleaching by sulfide oxidizers, direct contact between the cells and the sulfidic solids is believed to facilitate electron transfer, resulting in more efficient acid production (Vera et al., 2013). One-step bioleaching may also provide better leaching conditions if the cells adhere to the surface of the feedstock, since lixiviant would be produced directly proximate to the solid surface. Fathollahzadeh et al. reported that monazite dissolution and REE leaching increased in the presence of *Enterobacter aerogenes*, and microscopic imaging showed microbial attachment to the monazite surface suggesting that cell contact improved the bioleaching (Fathollahzadeh et al., 2018). However, toxins that are released during leaching can inhibit the desired lixiviant production. Qu and Lian suggested this as one reason that greater REE leaching was observed with two-step leaching of red mud compared to one-step leaching at pulp densities greater than 5% (w/w), although at <2% (w/w), the one-step process was more effective (Qu & Lian, 2013).

An additional consideration for one-step bioleaching is the effective mass transfer of nutrients for the microbes in the presence of the solid feedstocks. For aerobic organisms, provision of oxygen can be particularly challenging, because oxygen's limited solubility in water generally leads to the requirement for stirring and or sparging with air. This may limit the amount of REE feedstock that can be leached at one time due to the difficulty of aerating large volumes of saturated or suspended solids. In addition, biomass growth can also lead to clogging, between REE feedstock particles as well as inside particles (intra-porosity) if the particles are porous.

The choice of one-step or two-step bioleaching will depend on all of the factors mentioned and the downstream processing steps required. The additional unit operation required for a two-step process compared to a one-step imposes an additional cost factor. On the other hand, the size of unit operations for a single-step process may be prohibitively large to leach the quantity of REE feedstocks required to be processed for profitability. Following bioleaching by either approach, downstream separations will be necessary, since the leachate will be a complex mixture of REEs and other metals, as well as microbial products. If the presence of cells interferes with downstream operations, it may be necessary to remove cells prior to REE concentration and purification.

Another consideration for one-step bioleaching is the presence of other microorganisms in the solid feedstock, aside from those desired for leaching. Recycled feedstocks will not be sterile, and consequently microorganisms "carried along" with the solids may compete with organisms added expressly for bioleaching. If indigenous organisms use the growth substrates provided to the bioleaching organisms, and they do not produce compounds useful for bioleaching, this would reduce the efficiency of the process. However, heterotrophs are common in the environment, and indigenous mixed cultures can be effective for REE leaching (Barnett et al., 2018).

Indeed, use of mixed cultures may in fact be advantageous compared to use of single strains (Hopfe et al., 2017). In order to evaluate the potential for biological contamination and impacts on leaching, tests with real solid feedstocks and under representative conditions will be necessary.

8.3. Technical Considerations for Bioseparation

The chemical and physical characteristics of the solid feedstock bound the bioleaching mechanism and together with the lixiviant constituents bound the composition of the leachate. In bioseparation, the chemistry of the immobilized biological ligands together with the chemical composition of the leachate and the stripping agent chemistry define the degree of target metal recovery and product purity. These factors are discussed in the following sections.

8.3.1. Biological Ligand Chemistry

Microbially mediated surface adsorption (biosorption) represents a potentially cost-effective and ecofriendly approach for metal recovery (Fein et al., 2001; Mejáre & Bülow, 2001; Zhuang et al., 2015). Microorganisms have a high surface area per unit weight and exhibit high metal-adsorption capacities owing to an abundance of cell surface functional groups with metal coordination functionality (Moriwaki & Yamamoto, 2013). The organic functional groups generally considered most likely to play a role in chelation of metal ions are carboxylic, amino, hydroxyl, phosphoryl, and sulfhydryl groups (Beveridge & Murray, 1980; Cox et al., 1999; Yu & Fein, 2016). Microorganisms produce these moieties naturally as parts of proteins, nucleic acids, and lipids, and therefore the genetic coding for biosynthesis is already present in the organisms.

Owing to the chemical properties of the surface functional groups, a number of reports suggest that biosorption could be particularly amenable to REE extraction (Das & Das, 2013; Gupta et al., 2018). In particular, the selective extraction of REEs over non-REEs by microbial cell surfaces has been reported. For example, Tm^{3+} sorption was favored over Fe^{2+} sorption on the surface of *Bacillus subtilis* (Moriwaki et al., 2013), while Sm^{3+} adsorbed more tightly than Cu^{2+}, Mn^{2+}, Co^{2+}, Ni^{2+}, Zn^{2+}, and Cd^{2+} onto *Arthrobacter nicotianae* (Tsuruta, 2006). Additionally, the stability constant (log K) for the interaction of *B. subtilis* with Nd was stronger than for all tested metal cations, except the uranyl oxycation (Fein et al., 2001). In contrast, the difference in stability constants among individual REEs for cell surface sites is quite small with the exception of a strong preference for scandium (Sc) over lanthanides (Dong, Deblonde, Middleton, et al., 2021; Hosomomi et al., 2013; Karavaiko et al., 1996; Moriwaki et al., 2016; Ngwenya et al., 2010). Thus, while a moderate biosorption preference for middle/heavy REEs over light REEs has been observed (Park et al., 2017; Takahashi et al., 2005), native microbial surfaces are better suited for grouped lanthanide extraction rather than individual REE separation. Nevertheless, a recent study reported a scheme for preprotonation of cell surface functional groups followed by pH-dependent desorption that achieved separation factors for certain REE pairs that exceeded solvent extraction standards (Bonificio & Clarke, 2016). As such, pH modulation during the adsorption and desorption steps may prove an effective strategy for the selective enrichment of specific REEs from feedstock leachates.

In addition to moieties that are synthesized and expressed naturally, organisms may be genetically engineered to produce selective metal-binding peptides or proteins and express them at high density on the cell surface (Hennebel et al., 2015; Park et al., 2016; Wei et al., 2014; Z. Xu et al., 2010; Zhuang et al., 2015). As noted earlier, Park et al. reported the construction of recombinant strains of *Caulobacter crescentus* and *E. coli* with LBTs inserted at permissive sites of surface proteins, anchoring the LBTs to the cell surface (Park et al., 2016, 2017). LBTs are short peptides optimized to bind lanthanide ions with high affinity and specificity and were originally developed as fusion proteins for biological applications such as X-ray crystallography, nuclear magnetic resonance, and various luminescent visualization techniques (Allen & Imperiali, 2010; Daughtry et al., 2012; Nitz et al., 2003). The short-peptide-based nature of LBTs and high REE-binding affinity make them ideal for cell surface display to increase specific REE adsorption. Indeed, REE recovery performance with the LBT-displayed strains revealed improved adsorption capacity and selectivity, and enabled effective REE recovery from a variety of feedstocks (Park et al., 2016, 2017). In addition to LBTs, the more recently discovered lanmodulin protein from *Methylobacterium extorquens* also shows great promise as a selective binding agent for lanthanides (Cotruvo et al., 2018; Deblonde et al., 2020; Dong, Mattocks, Deblonde, et al., 2021; Mattocks et al., 2019). In recent work, the fungus *Yarrowia lipolytica* was genetically engineered to display lanmodulin on its cell surface for increased REE adsorption (Xie et al., 2022). Additionally, the acidophile *Acidithiobacillus ferrooxidans* was engineered for intracellular lanmodulin expression, resulting in a 4-fold increase in REE uptake under low pH conditions (pH 1.8) (Jung et al., 2023).

8.3.2. Effect of Leachate Chemical Composition on Biosorption

Biosorption performance depends on several factors, including the identity of the adsorbents, the solution chemistry of the leachate, and conditions such as temperature and the amount of sorbent relative to the target solutes. Among all of the factors, the pH of the leachate is perhaps the most critical driver of REE biosorption efficacy, here defined as the ability both to recover a high percentage of the total REE content and to do so preferentially over non-REE metals. The primary native cell surface functional groups involved in REE adsorption are carboxyl (pK_a, 4.3 and 5.5) and phosphoryl (pK_a, 2.2 and 6.9) groups (Markai et al., 2003; Martinez et al., 2014; Ngwenya et al., 2010; Texier et al., 2000). Given the protonation of these cell surface functional groups under acid conditions, adsorption efficacy at pH lower than 4 is generally poor for most REEs with the exception of Sc (Hosomomi et al., 2013; Moriwaki et al., 2016), which benefits from its smaller ionic radius and stronger Lewis acid characteristics compared to the lanthanides (Giret et al., 2018; Moriwaki et al., 2016). Since the REE feedstock is usually acidic following the (bio)leaching process, a pH adjustment step (e.g., NaOH addition) is necessary prior to sorption. REE-adsorption capacity generally increases with increasing pH (Das & Das, 2013; Kucuker et al., 2017; Park et al., 2016). However, REE solubility, a necessary condition for sorption, is highly pH-dependent and generally decreases with increasing pH, as presented in Section 8.2.2. To avoid the formation of REE hydroxide and or other REE mineral precipitates (depending on solution chemistry), REE biosorption studies are generally limited to pH $\leq$ 6. Consequently, the effective pH range for biosorption is largely constrained to between pH 4 and 6. For the coal fly ash leachates described in Section 8.2.2, poor REE solubility at pH 6 constrains the biosorption process to a pH range of 4–5. Given the reduction in adsorption capacity with decreasing pH, a higher biomass-to-REE ratio may be necessary to achieve high REE extraction efficiency. In addition, efforts to improve biosorption performance under low pH conditions using genetic engineering to produce binding motifs with higher REE affinity or by identifying organisms or macromolecules, such as lanmodulin (Deblonde et al., 2020), that are naturally efficient at REE sorption under acidic conditions may be warranted. For example, in simple synthetic solutions, the gram-positive bacterium *B. subtilis* was found to maintain high REE-adsorption capacity (relative to *E. coli*) down to a pH of 3 (Martinez et al., 2014; Takahashi et al., 2005). Testing with solutions more representative of real-world feedstock leachates will provide additional insight into the utility of *B. subtilis* and related bacteria for REE recovery via biosorption.

Within the permissible pH range, the identity and concentration of non-REEs are the other critical factors determining REE biorecovery. Although cell surface functional groups exhibit a general selectivity for REEs, many relevant feedstock leachates contain significant concentrations of non-REE metal cations and usually at concentrations much higher than that of REEs. High concentrations of the major elements (e.g., Na^+, K^+, Li^+, Mg^{2+}, Ca^{2+}, Ba^{2+}) are found in many REE feedstock leachates, but generally have little impact on REE extraction efficacy (Moriwaki & Yamamoto, 2013; Ngwenya et al., 2009; Park et al., 2017; Texier et al., 1999). For example, in our laboratories' studies with LBT-engineered *E. coli*, 95.1 ± 5.6% Tb adsorption was achieved with a synthetic solution mimicking the Great Salt Lake that contained 165,000 ppm major elements and 100 ppb Tb. Less than 1% of the major elements were adsorbed (Brewer, Chang, et al., 2019). Minimal adsorption of major elements was also observed in leachates of REE ore and low-grade mine tailings (Park et al., 2017). One notable exception is the elevated Ca^{2+} and Mg^{2+} recovery observed for native *A. nicotianae* and LBT-displayed *E. coli* (Park et al., 2020) in Powder River Basin fly ash leachate, which contains Ca^{2+} and Mg^{2+} at concentrations that are three orders of magnitude higher than total REEs. Similar to major elements, most transition metals such as Zn, Mn, and Ni also pose minimal impediments to REE-adsorption efficacy (Park et al., 2017, 2020), particularly when the biomass concentration is properly tailored to the feedstock (see in the following text). In contrast, studies with several bacteria have revealed that Cu, Al, Fe(III), Pb, and uranyl cations are the most effective competitors for REE binding to cell surface sites (Fein et al., 2001; Park et al., 2017, 2020; Texier et al., 1999), and will be coextracted with REEs if present at elevated concentrations. For example, Al^{3+} was found to be a strong inhibitor for REE binding to *Pseudomonas aeruginosa* (Texier et al., 1999). In addition, two-element competition experiments in our laboratories using LBT-displayed *E. coli* indicated that UO_2^{2+} and Al^{3+} were the most competitive non-REE cations, decreasing Tb-adsorption efficiency by 25% when present at concentrations fourfold higher than Tb (Brewer, Chang, et al., 2019). Lead and Cu reduced Tb-adsorption efficiency to a similar degree when present at concentrations tenfold and sixtyfold higher than Tb, respectively. For comparison, 1,400-fold and 33,000-fold higher Mg and Na concentrations, respectively, were required to similarly impact Tb adsorption.

The impact of non-REEs on REE-adsorption efficacy can be partially alleviated through solution conditioning and experimental optimization steps. For example, functionalization of the *E. coli* cell surface with LBTs enhanced the cell surface selectivity relative to native *E. coli* cells for Nd over all non-REEs except Cu (Park et al., 2017). Further genetic engineering efforts to increase the ratio of LBT to non-LBT sites or through the use of ligands that are even more selective, for example, lanmodulin (Cotruvo et al., 2018), can be employed to achieve even greater REE selectivity (Xie et al., 2022). In addition, the soluble concentration of many competing metals can be reduced through pH modulation. We have found that soluble Al, Pb, U, and Fe concentrations are significantly reduced in feedstock leachates upon pH adjustment to the 5–6 range, which is critical for high-purity extraction of REEs from mine tailings and coal byproducts (Park et al., 2017, 2020). Finally, optimizing the adsorbent-to-REE ratio for a given leachate can also minimize the effect of non-REE sorption on REE recovery. For example, biosorption experiments in our laboratories with synthetic geothermal fluids containing high Al content relative to REEs revealed that high REE extraction efficiency and REE purity can be obtained by adjusting adsorbent concentration relative to the REE concentrations in the leachate (Brewer, Chang, et al., 2019).

Compared to the cation composition, there has been less experimental investigation of how the anion content of leachates affects REE-adsorption efficacy by biosorbents. Many REE-containing leachates contain high concentrations of chloride, sulfate, or nitrate anions as a result of a mineral-acid leaching step. However, these anions are weak REE complexants compared to biological ligands and thus their effects on REE-adsorption efficacy should be minimal. Although carbonate forms stronger REE complexes, conducting the biosorption step at below circumneutral pH limits the free carbonate concentration. In contrast, phosphate is a strong complexant and REE phosphates exhibit extremely low solubility across the pH range relevant for biosorption. Therefore, prior to the application of biosorption to leachates that may contain significant quantities of phosphate, geochemical modeling as described in Section 8.2.2 is advised to guide selection of leaching conditions.

8.3.3. Stripping Agent Chemistry

After REEs are selectively adsorbed, the residual leachate solution is removed, and then a stripping agent is applied to release the captured metals and regenerate the binding sites. The stripping solution should be free of the nontarget components of the original leachate, and by using a smaller volume than used for sorption, increased concentration of the REEs as well as enrichment relative to the other ions in the original solution can be achieved. Use of a stripping agent that outcompetes the biological ligand but does not harm the structural architecture (cell structure) of the biosorbent allows the latter to be reused multiple times (Figure 8.10). Techno-economic analyses (TEA) of biosorption for recovery of REEs

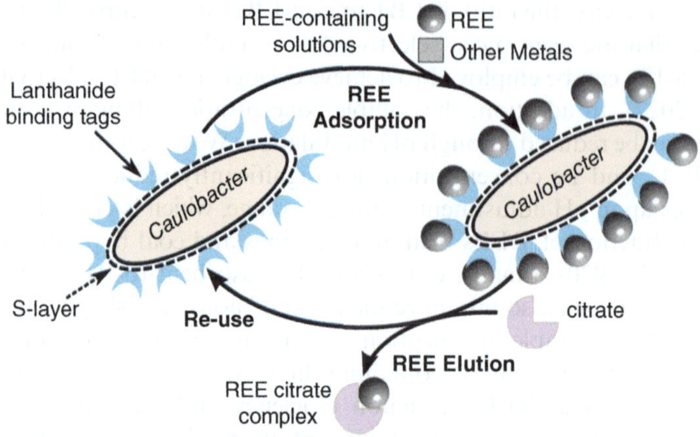

Figure 8.10 Process for REE recovery based on biosorption and desorption. REE bound to the Caulobacter cell surface can be stripped with citrate. Cells are reused for multiple cycles of sorption/desorption. Reproduced from Park et al. (2016), / American Chemical Society.

determined that biomass recycling, or regeneration and reuse of the sorbent with minimal loss of sorption capacity, can have a significant beneficial impact on cost (Alipanah et al., 2020; Jin et al., 2017). In most reported examples, stripping is performed in a nonselective manner, resulting in desorption of all adsorbed metals (Arunraj et al., 2019; Park et al., 2016, 2017; Texier et al., 2002). The selectivity of the extraction process is therefore driven almost exclusively by the biosorption step. Commonly employed stripping agents include organic and mineral acids. Citrate and ethylenediaminetetraacetic acid (EDTA) have been shown to be particularly effective for quantitative desorption given their strong REE complexation and minimal impact on subsequent biosorption capacity (Arunraj et al., 2019; Park et al., 2016, 2017; Texier et al., 2002).

Coupling the adsorption step with a selective desorption process may further improve the REE versus non-REE selectivity of the biorecovery process and potentially enable separation of certain REEs. For example, Bonificio and Clarke (2016) examined REE binding onto preprotonated non-engineered (native) bacterial cells, and reported that using different concentrations of nitric acid (pH manipulation) for desorption achieved separation factors for certain REE pairs that exceeded solvent extraction standards (Bonificio & Clarke, 2016). Ultimately, the possibilities to tune the selectivity of sorption and desorption, coupled with the fast kinetics of sorption and desorption, make bioseparation based on biosorption/desorption a promising platform for efficient REE recovery (Hennebel et al., 2015; Park et al., 2016; Takahashi et al., 2005; Texier et al., 1999).

8.3.4. Cell Immobilization for Biosorption

For an industrial-scale biosorption application, the biosorbent should be amenable to high-throughput contact with the REE-containing feedstock leachate, as well as with the subsequent stripping solution. Separation of the aqueous phase from the cells must be simple and robust. One approach that has been utilized for metal biosorption applications is encapsulation of planktonic cells within polymer matrices (Dodson et al., 2015). The goal is to provide a stable, permeable material that permits REE-specific adsorption by the immobilized microbes in a continuous flow extraction system. To maintain the high selectivity of the adsorbent, the use of minimally adsorptive carrier materials is preferable given the high concentrations of non-REE metal contaminants present in most feedstocks. Although most biosorption applications that have utilized encapsulation have focused on bioremediation of heavy metals (e.g., Pb, Cu, Zn, etc.), promising initial examples for REE extraction include the encapsulation of *P. aeruginosa* in polyacrylamide cubes for REE extraction from synthetic solutions (Texier et al., 2002) and the use of cellulose-embedded yeast for REE extraction from phosphor powder leachates (Arunraj et al., 2019). In our own work, we have pursued another promising approach, the encapsulation of LBT-displayed cells within polyethylene glycol diacrylate (PEGDA) hydrogel beads, for use in packed bed columns (Brewer, Dohnalkova, et al., 2019). In each case, REE extraction and stripping were performed in a continuous flow packed bed column and multiple reuse cycles were demonstrated. Additional work to improve the biomass loading capacity and mass transfer kinetics will be required to increase feedstock throughput. However, mass transfer kinetics and cell surface accessibility will likely be diminished above a certain cell loading threshold. Indeed, we have found that REE-adsorption capacity is no longer proportional to cell loading when immobilizing dense cell slurries (2 g wet weight cells/mL) using a Si sol–gel molding process. Coupled with appropriate cell immobilization strategies, a variety of continuous bioreactor configurations can be used for applying various REE-adsorbing microbes for REE recovery. The bioreactors should maximize contact between the biological agents and REE-containing fluids. In addition to the packed bed configurations described in the previously cited literature, fluidized bed bioreactors have also been proposed for various metal-adsorption applications (Ilamathi et al., 2014; Monge-Amaya et al., 2015). Efficient extraction of high-purity

REE will require establishing a cost-effective bioreactor design, an efficient workflow, and optimized operational conditions. The experimental burden of such optimization steps can be reduced by the development and implementation of reactive transport models that predict the optimal chemical and operational parameters for bioseparation operations.

8.4. General Techno-Economics of Biorecovery

Promising bench-scale technologies often falter during scale-up because the economics simply do not work at higher scales (Dunbar, 2017; Kappes, 2002). TEA is a tool used to project promising concepts to larger scale and identify areas where the technology needs improvement in order to be economically feasible. In the following text, we discuss economic factors relevant to bioleaching and bioseparation of REEs.

8.4.1. REE Bioleaching Economics

Bioleaching in the mining industry has been considered a low-cost option compared to more intensive metal recovery processes such as hydrometallurgical, pyrometallurgical, and electrometallurgical approaches, since in principle it only requires construction of an ore or tailings heap, stimulation of microbial activity, and time (Brierley & Brierley, 2013). It should be noted that bioleaching processes are inherently slower than other metallurgical processes occurring on the scale of months instead of hours to days, and this could potentially impact the economics as well. However, this consideration can be mitigated by developing continuous bioleaching processes that would have a steady output after an initial start-up time. We recently conducted a TEA of our REE bioleaching process using FCC catalysts as the REE waste stream (Thompson et al., 2018).

In the analysis, we assumed that we would construct a plant to bioleach 10% of the available FCC catalyst in the United States (19,000 tons/year), that the plant would be located next to a petroleum refinery to minimize transportation costs, and that leaching and biolixiviant production efficiencies would be the same as measured experimentally in our laboratory. The TEA identified the significant impact of the pulp density (mass of solid leached/mass of lixiviant applied) on the economics. Our laboratory research showed that lower pulp densities can provide better leaching efficiencies; approximately 50% of REE was recovered following leaching at 1.5% pulp density, versus about 28% recovery at 50% pulp density, in batch leaching studies of FCC catalyst (Reed et al., 2016). It would seem logical to assume that the more efficient recovery of REE's would be preferred. However, this assumption does not take into account the cost of the bioreactor required to produce the lixiviant. To leach 1,000 kg of FCC catalyst at 1.5% pulp density requires more than 65,000 kg of lixiviant, while at 50% pulp density only 2,000 kg is needed. The fixed capital costs (bioreactor, air compressor, sterilizer, heat exchanger, etc.) for leaching at 1.5% pulp density are 7 times higher than the capital costs for 50% pulp density, and the operating costs (nutrients, electricity, steam, water, and labor) are 27 times higher for 1.5% compared to 50% pulp density (Thompson et al., 2018). Consequently, constructing a plant for 50% pulp density leaching was more economical than the 1.5% pulp density case despite the lower leaching efficiency.

Costs were also sensitive to the amount of REEs in the leached feedstocks and the type of REEs present. REE content in FCC can range between 1.5% and 5% (Goonan, 2011), and leaching FCC with 5% REE content versus 1.5% REE content in our study would increase profits by tenfold. The REEs present in FCC catalyst are primarily La and Ce, and each was worth about \$2/kg in 2016 (Shanghai Metals Market). Sourcing REE feedstocks with higher values would improve economics. Among the most valuable REEs are Eu at \$150/kg, Tb at \$400/kg, Pr at \$52/kg, and Nd at \$42/kg. The analysis revealed that the bioleaching process by itself was profitable at REE prices in 2011 (a high point due to Chinese restrictions on exports), but was not profitable at REE prices in 2016. However, the process could be made profitable by including a tipping fee for acceptance of the FCC catalyst. Currently, in the petroleum industry, FCC catalyst is considered a hazardous waste and must be disposed in hazardous waste landfills. This costs petroleum refineries as much as \$200/ton for disposal (Marafi & Stanislaus, 2008). If a bioleaching company employing our process could recover this fee and render the waste nonhazardous (or reduce the volume of waste classified as hazardous), bioleaching of REEs became economical.

Another opportunity for improvement identified during the TEA was nutrient costs, in particular the glucose cost. Of the total costs for the plant, glucose was responsible for 44% of the overall costs. This finding has led us to focus on alternative carbon sources such as large volume agricultural and food processing wastes for bacterial growth and production of biolixiviant (Jin et al., 2019). Experiments indicated that total REE-leaching efficiency remained largely unchanged when corn stover and potato wastewater were used as carbon sources for microbial biolixiviant production. Although new costs would be associated with transport and processing of the complex alternative carbon to simple sugars for microbial growth, analyses suggested that the overall processing costs and environmental impacts would be greatly reduced. Indeed, compared to a glucose-based bioleaching plant, even with different REE-recovery income and process expenses, the predicted net profits quadrupled for the corn-stover-based plant and doubled for the potato-wastewater-based plant.

8.4.2. REE Bioseparation Economics

Many of the same economic drivers apply to bioseparation processes. An assessment of the techno-economics for integrating a bioseparation approach using LBT-displayed cells into a large-scale process for producing salable total rare earth oxides (TREOs) from 11 diverse REE feedstocks was recently conducted (Jin et al., 2017). For this study, the TEA assumed that the cells were immobilized onto biofilm carrier disks and biosorption was performed in an airlift bioreactor. Elution was performed in a separate column to which the loaded disks were transferred. Preprocessing steps for feedstocks (i.e., beneficiation and leaching) represented 70%–80% of the total cost of the REE-recovery operation. Thus, using feedstocks that require minimal preprocessing (e.g., geothermal brines, acid mine drainage) can provide a significant economic advantage. As with bioleaching, the feedstock REE loading and composition heavily influence the profitability of the operation. For example, Appalachian Basin coal ash, which contains a significant amount of Sc ($4,200/kg; MineralPrices.com; accessed on 31st March 2017), was the only profitable feedstock at the 2017 REE prices. Feedstocks with higher total REE concentrations, but predominantly comprising low-value light rare earths, were not profitable for recovery schemes including biosorption. Geothermal brine, despite not needing beneficiation or leaching, had such low REE concentrations that it was not profitable either. For the economics of the bioseparation module itself, a sensitivity analysis highlighted the importance of biomass reusability; 10 reuse cycles were sufficient to reduce the cost of the bioseparation steps by fourfold. Additional economic considerations for scaling a biosorption process include the costs associated with the specific immobilization platform and the adsorption capacity of the immobilized biosorbent. Maximizing cell-adsorption capacity and immobilized biomass density is critical to the economic viability of bioseparation technologies for REEs.

8.5. Conclusions and Directions for Future Research

The evaluation of biologically based approaches for recovery of REEs is still in its infancy, but research to date indicates their potential to play a role in the development of a more environmentally sustainable and robust supply chain for these metals so critical to modern society. Most likely, new bioleaching or bioseparation approaches for REE recovery will be applied initially to nontraditional REE sources, given the high bar to entry of new technologies into established industries. Potential such sources from which REEs might be recovered include ore tailings, bauxite mine residues, phosphogypsum wastes, incinerator ash, metallurgy slags, acid mine drainage, and industrial and municipal wastewaters and sludges (Binnemans, Jones, et al., 2013; Hennebel et al., 2015; Mulchandani & Westerhoff, 2016; Zhuang et al., 2015). These types of materials have the advantage of already being accessible and collected in large quantities, and in fact further processing to recover valuable components could mitigate some of the environmental liabilities that they pose. In addition, mine tailings and ash are already present in particle sizes that are amenable to leaching (Johnson, 2013) while mine drainage and wastewaters could potentially be directly processed by biosorption/desorption. Post-consumer electronics wastes and industrial wastes such as the FCC catalysts and lighting phosphors also represent potential REE feedstocks, although barriers to using these types of wastes may include the high costs (or absence) of systems to collect sufficient volumes for cost-effective recycling (Binnemans, Jones, et al., 2013).

Once a potentially viable feedstock is identified, rigorous characterization of its chemical composition is necessary for application of both bioleaching and biosorption, in order to ascertain compatibility with microbial cultures and to predict the effectiveness of the biohydrometallurgical processes. Empirical testing will inevitably be necessary, but the scope of such testing can be greatly narrowed by judicious use of geochemical equilibrium modeling. Such modeling is indispensable for interpreting experimental results and for guiding the selection of optimal leaching and sorption/desorption conditions. Availability of appropriate data for parameterization of the models is however still a challenge; robust thermodynamic data for complexation of REEs as well as other potentially competing metals by biologically produced organic ligands are a particular need. Understanding whether constituents in the REE feedstock (and leachate) pose a toxicity concern to bioleaching microbes is also critical, since the selection of a two-step versus one-step bioleaching approach can have a significant impact on costs. In this case, given the diversity of microbial species, testing of the specific feedstocks with the particular microbial culture is likely to be necessary at least for the near future. A better fundamental understanding of microbial interactions with REEs and metals in general is critical to accelerate the advancement of biotechnological approaches to REE recovery (Hennebel et al., 2015).

Of course, there are also substantial opportunities to "improve" upon nature, for example, by increasing microbial tolerance to potential inhibitors, optimizing production of leaching compounds (acids, or chelators), or designing biological ligands with higher affinity for targeted metals, by using synthetic biology techniques (Adesina et al., 2017; Dunbar, 2017). These opportunities are tremendously exciting, but use of genetically engineered organisms in large-scale REEs or other metal biorecovery operations will require additional studies on their interactions with indigenous microorganisms and native ecosystems, to develop deployment approaches that ensure both the survival of the desired species or traits and the mitigation of unwanted and unforeseen negative consequences.

Progress in the development of biotechnological approaches for REE recovery will require concerted effort on the part of microbiologists, molecular biologists, ecologists, geochemists, and engineers. Demonstrating performance at pilot scale will be critical to eliciting commercial interest, although the long and successful history of biotechnology in other industrial applications (e.g., food processing, pharmaceuticals, etc.) as well as in biomining should bolster confidence with respect to technical feasibility. Of course, economic factors must also be favorable for biotechnology integration into the REE supply chain. If REE demand continues to rise and the opportunity space for new approaches grows, the environmental benefits of biotechnology may become an increasingly important differentiating factor.

Acknowledgments

We thank David Combs (INL) for preparing the Figure 8.1 graphic and Debra Bruhn (INL) for assistance in collection of microorganisms from Yellowstone National Park and laboratory enrichment and bioleaching. We also thank the R. Donahoe lab at University of Alabama and the F. McLaughlin lab at University of Wyoming for generously sharing coal fly ash samples. The original research work by the authors described herein on REE recovery from coal or coal byproducts was supported by the DOE Office of Fossil Energy, DE-NETL Rare Earth Program, through awards DE-FWP-INL-B000-17-015 and DE-FWP-LLNL-18-FEW0239. Support for all other original research work first reported here (i.e., data in Figure 8.4b and 8.4c) and for the preparation of this manuscript was provided by the Critical Materials Innovation Hub funded by the U.S. Department of Energy, Office of Energy Efficiency and Renewable Energy, Advanced Materials and Manufacturing Technologies Office. Work was conducted under DOE Idaho Operations Office Contract DE-AC07-05ID14517 and Lawrence Livermore National Laboratory Contract DE-AC52-07NA27344. Accordingly, the U.S. Government retains a nonexclusive, royalty-free license to publish or reproduce the published form of this contribution, and allows others to do so, for U.S. Government purposes.

References

Abramov, S., He, J., Wimmer, D., Lemloh, M.-L., Muehe, E. M., Gann, B., et al. (2018). Heavy metal mobility and valuable contents of processed municipal solid waste incineration residues from Southwestern Germany. *Waste Management*, *79*, 735–743.

Adesina, O., Anzai, I. A., Avalos, J. L., & Barstow, B. (2017). Embracing Biological Solutions to the Sustainable Energy Challenge. *Chem*, *2*(1), 20–51.

Alipanah, M., Park, D. M., Middleton, A., Dong, Z., Hsu-Kim, H., Jiao, Y., & Jin, H. (2020). Techno-Economic and Life Cycle Assessments for Sustainable Rare Earth Recovery from Coal Byproducts using Biosorption. *ACS Sustainable Chemistry & Engineering*, *8*(49), 17914–17922. https://doi.org/10.1021/acssuschemeng.0c04415

Allen, K. N., & Imperiali, B. (2010). Lanthanide-tagged proteins—An illuminating partnership. *Current Opinion in Chemical Biology*, *14*(2), 247–254. http://www.sciencedirect.com/science/article/pii/S1367593110000037

Amin, M. M., El-Aassy, I. E., El-Feky, M. G., Sallam, A. M., El-Sayed, E. M., Nada, A. A., & Harpy, N. M. (2014). Fungal leaching of rare earth elements from lower carboniferous carbonaceous shales, southwestern Sinai, Egypt. *Romanian Journal of Biophysics*, *24*(1), 25–41.

Andrès, Y., Texier, A. C., & Le Cloirec, P. (2003). Rare earth elements removal by microbial biosorption: A review. *Environmental Technology*, *24*(11), 1367–1375. http://dx.doi.org/10.1080/09593330309385681

Antonick, P. J., Hu, Z., Fujita, Y., Reed, D. W., Das, G., Wu, L., et al. (2019). Bio- and mineral acid leaching of rare earth elements from synthetic phosphogypsum. *The Journal of Chemical Thermodynamics*, *132*, 491–496.

Arunraj, B., Sathvika, T., Rajesh, V., & Rajesh, N. (2019). Cellulose and Saccharomyces cerevisiae embark to recover europium from phosphor powder. *ACS Omega*, *4*(1), 940–952.

Ayora, C., Macías, F., Torres, E., Lozano, A., Carrero, S., Nieto, J.-M., et al. (2016). Recovery of rare earth elements and yttrium from passive-remediation systems of acid mine drainage. *Environmental Science & Technology*, *50*(15), 8255–8262. http://dx.doi.org/10.1021/acs.est.6b02084

Barmettler, F., Castelberg, C., Fabbri, C., & Brandl, H. (2016). Microbial mobilization of rare earth elements (REE) from mineral solids—A mini review. *AIMS Microbiology*, *2*(2), 190–204.

Barnett, M. J., Palumbo-Roe, B., & Gregory, S. P. (2018). Comparison of heterotrophic bioleaching and ammonium sulfate ion exchange leaching of rare earth elements from a Madagascan ion-adsorption clay. *Minerals*, *8*(6), 236. https://doi.org/10.3390/min8060236

Beveridge, T. J., & Murray, R. G. (1980). Sites of metal deposition in the cell wall of Bacillus subtilis. *Journal of Bacteriology*, *141*(2), 876–887.

Binnemans, K., Jones, P. T., Blanpain, B., Van Gerven, T., Yang, Y. X., Walton, A., & Buchert, M. (2013). Recycling of rare earths: A critical review. *Journal of Cleaner Production*, *51*, 1–22.

Binnemans, K., Pontikes, Y., Jones, P. T., Van Gerven, T., & Blanpain, B. (2013). Recovery of rare earths from industrial waste residues: A concise review. In *Proceedings of the 3rd International Slag Valorisation Symposium: The Transition to Sustainable Materials Management* (pp. 191–205).

Bonificio, W. D., & Clarke, D. R. (2016). Rare-earth separation using bacteria. *Environmental Science & Technology Letters*, *3*(4), 180–184. http://dx.doi.org/10.1021/acs.estlett.6b00064

Bosecker, K. (1997). Bioleaching: Metal solubilization by microorganisms. *FEMS Microbiology Reviews*, *20*(3–4), 591–604.

Bosshard, P. P., Bachofen, R., & Brandl, H. (1996). Metal leaching of fly ash from municipal waste incineration by Aspergillus niger. *Environmental Science & Technology*, *30*(10), 3066–3070. https://doi.org/10.1021/es960151v

Brewer, A., Chang, E., Park, D. M., Kou, T., Li, Y., Lammers, L. N., & Jiao, Y. (2019). Recovery of rare earth elements from geothermal fluids through bacterial cell surface adsorption. *Environmental Science & Technology*, *53*(13), 7714–7723. https://doi.org/10.1021/acs.est.9b00301

Brewer, A., Dohnalkova, A., Shutthanandan, V., Kovarik, L., Chang, E., Sawvel, A. M., et al. (2019). Microbe encapsulation for selective rare-earth recovery from electronic waste leachates. *Environmental Science & Technology*, *53*(23), 13888–13897. https://doi.org/10.1021/acs.est.9b04608

Brierley, C. L. (2016). Biological processing: Biological processing of sulfidic ores and concentrates—Integrating innovations. In *Innovative Process Development in Metallurgical Industry* (pp. 109–135). Switzerland: Springer International Publishing. https://doi.org/10.1007/978-3-319-21599-0_6

Brierley, C. L., & Brierley, J. A. (2013). Progress in bioleaching: Part B: Applications of microbial processes by the minerals industries. *Applied Microbiology and Biotechnology*, *97*(17), 7543–7552. Journal Article. https://doi.org/10.1007/s00253-013-5095-3

Brisson, V. L., Zhuang, W. Q., & Alvarez-Cohen, L. (2015). Bioleaching of rare earth elements from monazite sand. *Biotechnology and Bioengineering*, *113*(2), 339–348.

Calkins, W. H. (1994). The chemical forms of sulfur in coal: A review. *Fuel*, *73*(4), 475–484.

Campbell, P. G. C. (1995). Interactions between trace metals and aquatic organisms: A critique of the free-ion activity model,. In A. Tessier & D. R. Turner (Eds.), *Metal speciation and bioavailability in aquatic systems* (Vol. *3*, pp. 45–102). Chichester, UK: Wiley.

Cánovas, C. R., Pérez-López, R., Macías, F., Chapron, S., Nieto, J. M., & Pellet-Rostaing, S. (2017). Exploration of fertilizer industry wastes as potential source of critical raw materials. *Journal of Cleaner Production*, *143*, 497–505.

Chen, A., Shi, Q., Feng, J., Ouyang, Y., Chen, Y., & Tan, S. (2010). Dissociation of outer membrane for Escherichia coli cell caused by cerium nitrate. *Journal of Rare Earths*, *28*(2), 312–315.

Chen, A., Shi, Q., Ouyang, Y., & Chen, Y. (2012). Effect of Ce3+ on membrane permeability of Escherichia coli cell. *Journal of Rare Earths*, *30*(9), 947–951.

Corbett, M. K., Eksteen, J. J., Niu, X.-Z., Croue, J.-P., & Watkin, E. L. J. (2017). Interactions of phosphate solubilising microorganisms with natural rare-earth phosphate minerals: A study utilizing Western Australian monazite. *Bioprocess and Biosystems Engineering*, 1–14. http://dx.doi.org/10.1007/s00449-017-1757-3

Cotruvo, J. A., Featherston, E. R., Mattocks, J. A., Ho, J. V., & Laremore, T. N. (2018). Lanmodulin: A highly selective lanthanide-binding protein from a lanthanide-utilizing bacterium. *Journal of the American Chemical Society*, *140*(44), 15056–15061. https://doi.org/10.1021/jacs.8b09842

Cox, J. S., Smith, D. S., Warren, L. A., & Ferris, F. G. (1999). Characterizing heterogeneous bacterial surface functional groups using discrete affinity spectra for proton binding. *Environmental Science & Technology*, *33*(24), 4514–4521. https://doi.org/10.1021/es9906271

Cui, J., & Zhang, L. (2008). Metallurgical recovery of metals from electronic waste: A review. *Journal of Hazardous Materials*, *158*(2–3), 228–256.

Das, G., Lencka, M. M., Eslamimanesh, A., Anderko, A., & Riman, R. E. (2017). Rare-earth elements in aqueous chloride systems: Thermodynamic modeling of binary and multicomponent systems in wide concentration ranges. *Fluid Phase Equilibria*, *452*, 16–57.

Das, G., Lencka, M. M., Eslamimanesh, A., Wang, P., Anderko, A., Riman, R. E., & Navrotsky, A. (2019). Rare earth sulfates in aqueous systems: Thermodynamic modeling of binary and multicomponent systems over wide concentration and temperature ranges. *Journal of Chemical Thermodynamics*, *131*, 49–79.

Das, G., Lencka, M. M., Liu, J., Anderko, A., Riman, R. E., & Navrotsky, A. (2023). Modeling phase equilibria and speciation in aqueous solutions of rare earth elements with hydroxide and organic ligands. *The Journal of Chemical Thermodynamics*, *186*, 107125.

Das, N., & Das, D. (2013). Recovery of rare earth metals through biosorption: An overview. *Journal of Rare Earths*, *31*(10), 933–943.

Daughtry, K. D., Martin, L. J., Sarraju, A., Imperiali, B., & Allen, K. N. (2012). Tailoring encodable lanthanide-binding tags as MRI contrast agents. *ChemBioChem*, *13*(17), 2567–2574. http://dx.doi.org/10.1002/cbic.201200448

Deblonde, G. J.-P., Mattocks, J. A., Park, D. M., Reed, D. W., Cotruvo Jr, J. A., & Jiao, Y. (2020). Selective and efficient biomacromolecular extraction of rare-earth elements using lanmodulin. *Inorganic Chemistry*, *59*(17), 11855–11867.

Dehaine, Q., & Filippov, L. O. (2015). Rare earth (La, Ce, Nd) and rare metals (Sn, Nb, W) as by-product of kaolin production, Cornwall: Part1: Selection and characterisation of the valuable stream. *Minerals Engineering*, *76*, 141–153.

Di Piazza, S., Cecchi, G., Cardinale, A. M., Carbone, C., Mariotti, M. G., Giovine, M., & Zotti, M. (2017). Penicillium expansum Link strain for a biometallurgical method to recover REEs from WEEE. *Waste Management*, *60*, 596–600.

Dodson, J. R., Parker, H. L., Muñoz García, A., Hicken, A., Asemave, K., Farmer, T. J., et al. (2015). Bio-derived materials as a green route for precious & critical metal recovery and re-use. *Green Chemistry*, *17*(4), 1951–1965. http://dx.doi.org/10.1039/C4GC02483D

Dong, Z., Deblonde, G., Middleton, A., Hu, D., Dohnalkova, A., Kovarik, L., et al. (2021). Microbe-encapsulated silica gel biosorbents for selective extraction of scandium from coal byproducts. *Environmental Science & Technology*, *55*(9), 6320–6328.

Dong, Z., Mattocks, J. A., Deblonde, G. J.-P., Hu, D., Jiao, Y., Cotruvo, J. A., Jr., & Park, D. M. (2021). Bridging hydrometallurgy and biochemistry: A protein-based process for recovery and separation of rare earth elements. *ACS Central Science*, *7*(11), 1798–1808.

Dunbar, W. S. (2017). Biotechnology and the mine of tomorrow. *Trends in Biotechnology*, *35*, 79–89.

Emsbo, P., McLaughlin, P. I., Breit, G. N., du Bray, E. A., & Koenig, A. E. (2015). Rare earth elements in sedimentary phosphate deposits: Solution to the global REE crisis? *Gondwana Research*, *27*(2), 776–785.

Evangelou, V. P., & Zhang, Y. L. (1995). A review: Pyrite oxidation mechanisms and acid mine drainage prevention. *Critical Reviews in Environmental Science and Technology*, *25*(2), 141–199. https://doi.org/10.1080/10643389509388477

Fathollahzadeh, H., Becker, T., Eksteen, J. J., Kaksonen, A. H., & Watkin, E. L. J. (2018). Microbial contact enhances bioleaching of rare earth elements. *Bioresource Technology Reports*, *3*, 102–108.

Fathollahzadeh, H., Eksteen, J. J., Kaksonen, A. H., & Watkin, E. L. (2019). Role of microorganisms in bioleaching of rare earth elements from primary and secondary resources. *Applied Microbiology and Biotechnology*, *103*(3), 1043–1057.

Fein, J. B., Daughney, C. J., Yee, N., & Davis, T. A. (1997). A chemical equilibrium model for metal adsorption onto bacterial surfaces. *Geochimica et Cosmochimica Acta*, *61*(16), 3319–3328.

Fein, J. B., Martin, A. M., & Wightman, P. G. (2001). Metal adsorption onto bacterial surfaces: Development of a predictive approach. *Geochimica et Cosmochimica Acta*, *65*(23), 4267–4273.

Ferreira, D. M., Servulo, E. F. C., Silva, J. A. S., Dognini, J., & Oliveira, F. J. S. (2018). Bioleaching of Rare Earth Elements (REE) from spent cracking catalyst. In *Proceedings CRETE 2018, Sixth International Conference on Industrial & Hazardous Waste Management*, Chania, Crete, Greece.

Fonti, V., Amato, A., & Beolchini, F. (2015). Urban biomining: New challenges for a successful exploitation of WEEE by means of a biotechnological approach. In *Microbiology for minerals, metals, materials and the environment* (pp. 329–358). Boca Raton, FL: CRC Press. https://doi.org/10.1201/b18124

Fujita, Y., Barnes, J., Eslamimanesh, A., Lencka, M. M., Anderko, A., Riman, R. E., & Navrotsky, A. (2015). Effects of simulated rare earth recycling wastewaters on biological nitrification. *Environmental Science & Technology*, *49*(16), 9460–9468. http://dx.doi.org/10.1021/acs.est.5b01753

Fujita, Y., Walton, M., Das, G., Dohnalkova, A., Vanzin, G., & Anderko, A. (2020). Impacts of anthropogenic gadolinium on the activity of the ammonia oxidizing bacterium Nitrosomonas europaea. *Chemosphere*, *257*, 127250.

Fuma, S., Takeda, H., Miyamoto, K., Yanagisawa, K., Inoue, Y., Ishii, N., et al. (2001). Ecological evaluation of gadolinium toxicity compared with other heavy metals using an aquatic microcosm. *Bulletin of Environmental Contamination and Toxicology*, *66*(2), 231–238.

Funari, V., Bokhari, S. N. H., Vigliotti, L., Meisel, T., & Braga, R. (2016). The rare earth elements in municipal solid waste incinerators ash and promising tools for their prospecting. *Journal of Hazardous Materials*, *301*, 471–479.

Ganguli, R., & Cook, D. R. (2018). Rare earths: A review of the landscape. *MRS Energy & Sustainability*, *5*, 6.

Gaustad, G., Williams, E., & Leader, A. (2021). Rare earth metals from secondary sources: Review of potential supply from waste and byproducts. *Resources, Conservation and Recycling*, *167*, 105213. https://doi.org/10.1016/j.resconrec.2020.105213.

Giret, S., Hu, Y. M., Masoumifard, N., Boulanger, J. F., Juere, E., Kleitz, F., & Lariyiere, D. (2018). Selective separation and preconcentration of scandium with mesoporous silica. *Acs Applied Materials & Interfaces*, *10*(1), 448–457.

Glombitza, F., & Reichel, S. (2014). Metal-containing residues from industry and in the environment: Geobiotechnological urban mining. In A. Schippers, F. Glombitza, & W. Sand (Eds.), *Geobiotechnology I* (Vol. *141*, pp. 49–107): Berlin, Heidelberg: Springer. https://doi.org/10.1007/10_2013_254

González, V., Vignati, D. A., Leyval, C., & Giamberini, L. (2014). Environmental fate and ecotoxicity of lanthanides: Are they a uniform group beyond chemistry? *Environment International*, *71*, 148–157.

González, V., Vignati, D. A., Pons, M.-N., Montarges-Pelletier, E., Bojic, C., & Giamberini, L. (2015). Lanthanide ecotoxicity: First attempt to measure environmental risk for aquatic organisms. *Environmental Pollution*, *199*, 139–147.

Goonan, T. G. (2011). *Rare earth elements: End use and recyclability*. Reston: US Department of the Interior, US Geological Survey.

Gupta, N. K., Gupta, A., Ramteke, P., Sahoo, H., & Sengupta, A. (2018). Biosorption—A green method for the preconcentration of rare earth elements (REEs) from waste solutions: A review. *Journal of Molecular Liquids*, *274*, 148–164.

Hallberg, K. B., Grail, B. M., Plessis, C. A. D., & Johnson, D. B. (2011). Reductive dissolution of ferric iron minerals: A new approach for bio-processing nickel laterites. *Minerals Engineering*, *24*(7), 620–624.

Haque, N., Hughes, A., Lim, S., & Vernon, C. (2014). Rare earth elements: Overview of mining, mineralogy, uses, sustainability and environmental impact. *Resources*, *3*(4), 614–635.

Hassanien, W. A., Desouky, O. A., & Hussien, S. S. (2014). Bioleaching of some rare earth elements from Egyptian monazite using *Aspergillus ficuum* and *Pseudomonas aeruginosa*. *Walailak Journal of Science and Technology (WJST)*, *11*(6), 809–823.

Hennebel, T., Boon, N., Maes, S., & Lenz, M. (2015). Biotechnologies for critical raw material recovery from primary and secondary sources: R&D priorities and future perspectives. *New Biotechnology*, *32*(1), 121–127.

Herrmann, H., Nolde, J., Berger, S., & Heise, S. (2016). Aquatic ecotoxicity of lanthanum—A review and an attempt to derive water and sediment quality criteria. *Ecotoxicology and Environmental Safety*, *124*, 213–238.

Holmberg, O. (1907). Über die darstellung von reinem neodymoxyd und über zwei neue methoden zur trennung seltener Erden. *Zeitschrift für anorganische Chemie, 53*(1), 83–134.

Hopfe, S., Flemming, K., Lehmann, F., Möckel, R., Kutschke, S., & Pollmann, K. (2017). Leaching of rare earth elements from fluorescent powder using the tea fungus Kombucha. *Waste Management, 62*, 211–221.

Hosomomi, Y., Baba, Y., Kubota, F., Kamiya, N., & Goto, M. (2013). Biosorption of Rare Earth Elements by Escherichia coli. *Journal of Chemical Engineering of Japan, 46*(7), 450–454.

Ibrahim, H., & El-Sheikh, E. (2011). Bioleaching treatment of Abu Zeneima uraniferous gibbsite ore material for recovering U, REEs, Al and Zn. *Research Journal of Chemical Sciences, 1*(4), 55–66.

Ilamathi, R., Nirmala, G., & Muruganandam, L. (2014). Heavy metals biosorption in liquid solid fluidized bed by immobilized consortia in alginate beads. *International Journal of ChemTech Research, 6*, 652–662.

Jain, N., & Sharma, D. K. (2004). Biohydrometallurgy for nonsulfidic minerals—A review. *Geomicrobiology Journal, 21*(3), 135–144. http://dx.doi.org/10.1080/01490450490275271

Jha, M. K., Kumari, A., Panda, R., Rajesh Kumar, J., Yoo, K., & Lee, J. Y. (2016). Review on hydrometallurgical recovery of rare earth metals. *Hydrometallurgy, 161*, 77–101.

Jin, H., Park, D. M., Gupta, M., Brewer, A. W., Ho, L., Singer, S. L., et al. (2017). Techno-economic assessment for integrating biosorption into rare earth recovery process. *ACS Sustainable Chemistry & Engineering, 5*(11), 10148–10155.

Jin, H., Reed, D. W., Thompson, V. S., Fujita, Y., Jiao, Y., Crain-Zamora, M., et al. (2019). Sustainable bioleaching of rare earth elements from industrial waste materials using agricultural wastes. *ACS Sustainable Chemistry & Engineering, 7*(18), 15311–15319. https://doi.org/10.1021/acssuschemeng.9b02584

Johnson, D. B. (2012). Reductive dissolution of minerals and selective recovery of metals using acidophilic iron-and sulfate-reducing acidophiles. *Hydrometallurgy, 127*, 172–177.

Johnson, D. B. (2013). Development and application of biotechnologies in the metal mining industry. *Environmental Science and Pollution Research, 20*(11), 7768–7776.

Jung, H., Su, Z., Inaba, Y., West, A. C., & Banta, S. (2023). Genetic modification of Acidithiobacillus ferrooxidans for rare-earth element recovery under acidic conditions. *Environmental Science & Technology, 57*(48), 19902–19911.

Jyothi, R. K., Thenepalli, T., Ahn, J. W., Parhi, P. K., Chung, K. W., & Lee, J.-Y. (2020). Review of rare earth elements recovery from secondary resources for clean energy technologies: Grand opportunities to create wealth from waste. *Journal of Cleaner Production, 267*,122048.

Kaksonen, A. H., Boxall, N. J., Gumulya, Y., Khaleque, H. N., Morris, C., Bohu, T., et al. (2018). Recent progress in biohydrometallurgy and microbial characterisation. *Hydrometallurgy, 180*, 7–25.

Kappes, D. W. (2002). Precious metal heap leach design and practice. *Proceedings of the Mineral Processing Plant Design, Practice, and Control, 1*, 1606–1630.

Karavaiko, G. I., Kareva, A. S., Avakian, Z. A., Zakharova, V. I., & Korenevsky, A. A. (1996). Biosorption of scandium and yttrium from solutions. *Biotechnology Letters*, *18*(11), 1291–1296. https://doi.org/10.1007/BF00129957

Kim, P., Das, G., Lencka, M. M., Anderko, A., & Riman, R. E. (2020). Rare earth element recovery using monoethanolamine. *Journal of Materials Engineering and Performance*, *29*(9), 5564–5573. https://doi.org/10.1007/s11665-020-04887-7

Klauber, C., Gräfe, M., & Power, G. (2011). Bauxite residue issues: II. Options for residue utilization. *Hydrometallurgy*, *108*(1–2), 11–32.

Kucuker, M. A., Wieczorek, N., Kuchta, K., & Copty, N. K. (2017). Biosorption of neodymium on Chlorella vulgaris in aqueous solution obtained from hard disk drive magnets. *PLoS One*, *12*(4), e0175255.

Kurvet, I., Juganson, K., Vija, H., Sihtmäe, M., Blinova, I., Syvertsen-Wiig, G., & Kahru, A. (2017). Toxicity of Nine (Doped) Rare earth metal oxides and respective individual metals to aquatic microorganisms Vibrio fischeri and Tetrahymena thermophila. *Materials*, *10*(7), 754.

Loges, A., Migdisov, A. A., Wagner, T., Williams-Jones, A. E., & Markl, G. (2013). An experimental study of the aqueous solubility and speciation of Y (III) fluoride at temperatures up to 250 °C. *Geochimica et Cosmochimica Acta*, *123*, 403–415.

Løvik, A. N., Hagelüken, C., & Wäger, P. (2018). Improving supply security of critical metals: Current developments and research in the EU. *Sustainable Materials and Technologies*, *15*, 9–18.

Maes, S., Zhuang, W.-Q., Rabaey, K., Alvarez-Cohen, L., & Hennebel, T. (2017). Concomitant leaching and electrochemical extraction of rare earth elements from monazite. *Environmental Science & Technology*, *51*(3), 1654–1661. http://dx.doi.org/10.1021/acs.est.6b03675

Maleke, M., Valverde, A., Gomez-Arias, A., Cason, E. D., Vermeulen, J.-G., Coetsee-Hugo, L., et al. (2019). Anaerobic reduction of europium by a Clostridium strain as a strategy for rare earth biorecovery. *Scientific Reports*, *9*(1), 1–11.

Marafi, M., & Stanislaus, A. (2008). Spent hydroprocessing catalyst management: A review: Part II. Advances in metal recovery and safe disposal methods. *Resources, Conservation and Recycling*, *53*(1), 1–26.

Markai, S., Andres, Y., Montavon, G., & Grambow, B. (2003). Study of the interaction between europium (III) and Bacillus subtilis: Fixation sites, biosorption modeling and reversibility. *Journal of Colloid and Interface Science*, *262*(2), 351–361.

Martinez, R. E., Pourret, O., & Takahashi, Y. (2014). Modeling of rare earth element sorption to the Gram positive Bacillus subtilis bacteria surface. *Journal of Colloid and Interface Science*, *413*, 106–111.

Mattocks, J. A., Ho, J. V., & Cotruvo, J. J. A. (2019). A selective, protein-based fluorescent sensor with picomolar affinity for rare earth elements. *Journal of the American Chemical Society*, *141*(7), 2857–2861. https://doi.org/10.1021/jacs.8b12155

Mejáre, M., & Bülow, L. (2001). Metal-binding proteins and peptides in bioremediation and phytoremediation of heavy metals. *TRENDS in Biotechnology*, *19*(2), 67–73.

Migdisov, A. A., Reukov, V. V., & Williams-Jones, A. E. (2006). A spectrophotometric study of neodymium (III) complexation in sulfate solutions at elevated temperatures. *Geochimica et Cosmochimica Acta, 70*(4), 983–992.

Migdisov, A. A., & Williams-Jones, A. (2007). An experimental study of the solubility and speciation of neodymium (III) fluoride in F-bearing aqueous solutions. *Geochimica et Cosmochimica Acta, 71*(12), 3056–3069.

Migdisov, A. A., & Williams-Jones, A. E. (2002). A spectrophotometric study of neodymium (III) complexation in chloride solutions. *Geochimica et Cosmochimica Acta, 66*(24), 4311–4323.

Migdisov, A. A., & Williams-Jones, A. E. (2006). A spectrophotometric study of erbium (III) speciation in chloride solutions at elevated temperatures. *Chemical Geology, 234*(1–2), 17–27.

Migdisov, A. A., & Williams-Jones, A. E. (2008). A spectrophotometric study of Nd (III), Sm (III) and Er (III) complexation in sulfate-bearing solutions at elevated temperatures. *Geochimica et Cosmochimica Acta, 72*(21), 5291–5303.

Migdisov, A. A., Williams-Jones, A. E., Normand, C., & Wood, S. A. (2008). A spectrophotometric study of samarium (III) speciation in chloride solutions at elevated temperatures. *Geochimica et Cosmochimica Acta, 72*(6), 1611–1625.

Migdisov, A. A., Williams-Jones, A., & Wagner, T. (2009). An experimental study of the solubility and speciation of the Rare Earth Elements (III) in fluoride-and chloride-bearing aqueous solutions at temperatures up to 300 C. *Geochimica et Cosmochimica Acta, 73*(23), 7087–7109.

Mioduski, T. (1999). Identification of saturating solid phases in the system $Ce_2\ (SO_4)_3$-H_2O from the solubility data. *Journal of Thermal Analysis and Calorimetry, 55*(3), 751–763.

Mioduski, T., Gumiński, C., & Zeng, D. (2008). IUPAC-NIST solubility data series. 87. Rare earth metal chlorides in water and aqueous systems. Part 1. Scandium group (Sc, Y, La). *Journal of Physical and Chemical Reference Data, 37*(4), 1765–1853.

Mioduski, T., Gumiński, C., & Zeng, D. (2009a). IUPAC-NIST solubility data series. 87. Rare earth metal chlorides in water and aqueous systems. Part 2. Light lanthanides (Ce–Eu). *Journal of Physical and Chemical Reference Data, 38*(2), 441–562.

Mioduski, T., Gumiński, C., & Zeng, D. (2009b). IUPAC-NIST solubility data series. 87. Rare earth metal chlorides in water and aqueous systems. Part 3. Heavy lanthanides (Gd–Lu). *Journal of Physical and Chemical Reference Data, 38*(4), 925–1011.

Mioduski, T., Gumiński, C., & Zeng, D. (2012). IUPAC-NIST solubility data series. 94. Rare earth metal iodides and bromides in water and aqueous systems. Part 1. Iodides. *Journal of Physical and Chemical Reference Data, 41*(1), 013104-013104.

Mioduski, T., Gumiński, C., & Zeng, D. (2014). IUPAC-NIST solubility data series. 100. Rare earth metal fluorides in water and aqueous systems. Part 1. Scandium group (Sc, Y, La). *Journal of Physical and Chemical Reference Data, 43*(1), 013105.

Mioduski, T., Gumiński, C., & Zeng, D. (2015a). IUPAC-NIST solubility data series. 100. Rare earth metal fluorides in water and aqueous systems. Part 2. Light lanthanides (Ce–Eu). *Journal of Physical and Chemical Reference Data, 44*(1), 013102.

Mioduski, T., Gumiński, C., & Zeng, D. (2015b). IUPAC-NIST solubility data series. 100. Rare earth metal fluorides in water and aqueous systems. Part 3. Heavy lanthanides (Gd–Lu). *Journal of Physical and Chemical Reference Data*, *44*(2), 023102.

Mioduski, T., Gumiński, C., Zeng, D., & Voigt, H. (2013). IUPAC-NIST solubility data series. 94. Rare earth metal iodides and bromides in water and aqueous systems. Part 2. Bromides. *Journal of Physical and Chemical Reference Data*, *42*(1), 013101.

Monge-Amaya, O., Certucha-Barragán, M. T., Almendariz-Tapia, F. J., & Figueroa-Torres, G. M. (2015). Removal of heavy metals from aqueous solutions by aerobic and anaerobic biomass. *Biomass Production and Uses*, *31*, 31–54.

Morel, F. M., & Hering, J. (1993). *Principles and applications of aquatic chemistry*. New York: John Wiley & Sons.

Moriwaki, H., Koide, R., Yoshikawa, R., Warabino, Y., & Yamamoto, H. (2013). Adsorption of rare earth ions onto the cell walls of wild-type and lipoteichoic acid-defective strains of Bacillus subtilis. *Applied Microbiology and Biotechnology*, *97*(8), 3721–3728.

Moriwaki, H., Masuda, R., Yamazaki, Y., Horiuchi, K., Miyashita, M., Kasahara, J., et al. (2016). Application of freeze-dried powders of genetically engineered microbial strains as adsorbents for rare earth metal ions. *ACS Applied Materials & Interfaces*, *8*(40), 26524–26531.

Moriwaki, H., & Yamamoto, H. (2013). Interactions of microorganisms with rare earth ions and their utilization for separation and environmental technology. *Applied Microbiology and Biotechnology*, *97*(1), 1–8.

Mulchandani, A., & Westerhoff, P. (2016). Recovery opportunities for metals and energy from sewage sludges. *Bioresource Technology*, *215*, 215–226.

Nassar, N. T., Du, X., & Graedel, T. E. (2015). Criticality of the rare earth elements. *Journal of Industrial Ecology*, *19*(6), 1044–1054. http://dx.doi.org/10.1111/jiec.12237

Ngwenya, B. T., Magennis, M., Olive, V., Mosselmans, J. F. W., & Ellam, R. M. (2010). Discrete site surface complexation constants for lanthanide adsorption to bacteria as determined by experiments and linear free energy relationships. *Environmental Science & Technology*, *44*(2), 650–656.

Ngwenya, B. T., Mosselmans, J. F. W., Magennis, M., Atkinson, K. D., Tourney, J., Olive, V., & Ellam, R. M. (2009). Macroscopic and spectroscopic analysis of lanthanide adsorption to bacterial cells. *Geochimica Et Cosmochimica Acta*, *73*(11), 3134–3147.

Nitz, M., Franz, K. J., Maglathlin, R. L., & Imperiali, B. (2003). A Powerful Combinatorial Screen to Identify High-Affinity Terbium (III)-Binding Peptides. *ChemBioChem*, *4*(4), 272–276.

Olson, G. J., Brierley, J. A., & Brierley, C. L. (2003). Bioleaching review part B. *Applied Microbiology and Biotechnology*, *63*(3), 249–257. https://doi.org/10.1007/s00253-003-1404-6

Park, D., Brewer, A., Reed, D. W., Lammers, L. N., & Jiao, Y. (2017). Recovery of rare earth elements from low-grade feedstock leachates using engineered bacteria. *Environmental Science & Technology*, *51*(22), 13471–13480. http://dx.doi.org/10.1021/acs.est.7b02414

Park, D., Middleton, A., Smith, R., Deblonde, G., Laudal, D., Theaker, N., et al. (2020). A biosorption-based approach for selective extraction of rare earth elements from coal byproducts. *Separation and Purification Technology*, *241*, 116726.

Park, D., Reed, D. W., Yung, M., Eslamimanesh, A., Lencka, M. M., Anderko, A., et al. (2016). Bioadsorption of rare earth elements through cell surface display of lanthanide binding tags. *Environmental Science & Technology*, *50*(5), 2735–2742. http://dx.doi.org/10.1021/acs.est.5b06129

Peng, L., Weiying, Z., Xi, L., & Yi, L. (2007). Structural basis for the biological effects of Pr (III) ions: Alteration of cell membrane permeability. *Biological Trace Element Research*, *120*(1–3), 141–147.

Peng, L., Yi, L., Zhexue, L., Juncheng, Z., Jiaxin, D., Daiwen, P., et al. (2004). Study on biological effect of La^{3+} on *Escherichia coli* by atomic force microscopy. *Journal of Inorganic Biochemistry*, *98*(1), 68–72.

Pollmann, K., Kutschke, S., Matys, S., Kostudis, S., Hopfe, S., & Raff, J. (2016). Novel biotechnological approaches for the recovery of metals from primary and secondary resources. *Minerals*, *6*(2), 54.

Qu, Y., & Lian, B. (2013). Bioleaching of rare earth and radioactive elements from red mud using *Penicillium tricolor* RM-10. *Bioresource Technology*, *136*(0), 16–23.

Rao, L., Rai, D., & Felmy, A. R. (1996). Solubility of Nd (OH) 3 (c) in 0.1 M NaCl aqueous solution at 25 C and 90 C. *Radiochimica Acta*, *72*(3), 151–156.

Rasoulnia, P., Barthen, R., & Lakaniemi, A.-M. (2020). A critical review of bioleaching of rare earth elements: The mechanisms and effect of process parameters. *Critical Reviews in Environmental Science and Technology*, *51*(4), 1–50.

Rawlings, D. E. (2005). Characteristics and adaptability of iron- and sulfur-oxidizing microorganisms used for the recovery of metals from minerals and their concentrates. *Microbial Cell Factories*, *4*, 13.

Reed, D. W., Fujita, Y., Daubaras, D. L., Jiao, Y., & Thompson, V. S. (2016). Bioleaching of rare earth elements from waste phosphors and cracking catalysts. *Hydrometallurgy*, *166*, 34–40.

Rimbach, E., & Schubert, A. (1909). Löslichkeit einiger schwerlöslicher Salze seltener Erdmetalle. *Zeitschrift für Physikalische Chemie*, *67*(1), 183–200.

Roberto, F. F. (2017). Copper heap bioleach microbiology—Progress and challenges. *Solid State Phenomena*, *262*, 250–254.

Shin, D., Kim, J., Kim, B. S., Jeong, J., & Lee, J. C. (2015). Use of phosphate solubilizing bacteria to leach rare earth elements from monazite-bearing ore. *Minerals*, *5*(2), 189–202.

Silva, R. (1982). *The solubilities of crystalline neodymium and americium trihydroxides*. LBL-15055, Lawrence Berkeley National Laboratory, Berkeley, California.

Skorik, N. A., Kumok, V. N., Perov, E. I., Avgustan, K. P., & Serebrennikov, V. V. (1965). Citrato-complexes of rare-earth elements in acid solutions. *Russian Journal of Inorganic Chemistry*, *10*(3), 351–353.

Skorik, N. A., & Serebrennikov, V. V. (1965). Dependence of the solubility of the citrates of some rare-earth elements on pH. *Russian Journal of Inorganic Chemistry*, *10*(2), 219–220.

Sneller, F., Kalf, D., Weltje, L., & Van Wezel, A. (2000). Maximum permissible concentrations and negligible concentrations for rare earth elements (REEs). RIVM--601501011. National Institute of Public Health and the Environment, Bilthoven, Netherlands.

Sonesson, A. (1958). On the complex chemistry of the tervalent rare earth ions. *Acta Chemica Scandinavica*, *12*(2), 165–181.

Sovacool, B. K., Ali, S. H., Bazilian, M., Radley, B., Nemery, B., Okatz, J., & Mulvaney, D. (2020). Sustainable minerals and metals for a low-carbon future. *Science*, *367*(6473), 30–33.

Stewart, B. W., Capo, R. C., Hedin, B. C., & Hedin, R. S. (2017). Rare earth element resources in coal mine drainage and treatment precipitates in the Appalachian Basin, USA. *International Journal of Coal Geology*, *169*, 28–39.

Svoronos, D. R., Boulhassa, S., Guillaumont, R., & Quarton, M. (1981). Citric complexes and neodymium citrate: $NdCit \cdot 3H_2O$. *Journal of Inorganic and Nuclear Chemistry*, *43*(7), 1541–1545.

Takahashi, Y., Chatellier, X., Hattori, K. H., Kato, K., & Fortin, D. (2005). Adsorption of rare earth elements onto bacterial cell walls and its implication for REE sorption onto natural microbial mats. *Chemical Geology*, *219*(1–4), 53–67.

Takahashi, Y., Yamamoto, M., Yamamoto, Y., & Tanaka, K. (2010). EXAFS study on the cause of enrichment of heavy REEs on bacterial cell surfaces. *Geochimica et Cosmochimica Acta*, *74*(19), 5443–5462.

Talburt, D. E., & Johnson, G. T. (1967). Some effects of rare earth elements and yttrium on microbial growth. *Mycologia*, *59*, 493–503.

Texier, A. C., Andres, Y., Faur-Brasquet, C., & Le Cloirec, P. (2002). Fixed-bed study for lanthanide (La, Eu, Yb) ions removal from aqueous solutions by immobilized Pseudomonas aeruginosa: Experimental data and modelization. *Chemosphere*, *47*(3), 333–342.

Texier, A. C., Andres, Y., Illemassene, M., & Le Cloirec, P. (2000). Characterization of lanthanide ions binding sites in the cell wall of Pseudomonas aeruginosa. *Environmental Science & Technology*, *34*(4), 610–615.

Texier, A. C., Andres, Y., & Le Cloirec, P. (1999). Selective biosorption of lanthanide (La, Eu, Yb) ions by Pseudomonas aeruginosa. *Environmental Science & Technology*, *33*(3), 489–495.

Thompson, V. S., Gupta, M., Jin, H., Vahidi, E., Yim, M., Jindra, M. A., et al. (2018). Techno-economic and life cycle analysis for bioleaching rare-earth elements from waste materials. *ACS Sustainable Chemistry & Engineering*, *6*, 1602–1609. http://dx.doi.org/10.1021/acssuschemeng.7b02771

Tsuruta, T. (2006). Selective accumulation of light or heavy rare earth elements using gram-positive bacteria. *Colloids and Surfaces B: Biointerfaces*, *52*(2), 117–122.

Tunsu, C., Petranikova, M., Gergorić, M., Ekberg, C., & Retegan, T. (2015). Reclaiming rare earth elements from end-of-life products: A review of the perspectives for urban mining using hydrometallurgical unit operations. *Hydrometallurgy*, *156*, 239–258. http://www.sciencedirect.com/science/article/pii/S0304386X15300311

Van Gosen, B. S., Verplanck, P. L., Seal, R. R., II, Long, K. R., & Gambogi, J. (2017). *Rare-earth elements* (No. 1802-O). US Geological Survey.

Vera, M., Schippers, A., & Sand, W. (2013). Progress in bioleaching: Fundamentals and mechanisms of bacterial metal sulfide oxidation—Part A. *Applied Microbiology and Biotechnology*, *97*(17), 7529–7541.

Wang, P., Anderko, A., Springer, R., & Young, R. (2006). Modeling phase equilibria and speciation in mixed-solvent electrolyte systems: II. Liquid–liquid equilibria and properties of associating electrolyte solutions. *Journal of Molecular Liquids*, *125*(1), 37–44.

Wang, P., Anderko, A., & Young, R. D. (2002). A speciation-based model for mixed-solvent electrolyte systems. *Fluid Phase Equilibria*, *203*(1), 141–176.

Wang, P., Springer, R., Anderko, A., & Young, R. (2004). Modeling phase equilibria and speciation in mixed-solvent electrolyte systems. *Fluid Phase Equilibria*, *222*, 11–17.

Weber, R. J., & Reisman, D. J. (2012). Rare earth elements: A review of production, processing, recycling, and associated environmental issues. *US EPA Region 7*.

Wei, W., Liu, X., Sun, P., Wang, X., Zhu, H., Hong, M., et al. (2014). Simple whole-cell biodetection and bioremediation of heavy metals based on an engineered lead-specific operon. *Environmental Science & Technology*, *48*(6), 3363–3371.

Weltje, L., Verhoof, L. R. C. W., Verweij, W., & Hamers, T. (2004). Lutetium speciation and toxicity in a microbial bioassay: Testing the free-ion model for lanthanides. *Environmental Science & Technology*, *38*(24), 6597–6604. http://dx.doi.org/10.1021/es049916m

Wurm, M. (1951). The effect of lanthanum on growth and metabolism of Streptococcus faecalis R. *Journal of Biological Chemistry*, *192*(2), 707–714.

Xie, X., Yang, K., Lu, Y., Li, Y., Yan, J., Huang, J., et al. (2022). Broad-spectrum and effective rare earth enriching via Lanmodulin-displayed Yarrowia lipolytica. *Journal of Hazardous Materials*, *438*, 129561.

Xu, T. J., Ramanathan, T., & Ting, Y. P. (2014). Bioleaching of incineration fly ash by Aspergillus niger—Precipitation of metallic salt crystals and morphological alteration of the fungus. *Biotechnology Reports*, *3*, 8–14.

Xu, Z., Lei, Y., & Patel, J. (2010). Bioremediation of soluble heavy metals with recombinant Caulobacter crescentus. *Bioengineered Bugs*, *1*(3), 207–212.

Yoshida, T., Ozaki, T., Ohnuki, T., & Francis, A. J. (2004). Adsorption of rare earth elements by γ-Al2O3 and Pseudomonas fluorescens cells in the presence of desferrioxamine B: Implication of siderophores for the Ce anomaly. *Chemical Geology*, *212*(3–4), 239–246.

Yu, Q., & Fein, J. B. (2016). Sulfhydryl binding sites within bacterial extracellular polymeric substances. *Environmental Science & Technology*. http://dx.doi.org/10.1021/acs.est.6b00347

Zhang, L., Dong, H., Liu, Y., Bian, L., Wang, X., Zhou, Z., & Huang, Y. (2018). Bioleaching of rare earth elements from bastnaesite-bearing rock by actinobacteria. *Chemical Geology*, *483*, 544–557.

Zhang, W., Noble, A., Yang, X., & Honaker, R. (2020). A comprehensive review of rare earth elements recovery from coal-related materials. *Minerals*, *10*(5), 451.

Zhao, F., Cong, Z., Sun, H., & Ren, D. (2007). The geochemistry of rare earth elements (REE) in acid mine drainage from the Sitai coal mine, Shanxi Province, North China. *International Journal of Coal Geology*, *70*(1), 184–192.

Zhuang, W.-Q., Fitts, J. P., Ajo-Franklin, C. M., Maes, S., Alvarez-Cohen, L., & Hennebel, T. (2015). Recovery of critical metals using biometallurgy. *Current Opinion in Biotechnology*, *33*, 327–335.

9

ADSORPTION-BASED SEPARATION AND RECOVERY OF RARE EARTH ELEMENTS

Madhav Patel and Athanasios K. Karamalidis

Department of Energy and Mineral Engineering, Pennsylvania State University, State College, Pennsylvania, USA

Abstract

This chapter examines adsorption-based separation and recovery of rare earth elements (REEs) from aqueous solutions, covering a range of sorbents including ion exchangers, inorganic adsorbents, impregnated adsorbents, and ligand-grafted adsorbents. It discusses the diverse functional groups involved in REE adsorption and their roles in various separation techniques. The review assesses different ion-exchange resins and their efficacy in REE separation, and discusses novel inorganic adsorbents. Adsorbents known as extractant-immobilized materials synthesized by physically immobilizing extractant onto a solid support provide advantage over solvent extraction but suffer from stability issue. Surface-functionalized adsorbents with ligands chemically grafted onto solid support provide better stability and have been extensively researched for REE separation with silica- and carbon-based support being the major solid support for surface functionalization. Innovative approaches include using

Rare Earth Elements: Sustainable Recovery, Processing, and Purification, Special Publications 79, First Edition. Edited by Athanasios K. Karamalidis and Roderick Eggert.

DOI: 10.1002/9781119515005.ch9

magnetic nanomaterials for easy magnetic separation and ion-imprinted polymers for enhanced selectivity. Desorption typically involves mineral acids, though chromatographic separation with complexing ligands offers improved selectivity. While some studies have applied these adsorbents to real samples, there's a lack of examples demonstrating industrial-scale applications, and further research is needed to bridge this gap and advance practical REE-recovery solutions.

9.1. Introduction

This chapter briefly covers different technologies involved in the adsorption-based separation of rare earth elements (REEs), the 15 lanthanide elements of the periodic table with atomic numbers Z = 57–71, plus scandium (Z = 21) and yttrium (Z = 39). The separation technologies include a wide range of materials from ion-exchange resins (IERs), metal (hydro)oxide adsorbents, and solvent-impregnated resins (SIRs) that have the extractant impregnated into solid support to the recent technologies of surface-functionalized adsorbents with complexing ligand chemically anchored onto the solid support and ion-imprinted polymers with specific recognition sites for the desired REEs.

The adsorption sites in these adsorbents are various surface functional groups or ligands that can adsorb REEs through different mechanisms: physical adsorption, electrostatic interaction, and/or surface complexation. The adsorbent's behavior depends on the chemistry and chemical properties of these functional groups or ligands (discussed in Section 9.2). Additionally, the interaction of the surface sites and the REEs is controlled by the chemical properties of REEs, such as charge, ionic radius, coordination number, aqueous speciation, and the ability to form complexes with the ligand.

REEs occur dominantly in trivalent oxidation state, have variable coordination numbers (CN 8–12 (Cotton & Harrowfield, 2012)), and have similar ionic radii between adjacent REEs (which decrease with increasing atomic number among lanthanides) (Nash, 1993). They are considered hard acid cations and interact strongly with hard anions such as hydroxide, alkoxide, carbonates, and phosphates, and have strong complexes with organic ligands containing carboxylates and phosphonates (Johannesson et al., 1995; Noack et al., 2016; Pearson, 1963; Xie et al., 2014). More stable REE–ligand complexes are obtained with multidentate ligands such as amino-poly(carboxylic acids) due to the high coordination number of REEs with these ligands (Noack et al., 2016; Xie et al., 2014). These characteristics of REEs affect the extent, kinetics, and mechanism of the adsorption. In aqueous solutions, the pH of the solution controls the REE speciation. REEs are predominantly cationic REE(III) in an aqueous solution at acidic pH and REE hydroxides at alkaline pH (Callura, 2018; Ramasamy, Repo, et al., 2017). At high pH, the primary mechanism of REEs uptake can be the surface precipitation of REEs as hydroxides instead of adsorption (Dardenne et al., 2002; Farley et al., 1985; Iftekhar, Ramasamy, et al., 2018; Piasecki & Sverjensky, 2008).

A measure of the adsorbent's effectiveness for separation and its performance can be obtained by conducting batch or column adsorption experiments. In batch adsorption, the frequently used characteristics are equilibrium adsorption capacity (q_e) (Equation (1a)), maximum adsorption capacity calculated using Langmuir isotherm (q_m) (Equation (1e)), percentage adsorption (Equation (1b)), solid–liquid distribution (partitioning) coefficient (K_d) (Equation (1c)), and selectivity factor (SF) between elements A and B (Equation (1d)) in case of competitive adsorption or adsorption from a multi-element solution. These parameters are defined as follows:

$$q_e\left(\frac{mg}{g}\right) = (C_0 - C_e) \times V / m, \tag{1a}$$

$$Adsorption(\%) = \frac{C_0 - C_e}{C_0} \times 100, \tag{1b}$$

$$K_d\,(\mathrm{mL/g}) = \frac{C_0 - C_e}{C_e} \times \frac{V}{m}, \tag{1c}$$

$$\mathrm{SF(A/B)} = \frac{Kd_A}{Kd_B}, \tag{1d}$$

$$q_e\left(\frac{\mathrm{mg}}{\mathrm{g}}\right) = q_m \frac{K_L C_e}{1 + K_L C_e}, \tag{1e}$$

where C_o and C_e are initial and equilibrium concentrations of the adsorbate in the solution in batch adsorption (mg/L), m is the mass of the adsorbent used (g), V is the total volume of the solution (L), q is the adsorption (in mg/g), and K_L is Langmuir isotherm constant (L/mg).

9.2. Surface Functional Groups or Ligands

The functional groups on the surface of an ion-exchange resin or adsorbent are the active components that interact and bind the aqueous REEs species. These functional groups can be active molecules of the polymeric ion-exchange resins or ligands grafted on the solid support in functionalized adsorbents (shown in Figure 9.1) or the existing functional groups of a naturally occurring surface, such as –O/–OH on the surface of a mineral oxide/hydroxide. They are involved in the uptake of REEs from solution through electrostatic interaction or surface complexation. The different functional groups can be classified based on the heteroatom bonded to the adsorbent surface (Yang et al., 2019). The common groups are O-containing functional groups (e.g., –OH, –COOH, –C=O, and –C–O), S-containing functional groups (e.g., $-SO_3H$, C–S, C=S, and S=O), P-containing functional groups (e.g., $-PO_2H_2$, $-PO(OH)_2$), and N-containing functional groups (e.g., $-NH_2$, –NH, –C=N, and –C–N) (Figure 9.1). Additionally, ligands containing multiple types of functional groups are common, especially in chelating resins and functionalized adsorbents (Figure 9.1), resulting in wide variations in REE sorption chemistry (Hérès et al., 2018; Page et al., 2017; Trochimczuk & Alexandratos, 1994). Polyfunctional ligands can provide additional affinities (Alexandratos & Smith, 2004a). Also, the same functional group can have a higher or lower affinity for REEs based on its side group (Martell & Hancock, 1996; Zhu & Alexandratos, 2015).

The most common functional groups used in ion exchange and adsorption of REEs are carboxyl, found in both cation exchangers (Bezzina et al., 2018; Shu et al., 2007; Xiong et al., 2011; Xiong, 2008) and various functionalized adsorbents (Figure 9.1) (Ahmed, Lee, et al., 2019; Callura et al., 2018; Noack et al., 2016; Polido Legaria, Rocha, et al., 2017; Ravi, Zhang, et al., 2018). The mechanism of REE uptake by carboxyl-based adsorbents can be ion-exchange process in cation exchangers or surface complexation in the case of a chelating carboxyl-based surface ligand (e.g., complexation of REEs with diethylenetriaminepentaacetic acid (DTPA)-functionalized adsorbent (Noack et al., 2016)). Many commercially available carboxyl-based cation-exchange resins (CERs) have been used for REE extraction (Bezzina et al., 2018; Shu et al., 2007; Xiong et al., 2011; Xiong, 2008). Since carboxyl is a weak acid, its behavior is highly dependent on pH.

Sulfonic functional groups are the active component of the strongly acidic cation-exchange resins. The affinity of the counter ion for the ion exchanger depends on its charge and hydration radius (Harris & Lucy, 2015). REEs exist as triply charged positive ions in acidic aqueous

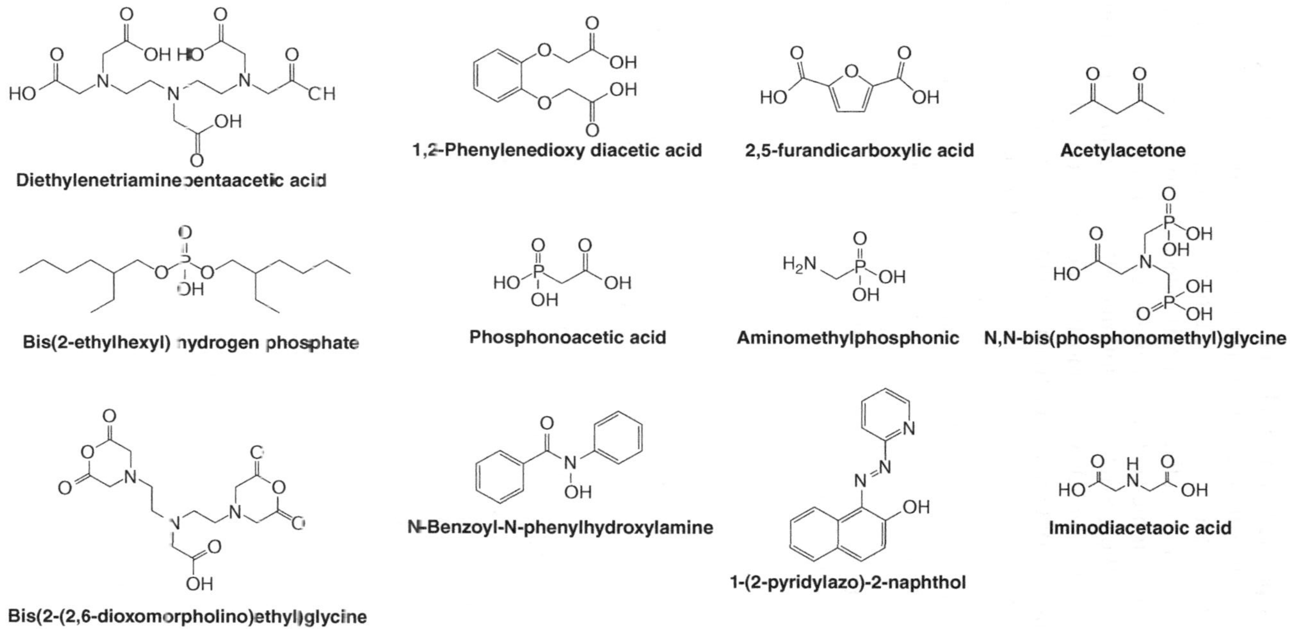

Figure 9.1 Different ligands used for REE adsorption in chelating ion-exchange resins or grafted onto solid supports.

solutions and hence have a high affinity to negatively charged sulfonic ions and can be separated from singly and doubly charged metal cations using interaction with sulfonic anion (Fritz & Garralda, 1963; Korkisch et al., 1967; Page et al., 2019; Strelow, 1960; Strelow et al., 1965). However, due to similar radii of adjacent Ln(III), selectivity for a particular Ln(III) is limited with sulfonic-acid cation exchangers (Boyd, 1978; Strelow, 1960; Strelow et al., 1965; Strelow & Bothma, 1964). The sulfonic group is a strong acid; hence, its behavior is mostly independent of pH.

Phosphorus-containing functional groups used for REE sorption are organophosphoric-acid-, phosphonic-acid-($-PO(OH)_2$)-, and phosphonic-ester-$PO(OR)_2$-based derivatives (Egawa et al., 1994; Ihara et al., 2001; Miklishanskii et al., 1968; Page et al., 2017). The ion-exchange-resin-containing phosphonic-acid-based groups are considered weakly acidic cation-exchange resins similar to carboxyl-based ion-exchange resins and are dependent on pH. Additionally, P-based chelating ligands have been used as functional groups in different chelating resin (Callura et al., 2019; Hérès et al., 2018) and functionalized adsorbents (Figure 9.1) (Artiushenko, Kostenko, et al., 2020; Bertelsen et al., 2019; Callura et al., 2018; de Decker et al., 2016; Kavun et al., 2021; Noack et al., 2016; Ravi, Lee, et al., 2018; Wang et al., 2002; Zheng et al., 2020).

Most common among N-containing functional groups is amine-based compounds. The REE sorption mechanism for amino-based sorbent is an anion-exchange process. These amino-based sorbents can be used to separate Th(IV) and U(IV), which form strong anionic compounds with nitrate and phosphate (Jackson, 1954). Among REEs, Ce(IV) can form anionic nitrates, which can be separated from other REEs using amino-based resins or adsorbents (Zhu & Chen, 2011). Moreover, N-based ligands are used for surface complexation of REEs in chelating resin (Niu et al., 2021; Page et al., 2019) and other adsorbents (Figure 9.1) (de Decker et al., 2016; Lee et al., 2019; Ramasamy, Puhakka, Repo, et al., 2018; Ramasamy, Repo, et al., 2017).

In naturally occurring oxidic adsorbents such as ferric hydroxide or silica oxide, the surface OH groups undergo protonation–deprotonation and produce surface charges, which lead to electrostatic attraction of cations and anions present in the solution. They can electrostatically attract REE cations, resulting in the adsorption of REEs.

9.3. Ion-Exchange Resins

Ion exchange in the context of solid-phase extraction (SPE) involves a reversible interchange of ions between a solid phase (or ion exchanger) and the solution through electrostatic sorption and desorption to and from the ion-exchange material. The ion exchangers are classified as cation or anion exchangers based on the mobile exchangeable ions (i.e., counterions). Acidic cation exchangers, introduced in 1944 by cross-linking polystyrene with 6%–8% divinylbenzene, have cations as counterions and attract and bind cations, while anion exchangers, introduced later in 1948, can attract and bind anions from solution. Based on their physical form, ion exchangers can be classified as membranes, fibers, hydrogels, or resins. Ion-exchange resins (IERs), which represent the major class of ion exchangers, are insoluble polymers with active-ion groups, commercially available as spherical beads or particles, and have been used to separate REEs mainly in chromatographic setup (Ehrlich & Lisichkin, 2017).

9.3.1. Cation-Exchange Resins

Strongly acidic cation-exchange resins: Strongly acidic cation-exchange resins are formed by copolymerization of polystyrene and divinylbenzene and have sulfonic (SO_3^-) groups introduced to the benzene rings. There are many sulfonic (SO_3^-) group-based resins available under various trade names (e.g., Amberlite IR-120, Dowex 50WX8), having different grain sizes, pore diameters, sorption capacities, degrees of cross-linking, and other characteristics (Ehrlich & Lisichkin, 2017). Sulfonic-acid-based cation exchangers have been used to separate REEs but are not very selective for a specific REE over the other.

In CERs, the strength of cation binding depends on the charge and radius of the hydrated ion (Harris & Lucy, 2015). The electrochemical attraction increases with decreasing ionic radius for ions with the same charge. Within Ln(III), the ionic radius decreases with the increasing atomic number from La(III) (ionic radius: 103 pm) to Lu(III) (ionic radius: 86.1 pm), and thus the binding strength to the cation exchanger increases with atomic number. However, the use of strongly acidic cation-exchange resins for the separation of individual REEs in mineral acids, such as HCl, H_2SO_4, and HNO_3, is not selective due to similar ionic radius of adjacent Ln(III) (Boyd, 1978; Strelow, 1960; Strelow et al., 1965; Strelow & Bothma, 1964). The similar ionic radii of neighboring Ln(III) ions lead to analogous electrostatic attraction, as indicated by the similar equilibrium constants observed in an ion-exchange reaction with a strongly acidic cation exchanger (Equation (2)). The ratio of the ion-exchange equilibrium constants of adjacent Ln(III) for a reaction similar to (Equation (2)) does not exceed 1.1 (Boyd, 1978):

$$Ce^{3+} + LnR_3 = CeR_3 + Ln^{3+} \qquad (2)$$

This limits the use of strongly acidic cation-exchange resins in individual REE separation from a solution containing a mix of different REEs. Still, these resins can be used for the separation of Ln(III) from other cations of lower or higher ionic charge (Fritz & Garralda, 1963; Korkisch et al., 1967; Page et al., 2019; Strelow, 1960; Strelow et al., 1965).

The separation of individual REEs increased with the use of complexing ligands as eluents for strongly acidic cation exchangers (Boyd, 1978; Chen et al., 2017; Spedding et al., 1947, 1954; Strelow & Victor, 1990). The REE ions form complexes with the ligand. These REE–ligand complexes have less affinity for the exchanger than corresponding REE ions, resulting in easier desorption of REE–ligand complex and a difference in relative REE desorption based on the stability of individual REE–ligand complex. With the same ligand, the relative concentration of different REE–ligand complexes depends on the ratio of their REE–ligand complex's stability constants. Since the ratio of stability constants of REE–ligand complexes (Karraker, 1961; Mackey, 1960; Schoeb, 1965; Suzuki et al., 1980; Wheelwright et al., 1953; Wood, 1993) of neighbor REEs is significantly higher than the ratio of ion-exchange constants (Boyd, 1978), better separation between REEs is obtained during elution (Chen et al., 2017). The stability of REE–ligand complexes increases with the increasing atomic number for most complexing ligands, which leads to an order of elution from Lu(III) to Ln(III) with complexing ligands as eluent (Karraker, 1961; Mackey, 1960; Schoeb, 1965; Suzuki et al., 1980; Wheelwright et al., 1953; Wood, 1993). The different complexing ligands used as eluent for REEs chromatographic separation using strong acidic cation exchangers are discussed later in the chapter (Section 9.8.2).

The sulfonic cation exchangers with complexing ligand eluents have been mainly used in the chromatographic separation of REEs (Ehrlich & Lisichkin, 2017). However, alternatively, in a batch adsorption process, Khawassek et al. (2019) used macroreticular sulfonic acid resin Dowex 50WX8 to adsorb REEs from effluents solution collected from the hydrometallurgical process for El-Erediya mineralization. The maximum total REE uptake from the leachate solution was 80 mg REEs/g. Similarly, strong acid resins Amberlite 200C Na and Amberlite 200C H showed high sorption efficiency (more than 99.8%) for La(III), Ce(III), and Nd(III) (Kołodyńska et al., 2019). Page et al. (2019) used strong acidic cation-exchange resin, Lewatit MonoPlus SP 112, for La(III), Sm(III), and Er(III) adsorption in sulfate and chloride media in the presence of competitive ions, such as Fe(III) and Al(III), and reported maximum adsorption capacity of 42 mg/g for Sm(III). The equilibrium constants for the ion-exchange reactions and the REE selectivities were higher in the chloride medium than in the sulfate medium. The ion-exchange equilibrium constants for competitive ions, Fe(III) and Al(III), were significantly lower in both media. A macroporous strongly acidic CER, SQS-6, adsorbed 97.6% La(III) and 90.36% Nd(III) in 4.0 M H_3PO_4 media with equilibrium capacities of 13.8 and 12.70 mg/g (in 8 M H_3PO_4), respectively (Abu Elgoud et al., 2019).

Carboxylic-acid-based cation-exchange resins: Carboxylic-acid-based cation-exchange resins can form complexes with REEs in addition to electrostatic interaction (Arnold & Hing, 1967; Kazantsev et al., 1974). The carboxylic-functionalized cation exchangers involve copolymerization of methacrylic acid or acrylic, or acrylonitrile, generally, with divinylbenzene. The type of carboxylic group on the cation exchanger affects the adsorption of REEs; for example, polyacrylic acid resin (Amberlite XE 89) showed lower selectivity for lanthanide than the polymethacrylic acid resin (Amberlite IRC 50) (Arnold & Hing, 1967).

Many macroporous carboxylic cation-exchange resins such as D113, D155, D152, D151, and SQD-85 have been used for REE adsorption in acetate medium. The macroporous carboxylic acid resin D113 was used for adsorption of La(III) (q_m = 273.3 mg/g (Shu et al., 2007)), Dy(III) (q_m = 292.7 mg/g (Wang & Gao, 2007)), Nd(III) (q_m = 232.56 mg/g (Xiong et al., 2011)), and Er(III) (q_m = 250 mg/g (Xiong et al., 2009)). The separation coefficients for La (III) adsorption onto D113 were 2.29, 3.64, 4.27, and 0.627 against Ce(III), Gd(III), Er(III), and Y(III), respectively (Shu et al., 2007). With another carboxylic acid resin, D152, the maximum adsorption capacities at pH 6.70 for different REEs were between 238 mg/g (for Er(III)) and 510 mg/g (for Sm(III)) (Xiong et al., 2008). Similarly, D155 was used for Ce(III) (q_m = 294 mg/g) and Gd(III) (q_m = 283 mg/g) (Xiong, 2008), SQD-85 for Yb(III) (q_m = 347.6 mg/g) (Xiong et al., 2011), D151 for Ce(III) (q_m = 392 mg/g) (Yao, 2010), and a gel-type weak acid resin (110) for Yb(III) (q_m = 265.8 mg/g) (Zheng & Xiong, 2011). Commercial resins Amberlite IRC86 (functional group: carboxylic acid) and Purolite S910 (functional group – amidoxime) were used in various buffer solutions (i.e., malic acid, formic acid, acetic acid, alanine, and lactic acid) for adsorption of REEs in the presence of competitive Fe(III) and Al(III) ions (Bezzina et al., 2018). The use of a carboxylate-buffered system allowed easy separation of REEs from Fe(III) and Al(III) ions. The carboxylic resin IRC86 showed a higher capacity for HREEs in acetic acid media at pH approximately 4.4, showing a preference order of Y(III) > Er(III) > Sm(III) > La(III).

The limitation of using carboxylic cation-exchange resins is that these resins operate in the weak acidic pH range (pH > 3.5), posing additional challenges due to hydrolysis of REE ions and occurrence of other competitive cations such as Fe(III) (Ehrlich & Lisichkin, 2017; Kazantsev et al., 1974; Wood, 1993).

P-containing cation-exchange resins: The phosphorus-based ligand-containing resins are considered part of the family of weakly acidic cation exchangers. Among different phosphorus-containing functional groups, phosphinic showed higher selectivity than phosphonic, with both showing higher selectivity than their respective oxidized derivatives (Egawa et al., 1994). Their Ln(III) distribution coefficient increased with increasing atomic number in general but with a plateau from Sm(III) to Ho(III). Macroreticular methylenephosphonic acid resin showed affinity series of Fe (III) ~ U(VI) ~ Mo(VI) > Bi(III) > AI(III) > Gd(III) > La(III) ~ V(V) > Pb(II) > Cd(II) > Cu(II) ≈ Ca(II) ≈ Ba(II) ≈ Zn(II) > Mg(II) ≈ Co(II) ≈ Ni(II) (Jyo et al., 1997); whereas macroreticular phosphonic acid resin had affinity series of Mo(VI) ≈ Fe(III) ≈ U(VI) > Bi(III) ≈ Lu(III) > Al(III) ≈ Gd(III) > La(III) > Cr(III) > Pb(II) > Mn(II) ≈ Cd(II) ≈ Cu(II) > Ca(II) ≈ Co(II) ≈ Zn(II) ≈ Ba(II) ≈ Sr(II) > Ni(II) ≈ Mg(II) (Ihara et al., 2001).

Many different P-based cation-exchange resins used for REE adsorption involve different side groups with phosphorous. The P-based resin containing P(R)(O)OH (R = Et, Bu) and –P(O)(OR)OH (R = Et, Pr, Bu) were used for chromatographic separation of La, Pm, and Eu with increasing K_d values with increase in atomic number (Miklishanskii et al., 1968). Among different adsorbents with phosphorus-containing ligands, the strength of metal ion binding is affected by electron density on the oxygen atom, the spacer connecting the phosphorous-containing ligand to the polymer matrix, and the hydrophilicity of its environment (Alexandratos & Hussain, 1998; Alexandratos & Zhu, 2005, 2008, 2015; Trochimczuk & Alexandratos, 1994; Zhu & Alexandratos, 2014, 2015). The affinity for REEs increased with the introduction of hydroxyl or polyol into the spacer due to their stronger interaction with REEs (Alexandratos & Zhu, 2008).

Adsorbents containing difunctional and polyfunctional groups in addition to P-based groups have been studied for REE separation (Alexandratos & Hussain, 1995; Alexandratos & Natesan, 1999; Alexandratos & Smith, 2004b; Shumilova et al., 2012; Trochimczuk, 2000). The addition of a second functional group can enhance the behavior of the adsorbent, as in phosphorylated adsorbents, where the introduction of highly hydrophilic sulfonic acid eliminated intra-ligand hydrogen bonding, which reduced intra-ligand cooperation and increased the accessibility of P-containing groups (Alexandratos & Smith, 2004b). Additionally, the sulfonic acid itself has an affinity for REEs, which can increase REE adsorption (Ehrlich & Lisichkin, 2017). Bonding sulfonic acid groups to the phosphinated polystyrene matrix increased the Eu(III) distribution coefficient from 61.8 to 220.5 in a 1 N HNO_3 solution (Alexandratos & Hussain, 1995). Hérès et al. (2018) used Monophos and Diphonix resins, which had sulfonic groups together with alkylphosphonic groups for adsorption of REEs. The extraction efficiency of Monophos and Diphonix resins increased with increasing ionic radius of REEs.

9.3.2. Anion-Exchange Resins

REE ions do not form anions with inorganic ligands, thus are poorly adsorbed on the anion-exchange resins (AERs) (Ehrlich & Lisichkin, 2017; Koodynska & Hubicki, 2012; Minczewski et al., 1982). Still, Th(IV) and U(VI) can be separated from solution containing REEs using AERs, as they form anion complexes with nitrate and phosphate (Ehrlich & Lisichkin, 2017; Koodynska & Hubicki, 2012). Among REEs, cerium(IV) can form negatively charged complexes with nitrate and can be adsorbed to the AERs, for example, anion exchanger containing N-methyl-imidazolium group (Zhu & Chen, 2011).

Anionic REE–ligand complexes can be formed by adding organic compounds (e.g., methanol (Faris & Warton, 1962), ethanol (Hubicki & Olszak, 1998), and propanol (Hubicki & Olszak, 2002)) to nitric acid. These anionic complexes can be separated using a strongly basic anion-exchange resin. Anionic REE–ligand complex electrostatically binds to the positive anion exchanger. Similarly, complexing ligands such as ethylenediaminetetraacetic acid (EDTA) and cyclohexane-1,2-diaminetetraacetic acid (CDTA) were used to form anionic REE–ligand complex for REE separation using anion-exchange resins (Hubicka & Hubicki, 1986; Hubicka & Kołodyńska, 2004, 2008; Minczewski & Dybczyński, 1962; Wódkiewicz & Dybczyński, 1968). Binary mixtures of Y–Nd, Sm–Tb, La–Nd, and Dy Er were separated using frontal chromatography with EDTA as eluent using different Dowex anion exchangers (Hubicka & Hubicki, 1986). Strongly basic anion resins Dowex 1 and Dowex 2 were used to separate the REE complex with N-hydroxyethylenediaminetriacetic acid (HEDTA) and with iminodiacetic acid (IDA) (Hubicka & Drobek, 1997, 1998, 1999). Similarly, REE complexes with trans-1,2-diaminocyclohexane-N,N,N′,N′-tetraacetic acid (DCTA) of the $Ln(dcta)^-$ type were chromatographically separated using different polyacrylate anion exchangers (Amberlite IRA 68, Amberlite IRA 458, and Amberlite IRA 958) and different polystyrene anion exchangers (Lewatit MonoPlus M 500, Lewatit MonoPlus M 600, Lewatit MonoPlus MP 64, Lewatit MonoPlus MP 500, and Lewatit MP 62) (Hubicka & Kołodyńska, 2004, 2008).

9.3.3. Chelating Ion-Exchange Resins

Chelating ion-exchange resins are equipped with ligands that form complexes with metal ions through their functional groups (Garg et al., 1999). Chelating ion-exchange resins used for REEs separation contain iminodiacetate, aminomethylphosphonic, hydroxamine, and diamide (Ehrlich & Lisichkin, 2017; Koodynska & Hubicki, 2012). Many commercially available iminodiacetate-based resins such as Amberlite IRC 718, Chelex 100, Dowex A-1 (Mathur & Khopkar, 1985; Schrobilgen & Lang, 1968), Duolite ES466, Lewatit TP 207 (Botelho Junior et al., 2021; Niu et al., 2021), and Purolite S930 Plus (Page et al., 2019) have been used for REEs separation (Ehrlich & Lisichkin, 2017; Koodynska & Hubicki, 2012).

Iminodiacetic acid resins are also characterized by a very high selectivity for REEs over the alkali and alkaline earth metals, often present in high concentrations in alternative REE resources (Ehrlich & Lisichkin, 2017; Koodynska & Hubicki, 2012; Page et al., 2017, 2019). An iminodiacetic acid resin, Purolite S930 Plus, showed higher selectivity for REEs in the presence of high Na and Ca concentrations than the sulfonic acid resin (SA), sulfonic–phosphonic resin (Purolite S957), and aminophosphonic resin (Purolite S950) (Page et al., 2017). However, the SA resin, Purolite S957, and Purolite S950 were tolerant toward REEs adsorption from acid solutions up to 0.5 M $[H^+]$, whereas iminodiacetic acid resins effectively adsorbed REEs only at $[H^+] < 0.001$ M. For the sulfonic acid resin, the selectivity was REE $\simeq$ Th > Fe $\simeq$ Al, whereas for other resins, it was Th $\simeq$ Fe $\gg$ REE $\simeq$ Al. In another study with Purolite S930 Plus, the resin selectively adsorbed La(III) over Mg(II) with a selectivity factor of 3,046 in 1 M $MgCl_2$, but the selectivity factor decreased to 453 and 240 in 0.5 M $MgCl_2$ + 0.5 M $MgSO_4$ and 1 M $MgSO_4$ medium, respectively (Page et al., 2019). Another iminodiacetic-acid-based resin, TP207, showed an adsorption capacity order of La(III) > Eu(III) > Y(III), with a capacity of 0.73 mmol/g (or 101.5 mg/g) for La(III) at pH 4.0 (Niu et al., 2021).

The REE extraction efficiency using different aminophosphonic resins (Tulsion CH-93, Purolite S940, Amberlite IRC-747, Lewatit TP-260) decreased with an increase in Z/IR (atomic number/ionic radius) from Sc up to a plateau of La, Nd, and Gd, and increased after that with Z/IR from Dy to Yb (Hérès et al., 2018). Concurrently, the extraction efficiency decreased with an increase in IR in general, with a slight spike at Nd(III) and Pr(III). Another chelating ligand diglycolamic-acid-containing polymeric adsorbent showed adsorption capacity (q_m) of 0.113 mmol Dy/g (or 18.3 mg/g) and selectively adsorbed REEs at low pH values (Shinozaki et al., 2018). Other chelating ligands diethylenetriaminepentaacetic dianhydride

(DTPADA), phosphonoacetic acid (PAA), and N,N-bis(phosphonomethyl) glycine (BPG) containing polymers had maximum adsorption capacities of 2.9, 5.0, and 3.0 mg/g, respectively, for a mixture of three REEs (Nd(III), Gd(III), and Ho(III) (Callura et al., 2019, 2021)). N-based chelating resin Dowex M 4195 with bis(2-pyridylmethyl)amine showed very low sorption (maximum 3.1%) for La(III), Ce(III), and Nd(III) (Kołodyńska et al., 2019).

9.4. Inorganic Adsorbents

Inorganic adsorbents used for REE adsorption include mineral oxides and hydroxides, clay minerals, and nanomaterials. Mineral oxides and hydroxides are adsorbents with reactive sites containing oxygens (O^{2-}) or hydroxides (–OH) that interact with cations and anions. The mineral adsorbents used for REE adsorption are silica, alumina, iron (hydro) oxides, Mn oxides, and TiO_2.

Silica has been studied as an adsorbent for Gd(III), Y(III) (Kosmulski, 1997b), and Yb(III) (Marmier et al., 1999). These investigations focused on surface complexation modeling and percentage uptake and did not report maximum uptake in mg/g, distribution (partitioning) coefficient, or separation factor. Alumina was studied for adsorption of Gd(III), Y(III) (Kosmulski, 1997a), Yb(III) (Marmier et al., 1997), and Eu(III) (Wang et al., 2006). TiO_2 was studied for the adsorption of Y(III) (Zhang et al., 2004), Nd(III) (Ridley et al., 2005; Zhang et al., 2004), and Eu(III) (Tan et al., 2009). The q_m for Eu(III) was 1.5 mg/g with TiO_2 and 2.7 mg/g in the presence of fulvic acid (Tan et al., 2009).

Other studies involving inorganic adsorbents are Yb(III) adsorption on hematite (Marmier et al., 1997), La(III) onto hematite (Marmier & Fromage, 1999), REEs onto Mn oxide, Fe oxy-hydroxide (Davranche et al., 2004, 2005; Ohta & Kawabe, 2001), and REEs on amorphous ferric hydroxide (Quinn et al., 2006, 2007). In addition to oxides and hydroxides, other minerals have been used as adsorbents as well. Adsorption of Nd(III) onto kaolinite (Aja, 1998), Ln(III) onto kaolinite and smectite (Coppin et al., 2002), Eu(III) onto montmorillonite (Kowal-Fouchard et al., 2004), REEs on kaolin (Xiao et al., 2016), REEs adsorption onto kaolinite (Feng et al., 2021), and Eu(III) on halloysite and kaolinite (Zhou et al., 2022) are such studies involving inorganic mineral adsorbents. Kaolin showed saturated adsorption capacities of 1.731, 1.587, and 0.971 mg/g for La(III), Nd(III), and Y(III), respectively (Xiao et al., 2016). Mosai et al. (2019) used natural zeolite and bentonite to recover REEs from aqueous solution and reported q_m values of 0.09–0.24 mg/g and 1.07–2.87 mg/g for adsorption of different REEs onto zeolite and bentonite, respectively.

Most of the research mentioned in the earlier text focuses on understanding the mechanism of REE adsorption onto the inorganic adsorbent and uses surface complexation models. The adsorption mechanism is required for a better understanding of natural and engineered systems. However, since the adsorption capacity of these unmodified adsorbents is low (i.e., around 0.01–0.1 mmol/g) (Ehrlich & Lisichkin, 2017), it is impractical to employ these adsorbents for REE separation at an industrial scale.

Nonetheless, some recent studies involving novel inorganic adsorbents show promising results. Multiple zirconium organophosphonates (ZrP) with different Zr:P ratios were prepared using low-temperature hydrothermal reactions of zirconium propoxide with aminotris(methylene phosphonic acid) (ATMP). The adsorbents showed selectivity for lanthanide elements over monovalent and divalent metal ions. The sorption capacities of 37, 56, and 60 mg Eu/g were reported for ZrP-0.65, ZrP-0.71, and ZrP-0.76 (legend 32, Figure 9.2), respectively (Veliscek-Carolan et al., 2014). In another study, Attallah et al. (2016) synthesized a number of aluminum silicotitanates (ATS 1–31) and obtained 99% La(III) and Eu(III) adsorption. Inorganic titanium phosphate showed Sc(III) uptake capacities of 26.1, 8.25, and 3.3 mg/g for amorphous TiP, α-TiP, and γ-TiP, respectively. In chromatographic separation using amorphous TiP, almost all Sc(III) were eluted using a mixed acid comprising nitric acid and phosphoric acid, with Sc(III)/Fe(II) and Sc(III)/Al(III) enrichment factor of 9 and 265, respectively (Zhang, Koivula, et al., 2017). Titanium phosphate prepared from acidic wastewater of titanium dioxide achieved a maximum Ce(III) adsorption capacity (q_m) of 158.0 mg/g and was selective for Ce in the presence of Na, K, Ca, and Mg ions (Liu et al., 2021). Another inorganic adsorbent, layered titanium (IV) butyl phosphate (TiP), showed Dy(III) uptake of more than 25 mg/g and a maximum separation factor of 2,065 for Dy(III)–Nd(III), 11.5 for Dy(III)–Tb(III), and around 25 for La(III)–Ce(III) in respective binary element solutions (Zhang et al., 2018). Adsorptions conducted at different pH resulted in selectivity for different Ln(III) with the highest distribution coefficient (K_d) of around 10^5 mL/g (legends 17–20, Figure 9.2). The adsorbent TiP showed selectivity for Yb(III) and Lu(III) at pH 2.6. The distribution coefficient (K_d) of different REEs measures the selectivity of the adsorbent for a particular REE in competitive adsorption (Figure 9.2).

With another inorganic adsorbent, amorphous zirconium phosphate (am-ZrP), maximum adsorption of 27.5 mg/g and 38.3 mg/g for Nd and Dy were obtained (Xu et al., 2018). The SF for Dy(III)/Co(II), Nd(III)/Co(II), and Dy(III)/Nd(III) were 1,811, 958, and 1.9, respectively (Xu et al., 2018). A microporous titanosilicate, ETS 10, had adsorption capacities (q_m) of 48.8, 68.1, 99.0, 124.7, and 122.7 mg/g, respectively, for Y(III),

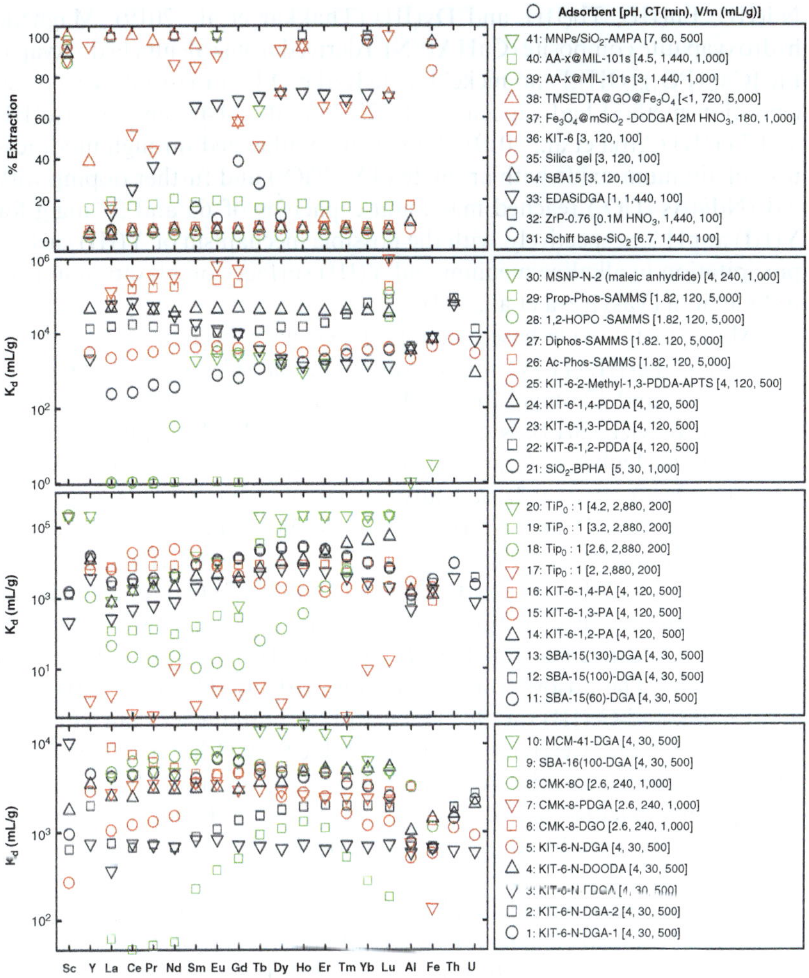

Figure 9.2 Distribution coefficient (K_d) for various adsorbents. The legends include adsorbent with the experimental condition (pH, contact time (CT), and V/m ratio (volume of solution/mass of adsorbent)). Source for data: adsorbents 1–2 (Florek et al., 2014), 3–5 (Florek et al., 2015), 6–8 (Perreault et al., 2017), 9–13 (Juère et al., 2016), 14–16 (Hu et al., 2017), 17–20 (pH is equilibrium pH) (Zhang et al., 2018), 21 (Artiushenko, Ávila, et al., 2020), 22–25 (Hu et al., 2019), 26–29 (Yantasee et al., 2009), 30 (Zheng et al., 2015), 31 (Amarasekara et al., 2009), 32 (Veliscek-Carolan et al., 2014), 33 (Ogata et al., 2015b), 34–36 (Giret et al., 2018), 37 (Li et al., 2018), 38 (Barrak et al., 2019), 39–40 (Lou et al., 2019), and 41 (Kostenko et al., 2019).

Nd(III), Eu(III), Tb(III), and Dy(III) (Thakkar et al., 2019). Magnetic hydroxyapatite composite (CaHAP/NF) derived from calcium hydroxyapatite [$Ca_{10}(PO_4)_6(OH)_2$] and nickel ferrite [$NiFe_2O_4$] synthesized by a coprecipitation method had capacities of 137.35 and 130.43 mg/g for Eu(III) and Tb(III) (Attia et al., 2019). Adsorbent synthesized through modification of titanium oxide with arsenate (4As–TiO_2) and further doping with Nd (Nd/4As–TiO_2) reached maximum capacities of 65 and 37 mg/g for Y(III) at pH 9, respectively, with the possible mechanism of Y(III) surface precipitation in alkaline medium and Y(III) surface complexation in neutral medium (Vasylyeva et al., 2021).

Many metal-oxide-based nanoparticles have been tested for REE adsorption. Fe(III)–Ti(IV) mixed oxide nanoparticles showed capacities of 229, 181, and 126 mg/g for Ce(III), Nd(III), and Gd(III), respectively (Metwally & Rizk, 2014). Nano maghemite showed Y(III) uptake capacity of 13.5 mg/g (Dubey & Grandhi, 2016). A magnetic nanoparticle (MNP) $Co_xMn_1{-}_xFe_2O_4$ (x = 0.2) had adsorption capacities of 125 mg/g and 25 mg/g for La(III) and Ce(III), while $Co_xMn_1{-}_xFe_2O_4$ (x = 0.8) had adsorption capacities of 189 mg/g and 294 mg/g for La(III) and Ce(III) (Ghobadi et al., 2017). In another study, MNP $MnFe_2O_4$ prepared via coprecipitation and further modified with graphene oxide ($MnFe_2O_4$-GO) showed high capacities for La(III) and Ce(III) adsorption (Ghobadi et al., 2018). The adsorption capacity of $MnFe_2O_4$-GO for La(III) was 1,001 mg/g, and Ce(III) was 982 mg/g at pH 7, both higher than $MnFe_2O_4$ (La(III) = 785 and Ce(III) = 770 mg/g) (Table 9.1). Other magnetic nano-$MnFe_2O_4$ composites with Al_2O_3 and SiO_2-chitosan showed similar high adsorption capacities (Liu et al., 2021) (Table 9.1). Other high-capacity REE adsorbents are listed in Table 9.1 (adsorbents are discussed in their respective sections).

Magnetic nanoparticle $CuFe_2O_4$ showed adsorption capacities of 51.02, 42.02, and 40.16 mg/g for Nd(III), La(III), and Ce(III), respectively (Tu & Johnston, 2018). Marwani et al. (2018) synthesized CeO_2–CdO nanomaterial and used it for adsorption of Y(III) from a solution containing Co(II), Cu(II), Cr(III), Cr(VI), Pb(II), W(VI), Y(III), and Zn(II). The distribution coefficient (K_d) for Y(III) reported was 17,743.90 mL/g with a capacity (q_m) of 84.67 mg/g. Other nano-adsorbents, nano-thorium (IV) oxide and nano-zirconium oxide, showed adsorption capacities of 10.5 and 18.0 mg/g for Y (III), respectively (Dubey & Grandhi, 2019). High adsorption capacities of 1,111 mg/g for La(III) and 1,000 mg/g for Ce(III) were obtained at pH 5 with transition metal oxide (TMO) nanowires synthesized via hydrothermal processing of a metal mix (Co, Fe, and Ni) followed by pyrolysis (Sarmadi et al., 2021).

Table 9.1 Some adsorbents with high adsorption capacity (q_m > 350 mg/g) for various REEs (III).

Adsorbent	Type of adsorbent	q_m (mg/g) for REE(III)	pH	Reference
D152	Carboxyl acid-CER	La: 486, Pr: 373, Nd: 453, Sm: 510, Gd: 391, Dy: 413, Lu: 401, Yb: 393, other REE > 238	6.7	(Xiong et al., 2008)
D151	Carboxyl acid-CER	Ce: 392	6.5	(Yao, 2010)
$MnFe_2O_4$	Inorganic adsorbent	La: 785, Ce: 770	7	(Ghobadi et al., 2018)
$MnFe_2O_4$-GO	Inorganic adsorbent/GO based	La: 1,001, Ce: 982	7	(Ghobadi et al., 2018)
TMO-nanowire	Inorganic adsorbent	La: 1,111, Ce: 1,000	5	(Sarmadi et al., 2021)
MnF_2O_4–Al_2O_3	Inorganic adsorbent	La: 885, Ce: 879	7	(Liu et al., 2021)
$MnFe_2O_4$@SiO_2-CS	Composite adsorbent	La: 1,030, Ce: 1,020	7	(Liu et al., 2021)
SH-MCM-41	OMS based	La: 560.56, Gd: 467.60, Yb: 546.68	5, 6 (Gd)	(Li et al., 2019)
APHMSN	Functionalized mesoporous silica	Gd: 387.3	5	(Yin et al., 2020)
Fe_3O_4/MnO_2/rGO	GO based	La: 1,016, Ce: 981	7	(Liu et al., 2020)
GO@TiP-Sponge	GO based	Dy: 576.17	Not clear	(Peng et al., 2020)
CNT/GO hybrid hydrosol	CNT-GO based	Gd: 534.76	5.9	(Guo et al., 2018)
ZIF-8	MOF	Dy: 430.4	7	(Abdel-Magied et al., 2019)
ZIF-8	MOF	La: 385	Not clear	(Jiang et al., 2016)
HKUST	MOF	Ce: 353	6	(Zhao et al., 2019)
ZnGA	MOF	Y: 377.02	6	(Mahmoud et al., 2019)
Mg-Fe-LDH-Cyanex272	Metal based	La: 480.8	1	(Gasser & Aly, 2013)
CMC-g-PAC hydrogel	Composite adsorbent	La: 381.72	5.58	(Zhu et al., 2016)

CER: cation-exchange resin, TMO: transition metal oxide, GO: graphene oxide, OMS: ordered mesoporous silica, APHMSN: amino-phosphonic functionalized hollow mesoporous silica nanoparticle, CNT: carbon nanotube, MOF: metal-organic framework, LDH: layered double hydroxide, CMC-g-PAC: carboxymethylcellulose grafted polyacrylic acid.

9.5. Extractant Immobilized Materials

Extractant immobilized materials (EIMs) are a class of adsorbents in which the active component or extractant used to bind the target analyte separating it from its original medium is physically immobilized onto a solid support (Hidayah & Abidin, 2017). The active component or the extractant is a compound capable of ionic or complex interaction with the target metal ions. The active component is attached to the solid support's surface by physical adsorption, or it resides in a thin solvent layer covering the support surface, or it can be part of the bulk solvent filling the pores of the solid support. In literature, solvent-impregnated resin (SIR) is used to refer to the adsorbents obtained when solvent extractants such as N,N-dimethylformamide (DMF), N,N,N′,N′-tetraoctyl diglycolamide (TODGA), and Cyanex-923 are infused onto solid support like silica or resin (Kabay et al., 2010; Sun et al., 2009; Zhu & Chen, 2011). Another similar term, supported ionic liquid (SIL), describes adsorbents where ionic liquids (ILs) are infused into the solid support (Bao et al., 2016; Zhu et al., 2012).

The separation processes involving EIMs have been termed as supported liquid extraction (SLE) and solid-supported liquid–liquid extraction (SLLE) (Florek et al., 2016; Hidayah & Abidin, 2017; Hu et al., 2018). SLE is analogous to traditional liquid–liquid extraction (LLE), where target species are separated based on their relative solubilities in two different immiscible liquids, usually water and an organic solvent containing the extractant, but instead of shaking the two immiscible phases together, the extractant is supported on a solid support and put in a column (or cartridge). The REE-containing solution is passed through, the REE ions get adsorbed, and then eluent is used to desorb the adsorbed REE species. In SLE, it is possible to achieve higher recovery due to the increase in surface area of interaction between the extractant and the target cations, and increasing the diffusion rate through control of the porous structure of the support material. Other advantages over traditional LLE are less use of organic compounds, waste reduction, and lack of third-phase formation (Florek et al., 2016; Hidayah & Abidin, 2017; Kabay et al., 2010).

The synthesis of EIMs (SIRs and SILs) involves modification of already synthesized polymers or other solid supports like silica or activated carbon using dry or wet methods. In the dry method, the active component is impregnated into support in excess using a volatile solvent, and then the excess amount is evaporated by heating, thus leaving the active component physically adsorbed to the support surface. In the wet method, support is impregnated with a solution of active component in a non-polar or weakly polar solvent followed by repeated washing with water, leaving a layer of the non-polar solution of the active component (Bao et al., 2016; Ehrlich & Lisichkin, 2017). Based on the types of solid support used for preparing, these adsorbents can be broadly divided into organic polymer-based supports and inorganic supports.

9.5.1. Organic Polymer-Based Supports

Many researchers have used organic polymers as the solid support for solvent impregnated resins (SIRs) used for REE uptake from solution. Relevant examples of polymer-based SIRs include TODGA- and N,N,N′N′-tetrakis-2-ethylhexyldiglycolamide (TEHDGA)-impregnated Amberchrom CG-71 for REE separation; tributyl-phosphate-infused styrene–divinylbenzene to separate copolymers Am and Eu (Louis & Duyckaerts, 1984, 1985); 2-ethylhexyl phosphonic acid mono-2-ethylhexyl ester (PC88A)-infused Amberlite XAD-7 for separation of Gd and Y mixtures (Wakui et al., 1988); 8-Quinolinol (oxine)- and 2-(2-(5-chloropyridyl azo)-5-dimethylamino)-phenol (5ClDMPAP)-immobilized Amberlite XAD-4 and XAD-7 each for preconcentration of Ce, La, and Pr (Masi & Olsina, 1993); Amberlite XAD-7 impregnated with binary mix of cobalt dicarbollide and dibutyl-N,N-diethylcarbamoylmethyl phosphonate (DBDECMP) for Eu sorption (Svoboda et al., 1997); bis(2,4,4-trimethyl pentyl)phosphinicacid(Cyanex272)(Wangetal.,1998),1-hexyl-4-ethyloctyl hydrogen isopropylphosphonic acid (HEOPA) (Wang et al., 2002), and bis(2,4,4-trimethylpentyl) monothiophosphinic acid (CL302, HL) (Jia et al., 2004) infused resin for HREE separation.

Additional examples include PC88A-infused styrene–divinylbenzene copolymer for Gd(III) and Tb(III) chromatographic separation (Park et al., 2005), octacarboxymethyl-C-methylcalix 4-resorcinarene impregnated Amberlite XAD-16 for La(III), Ce(III), and Y(III) preconcentration (Gok et al., 2007); XAD-4 infused with tri-n-octylmethylammonium chloride (Aliquat-336) in benzene for Ln(III) and Gd(III) (Elsofany, 2008); microporous polymer infused with phosphorus podands bearing two $Ph_2P(O)CH_2C(O)NH$ for REEs (Turanov et al., 2008); Cyanex272-2-ethylhexyl phosphoric acid-2-ethylhexyl ester (P507) impregnated resin for HREEs (Liao et al., 2010); TOPS-99-infused Amberlite XAD-4 to extracted Tb (Kumar et al., 2011), β-glycerophosphate impregnated Amberlite XAD 7 for La(III) (Gabor et al., 2016); di-(2-ethylhexyl)phosphinic acid (P227) impregnated XAD-7HP for HREEs (Yang et al., 2020); and bis(2-ethylhexyl) hydrogen phosphate (HDEHP) impregnated XAD-7 for REEs (Sert et al., 2021).

Lee et al. (2009) used trialkylphosphine-based Cyanex 923, PC88A, and HDEHP for selective separation of La from Ce, Pr, Nd, and Sm, as well as from other concomitant metals in chloride medium. HDEHP provided the best separation with separation factors of 2.57, 3.63, 4.72, and 32.06 for Ce, Pr, Nd, and Sm over La (Lee et al., 2009, 2010b). The same adsorbent was used, followed by anion exchange and oxalate precipitation to obtain La_2O_3 of more than 99.9998% purity (Lee et al., 2010a). Amberlite XAD-7 coated with β-glycerophosphate had q_m of 33.8 mg/g for La (Gabor et al., 2016). Mondal et al. (2019) used N,N,N′,N′-tetrakis-2-ethylhexyldiglycolamide (TEHDGA)-impregnated XAD-7 resin to selectively adsorb REEs from coal fly ash solution containing Fe, Ca, Al, Mg, and Si with K_d of 200–520 mL/g for REEs and 0.2–2.9 mL/g for impurities. Cyanex-272-impregnated Amberlite XAD-7 resin showed distribution coefficient in order of Gd > Eu > Sm > Nd > Pr > La with K_d value for Gd and Eu as 208.1 and 156.4 mL/g (İnan et al., 2018). The SIR-P227-infused XAD-7HP showed selectivity series of Fe > Lu > Tm > Zn > Mg > Ca > Ho > Co > Ni > Cu > Al (B. Yang et al., 2020). Another adsorbent, HDEHP-XAD-7, showed selectivity series Gd > Eu > Sm > Nd > Pr > La with separation factors achieved Gd/La separation factors of >140 (Sert et al., 2021).

The example SILs with organic polymer include 1-octyl-3-methylimidazolium hexafluorophosphate ($C8mim^+PF_6^-$) containing Cyanex-923 immobilized on XAD-7 for REE extraction (Sun et al., 2008), trialkylmethylammonium sec-nonylphenoxyacetate ([A336][CA-100]) impregnated on Amberlite XAD-7 (Sun et al., 2009), trihexyl(tetradecyl) phosphonium mono-(2-ethylhexyl) 2-ethylhexyl phosphonate ([P_{66614}][EHEHP]) and trioctylmethylammonium bis(2,4,4-trimethylpentyl) phosphonate ([N_{1888}][BTMPP]) impregnated Amberlite XAD-7 (Zhao et al., 2016), betainium sulfonyl(trifluoromethanesulfonylimide) poly(styrene-co-divinylbenzene) [Hbet–STFSI–PS–DVB] impregnated Amberlite XAD-16 (Avdibegović et al., 2017), 1-butyl-3-methylimidazoliumbis(tri fluoromethylsulfonyl)imide ([Bmim][NTf2]) combined with N,N-dioctyldiglycolamic acid (DODGAA) and impregnated into Amberlite XAD7HP (Friend et al., 2020), and trihexyl (tetradecyl) phosphine bis (2,4, 4-trimethyl-amyl)–phosphonate (Cyphos IL 104) impregnated polymer membrane (Wang et al., 2020).

9.5.2. Inorganic Supports

Inorganic supports mostly include silica-based polymers. In some cases, ordered mesoporous silica (OMS) and carbon nanotubes (CNTs) have been used. Extractant HDEHP-coated C18-hydrophobized silica was used for chromatographic separation of lanthanides (Sivaraman et al., 2002); 2-ethylhexylphosphonic acid mono-2-ethylhexyl ester impregnated silica-based urea–formaldehyde (SiO_2/UF) for Eu(III) and Nd(III) (Naser et al., 2015); HDEHP-impregnated polyethersulfone (PES) embedded with polyvinyl alcohol (PVA) and multiwalled carbon nanotubes (MWCNT) for Y(III) sorption (Yadav et al., 2015); and tetraoctyldiglycolamide-impregnated carbon inverse opals (C-IOP) for REEs (Turanov et al., 2015).

Cyanex 272 impregnated mesoporous silica MCM 41 was used for the extraction of lanthanides. The selectivity between light lanthanide (La(III)), medium lanthanide (Eu(III)), and heavy lanthanide (Lu(III)) followed the order of Lu(III) > Eu(III) > La(III) with separation factor of 61, 61, and 3504 for Lu(III)/Eu(III), Eu(III)/La(III), and Lu(III)/La(III), respectively (Mohammedi et al., 2019). Silica gel modified with 1-(-2-pyridylazo) naphthol (PAN) and acetylacetone (Acac) through solvent evaporation process was selective for Sc(III) against Fe(III) with selectivity factors of 5.79 and 8.95, respectively at pH 6 (Ramasamy, Puhakka, Repo, & Sillanpää, 2018). Magnetite nanoparticle was used as a support for coating di(2-ethylhexyl) phosphonate (HDEHP), bis(2,4,4-trimethylpentyl)phosphinic acid (CYANEX 272), and bis(2,4,4-trimethylpentyl)dithiophosphinic acid (CYANEX 301) (Molina et al., 2019). These coated magnetite nanoparticles had q_m of 6.6–8.3 mg/g for La(III), 7.7–8.7 mg/g for Pr(III), and 4.8–8.9 mg/g for Nd(III). The same extractant, HDEHP, was infused into styrene–divinylbenzene copolymer immobilized silica particles (HDEHP-SiO_2-P) for REE adsorption (Shu et al., 2017, 2018; Zhang et al., 2019). The adsorption capacity (q_m) of HDEHP–SiO_2–P was 38.95 mg/g for Ce(III) and 52.84 mg/g for Gd(III) (Shu et al., 2017). In a different study, the HDEHP–SiO_2–P had q_m of 14.1 mg/g for Sc(III) with Sc(III) selectivity factor of >50 over other REEs (Zhang et al., 2019).

Inorganic solid supports infused with ionic liquids are mesoporous silica doped with binary IL mixtures (C8mim+$PF6^-$/Cyphos IL 104 or C4mim+PF6−/Cyphos IL 104) (Liu et al., 2009), silica gel impregnated with N-phenacylPyrNTf2 ionic liquid (Marwani & Alsafrani, 2013), silica doped with bifunctional ionic liquid trioctylmethylammonium 1-phenyl-3-methyl-4-benzoylpyrazol-5-onate ([A336]+[L]−) (Turanov et al., 2016), and MCM-41 silica impregnated with [Hbet–STFSI–PS–DVB] (Avdibegović et al., 2017). Among the SILs, trihexyl(tetradecyl)phosphonium bis(2,4,4-trimethylpentyl)phosphinate (Cyphos) entrapped in silica reached maximum adsorption capacities of 14.7 mg/g for Nd(III) and 19.8 mg/g at pH 4.0 (Mohamed et al., 2017). Ionic liquid 1-butyl-3-methylimidazolium bromide ($[C4mim]^+[Br]^-$) impregnated metal-organic framework (MOF) UiO-66 had a maximum Gd(III) adsorption capacity of 65 mg/g at pH 6 with Gd(III) selectivity in the presence of other metal ions (Na(I), Ca(II), Ma(II), Al(III), and Fe(III)) (Ahmed, Adhikary, et al., 2019).

The main limitation of supported liquid extractions (SLEs) is the gradual washout of the active component during operation (Bao et al., 2016). Possible solutions include the selection of highly hydrophobic and low soluble active components (Muraviev et al., 1998; Sparfel & Cote, 2004) and/or thorough washing of weakly adsorbed active components after their application onto the solid surface (Muraviev, 1998; Muraviev et al., 1998). Another method to prevent the leaking of the extractant from the adsorbent grain is to coat a layer of polymer on the adsorbent, which acts as a semipermeable membrane or protective barrier for the extractant while allowing the metal ion diffusion (Bao et al., 2016; Muraviev et al., 1998; Nishihama et al., 2013; Trochimczuk et al., 2004).

A protective barrier around the SIR was formed using polysulfone in dimethylformamide (DMF); however, loss of the adsorbed extractant [bis(2-ethylhexyl) hydrogen phosphate, bis-(2-ethylhexyl) hydrogen dithiophosphate, or bis(3-propylphenyl) dithiophosphate] still occurred (Muraviev et al., 1998). Better results were obtained with the coating of water-soluble polymer poly(vinyl alcohol) (PVA) precipitate onto SIR (Trochimczuk et al., 2004). The stability of the coating layer can be increased by cross-linking PVA with vinyl sulfone (Trochimczuk et al. 2004), boric acid (Yuan et al., 2010), and glutaraldehyde (Nishihama et al., 2013).

In general, EIMs (SIRs and SILs) can improve the performance of extractants by avoiding drawbacks of LLE, such as poor contact between the organic and aqueous phase, requirement of a large amount of extractant, emulsion formation, and formation of the third phase. The issue with SLEs is the stability of the active layer, which can wash out during operation. Moreover, the extractant layer may also be not uniform throughout the support surface; for example, capillary forces may accumulate the extractant in narrow pores, a process enhanced in contact with an aqueous solution (Bao et al., 2016). In addition, the complex nature of real-life REE sources, such as very acidic solutions, can put extra pressure on these adsorbents, which can limit their industrial applications.

The adsorbents obtained through ligand functionalization to the surface of solid support using coupling agents provide an alternative to EIMs. Surface-functionalized adsorbents can be an alternative with better control of ligand distribution on the surface, better ligand and surface bonding, and thus better stability, but they can be more resource-intensive to produce.

9.6. Surface-Functionalized Adsorbents

In this class of adsorbents, the active component or the extractant is chemically immobilized onto the surface of the solid support through chemical bonds. The active component is usually a complexing ligand containing specific functional groups that can selectively bind to the target REE. These adsorbents have been used for SPE of various elements. SPE is similar in the process as the SLE; however, the extractant or the ligand is chemically immobilized onto the surface in SPE as opposed to physical immobilization in the case of SLEs (Florek et al., 2016; Hidayah & Abidin, 2017; Hu et al., 2018).

A typical ligand-functionalized adsorbent involves solid support (e.g., silica, polymer beads, or carbon nanomaterials), coupling agent (e.g., silane), ligand containing a specific functional group (i.e., binding unit), and spacer between the coupling agent and the ligand (Pallavicini et al., 2014). Since the ligand is chemically bound to the surface, it is more robust than EIMs, provides better control of surface ligand distribution, and can work under extreme solution conditions. The surface-functionalized adsorbents used for REE adsorption can be classified based on the type of solid support. Silica-based and carbon-based supports have been the primary focus for SPE of REEs; however, magnetic nanomaterials, MOFs, and other composite supports are gaining attention due to their various advantages.

9.6.1. Silica-Based Supports

Silica-based supports are popular for functionalized adsorbents used in SPE, as they fulfill the requirements of good solid support: (1) they can have a large specific surface area (silica gel: usually 100–750 m^2/g (Bhatnagar & Sillanpää, 2010), mesoporous silica > 1,000 m^2/g (Meynen et al., 2009)), (2) are robust and reusable, (3) are easy to modify with functional groups, (4) have the possibility of shape control for various uses, and (5) show negligible swelling (Florek et al., 2016; Hu et al., 2018). Limiting factors affecting the use of silica as solid support are its tendency to lose the integrity of its porous structure under strong acidic conditions (El Mourabit et al., 2012; Florek et al., 2016) and its high dissolution rate in alkaline solution (Croissant et al., 2017; Crundwell, 2017).

Silica gel is a commonly used silica-based support. The other type of silica support is nanoporous silica material, further classified as macroporous (pore size > 50 nm), mesoporous (pore size 2–50 nm), microporous (pore size < 2 nm), and hierarchically porous materials, which contain multiple levels of abovementioned porosities (Sun et al., 2016; Wan & Zhao, 2007). Mesoporous silica nanoparticles are gaining widespread attention in various fields due to uniform and tunable pore size, presence of internal and external pores, controlled morphology, and easy functionalization (Hu et al., 2018; Mehmood et al., 2017; Narayan et al., 2018).

Silica Gel

Silica gel is the porous amorphous form of silicon dioxide (SiO_2), composed of interconnected networks of microscopic pores. It can have a pore size of 2–25 nm. The surface silanol groups present in silica gel are used to attach organofunctional alkoxysilane through the silanization process. The organofunctional alkoxysilane is used to attach functional groups in a process known as grafting. The most commonly used silanes are aminopropyl alkoxy silanes (3-aminopropyl triethoxysilane (APTES) and 3-aminopropyl trimethoxysilane (APTMS)), as they allow easy attachment of amino-poly(carboxylic acid) and other functional groups (Noack et al., 2016; Ramasamy, Khan, et al., 2017). Both the silica gel and silica gel with attached organo-silanes (e.g., 3-aminopropyl silica gel) are available commercially under various brands and have been functionalized further for adsorption of REEs (Asadollahzadeh et al., 2020; Callura et al., 2018; Noack et al., 2016; Ogata et al., 2014, 2015a; Ramasamy, Khan, et al., 2017; Ramasamy, Puhakka, et al., 2017).

Instead of surface modification of already prepared silica gel or 3-aminopropyl silica gel, co-condensation of tetra-alkoxy silane $Si(OR)_4$ (generally, R = Me or Et) with silane coupling agent $(RO)_3Si(CH_2)_3X$ (where R = Me or Et, X = complexing ligand group) has been used to obtain the required adsorbent (El-Nahhal & El-Ashgar, 2007).

Functional ligands grafted on the silica gel for REE adsorption are 4-acylpyrazolone Schiff base (Amarasekara et al., 2009), diglycolamic acid (Ogata et al., 2014, 2015a, 2015b, 2016), 1-(-2-pyridylazo) naphthol (PAN) (Ramasamy, Puhakka, Repo, et al., 2018; Ramasamy, Repo, et al., 2017), acetylacetone (Ramasamy, Repo, et al., 2017), phosphonoacetic acid (PAA), N,N-bis(phosphonomethyl) glycine (BPG), diethylenetriamine-pentaacetic dianhydride (DTPADA) (Callura et al., 2018; Noack et al., 2016), diethylenetriaminepentaacetic acid (DTPA) (Noack et al., 2016), diglycolamide (Liu et al., 2019), N-benzoyl-N-phenylhydroxylamine (BPHA) (Artiushenko, Ávila, et al., 2020), amino-di(methylene-phosphonic) acid (Artiushenko, Kostenko, et al., 2020), N,N-dioctyldiglycolic acid (DODGA)(Li et al., 2020), polyhexamethylene guanidine and Arsenazo I or Arsenazo III (Losev et al., 2020), and bis(ethylhexyl)amido diethylenetriaminepentaacetic acid (Hovey et al., 2021).

The diglycolamic-acid-functionalized silica gel showed selectivity for HREEs with the highest adsorption capacity of 0.148 mmol/g (or 24.4 mg/g) for Ho(III) (Ogata et al., 2015a). Another adsorbent PAN-functionalized silica gel showed capacities of 82.72, 75.5, and 62.92 mg/g for La(III), Sc(III), and Y(III), respectively (Ramasamy, Puhakka, Repo, et al., 2018; Ramasamy, Repo, et al., 2017). The BPHA-functionalized

SiO_2 with adsorption capacities (q_m) of 6.7 and 8.3 mg/g for Eu(III) and Tb(III), respectively, was selective for HREEs with K_d values of 3,500 mL/g for Lu(III) and Yb(III) each (legend 21, Figure 9.2) (Artiushenko, Ávila, et al., 2020).

Ordered Mesoporous Silicas

Ordered mesoporous materials have pore sizes of 2–50 nm with ordered arrangements of pores. Common ordered mesoporous silica materials are available as groups, defined as MCM (Mobil Composition of Matter) (e.g., MCM-41 (2d hexagonal, p6m), MCM-48 (cubic Ia3d), MCM-50 (lamellar)) (Beck et al., 1992; Florek et al., 2016; Kumar et al., 2001), SBA (Santa Barbara Amorphous), KIT (Korean Advanced Institute of Science and Technology), and COK (Centre for Research Chemistry and Catalysis). Within each mentioned group, there are different types of OMS with different pore symmetry, pore sizes, and pore volume (Beck et al., 1992; Florek et al., 2016; Hu et al., 2018; Huo et al., 1996; Trewyn et al., 2007).

Similar to surface functionalization of silica gel, the silica mesoporous can be functionalized through the reaction of organosilanes with surface silanol group (grafting) or by co-condensation with tetra-alkoxy silane $Si(OR)_4$ (generally, R = Me or Et) and trialkoxyorganosilanes $(R'O)_3SiRX$ (X = ligand) in the presence of a surfactant template (one-pot synthesis) (Florek et al., 2016; Hoffmann et al., 2006; Hu et al., 2018).

The abundance of silanol groups on the surface of silica (1–2 Si-OH per nm^2 on average (Ide et al., 2013)) allows for facile and efficient functionalization through grafting. The limitation of grafting manifests in the case of small pore size or narrow entrance of support and/or with a bulky size of the functional group. In such cases, the entrance may have a higher number of functional groups, or some smaller pores may lack functional groups due to pore blocking, resulting in an overall nonhomogenous distribution (Hoffmann et al., 2006). The limitation of pore blocking is not faced in co-condensation synthesis, and organic units are more homogeneously distributed than in the grafting process (Hoffmann et al., 2006). However, in co-condensation, the concentration of ligands in the pore wall tends to be lesser than the starting $(R'O)_3SiRX$ in the reaction mixture. Additionally, the degree of mesoscopic order of the OMS decreases in co-condensation with increasing concentration of $(R'O)_3SiRX$ in the reaction mixture, which leads to disordered products (Hoffmann et al., 2006). The increasing concertation of $(R'O)_3SiRX$ in the reaction mix favors homo-condensation over co-condensation, thus reducing the homogeneity and mesoscopic order of the OMS. Another challenge with one-pot syntheses of OMS is removing the surfactant template after co-condensation without destroying the surface ligands (Hoffmann et al., 2006).

Mesoporous KIT-6 has been functionalized with EDTA (Ravi, Zhang, et al., 2018), phenylenedioxy diamide (PDDA) (Hu et al., 2019), bidentate phthaloyl diamide (PA) (Hu et al., 2017), diglycolamide-based (Florek et al., 2014, 2015, 2020), furan-2,4-diamido-propyltriethoxysilane (Florek et al., 2015, 2020), and 3,6-dioxaoctanedioic acid (Florek et al., 2015, 2020) for REE adsorption. Similarly, MCM-41 has been functionalized with titanium(IV) alkylphosphate (Wenzhong Zhang, Avdibegović, et al., 2017), diglycolylamide (Juère et al., 2016), and iminodiacetamide (Fryxell et al., 2011), and SBA15 has been functionalized with diglycolamide (DGA) (Juère et al., 2016), benzene-1,3,5-triamido-tetraphosphonic acid (BTATPA) (Ravi, Lee, et al., 2018), phosphoric acid (Zheng et al., 2020), and 1,4-phthaloyl diamido-propyltriethoxysilane (Ryu et al., 2021) for REE adsorption. In the case of unmodified SBA-15 and KIT-6, uptake of Sc(III) (KIT-6: q_m = 1.0 mg/g (legend 36, Figure 9.2), SBA-15: q_m = 1.1 mg/g (legend 34, Figure 9.2)) was achieved through adsorption to accessible surface silanols (Giret et al., 2018).

DGA was used as a common ligand for three different OMS solid supports, SBA-15, MCM-41, and SBA-16, to investigate the effect of silica support pore network structure (2D vs. 3D), pore shape (cylindrical vs. cage-like), and pore size on the adsorption of REEs from synthetic samples of REEs and REEs with Al, Fe, Th, and U ions (shown as legends 9–13, Figure 9.2) (Juère et al., 2016). The 2D hexagonal structure (SBA-15) has the advantage over the 3D cage-like structure of SBA-16, since the cage-like pore structure and narrow connectivity made SBA-16 more prone to pore blocking. The pore size of SBA-15 was tuned by varying aging temperatures during synthesis. The highest capacity of extraction was observed for SBA-15(80)–DGA (pore size, 5.2 nm), which was attributed to the presence of certain confinement of the targeted ions by SBA-15 material with a smaller pore (5–8 nm) (Juère et al., 2016). Functionalization of DGA onto MCM-41 (3.2 nm, 2D pore) resulted in the obstruction of the pore entrance during grafting due to smaller pore size, resulting in lower extraction capacity (Juère et al., 2016).

The ligand immobilization on the surface of solid support can lead to reduced flexibility, which results in a more stable bite angle (angle formed by chelate ligands) and yields enhanced selectivity toward particular REE cations (Hu et al., 2018). The ligand DGA was functionalized onto KIT-6 by anchoring on both ends (KIT-6-N-DGA-1) or single end (KIT-6-N-DGA-2) (Florek et al., 2014). KIT-6-N-DGA-1 had a higher extraction capacity for REEs (especially middle REEs (MREEs)) than KIT-6-N-DGA-2 (ligands 1–4, Figure 9.2). The higher extraction capacity of

KIT-6-N-DGA-1 was attributed to increased rigidity in the DGA ligand anchored on the surface at both ends, which increased the size-specific cavity (Hu et al., 2018). The presence of competitive ions did not affect K_d values (ligands 3 and 4, Figure 9.2). The separation factor between Eu(III) and Th(IV) was 6.3, and between Eu(III) and U(IV) was 3.4 for KIT-6-N-DGA-1 (Florek et al., 2014).

The effect of different bite angles of ligands on the selectivity of REEs was studied using DGA, 3,6-dioxaoctanediamidopropyl (DOODA), and furan-2,4-diamidopropyl (FDGA) functionalized KIT-6 (Florek et al., 2015, 2020). The DOODA ligand has a smaller bite angle than DGA (Figure 9.3), hence showed a preference for smaller Ln(III) (Ho–Lu, legends 4 and 5, Figure 9.2) (Florek et al., 2015; Hu et al., 2018). Hu et al. (2017) grafted phthaloyl diamide (PA) bearing different bite angles onto KIT-6 by varying the position (ortho, meta, and para) of the amide groups. The KIT-6-1,2–PA (ortho-position) with a smaller bite angle showed higher affinity for heavy Ln(III), which have smaller ionic radii (legend 14, Figure 9.2), whereas the KIT-6-1,3–PA showed selectivity for middle-size Ln(III) due to its larger bite angle (legend 15, Figure 9.2). KIT-6-1,4–PA did not show selectivity for a particular lanthanide (legend 16, Figure 9.2) due to the absence of any synergist action of the moieties (Hu

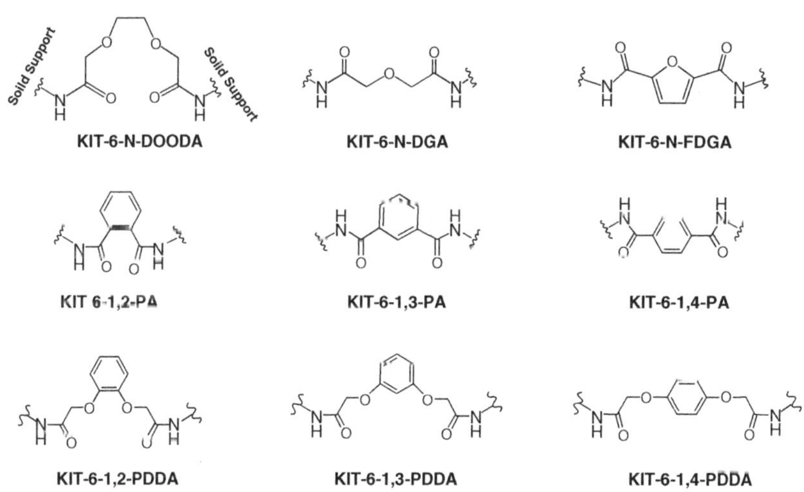

Figure 9.3 Different bite angles of the ligands grafted onto KIT-6 (modified after (Florek et al., 2015; Hu et al., 2017, 2019)). DOODA: 3,6-dioxaoctanediamidopropyl, DGA: diglycolylamide, FDGA: furan-2,4-diamidopropyl (Florek et al., 2015, 2020), PA: phthaloyl diamide (Hu et al., 2017, 2018), and PDDA: phenylenedioxy diamide (Hu et al., 2019).

et al., 2017, 2018). In another study, the bite angle of phenylenedioxy diamide (PDDA) functionalized modified KIT-6 was varied by using PDDA with amide at different positions (Hu et al., 2019). Results analogous to KIT-6–PA systems were obtained, as adsorbent KIT-6-1,2–PDDA having the smallest bite angle (Figure 9.3) showed high affinity for heavy lanthanides (Tm(III), Yb(III), and (Lu(III)) with smallest ionic radii (legend 22, Figure 9.2), and adsorbent KIT-6-1,3–PDDA with larger bite angle showed higher selectivity for light lanthanides (La(III), Ce(III), and Pr(III)) with larger ionic radii (legend 23, Figure 9.2). KIT-6-1,4–PDDA did not show selectivity (legend 24, Figure 9.2) and behaved similarly to KIT-6-1,4–PA.

In other examples of functionalized ordered mesoporous silica used for REE uptake, Dy(III) was selectively adsorbed from a solution of Cu(II), Dy(III), Fe(III), Nd(III), and Zn(II) using amino and carboxylic functionalized mesoporous silica having different pore sizes (3, 5, 12, and 22 nm) (Kaneko et al., 2018) and using amino, carboxylic, and diglycolic anhydride functionalized mesoporous silica with a sheet and spherical morphologies as well as nonporous Stöber silica (Kaneko et al., 2019). Zhang and their colleagues (Zhang, Avdibegović, et al., 2017) used titanium(IV) alkylphosphate grafted MCM41 for adsorption in the binary equimolar solutions of Sc–La and Nd–Dy. Ravi, Zhang, et al. (2018) functionalized KIT-6 and KCC-1 with EDTA and employed them for Nd(III) adsorption from a single-element solution (Ravi, Zhang, et al., 2018). The same team, in a different study, used benzene-1,3,5-triamidotetraphosphonic acid (BTATPA) modified SBA-15 for adsorption of Nd(III), Y(III), La(III), and Ce(III) from single element solution and reported K_d (mL/g) values of approximately 10^4, 10^4, 1.5×10^5, and 1.04×10^5, respectively (Ravi, Lee, et al., 2018). In other studies, SBA-15 functionalized with phosphorus acid achieved a maximum Gd(II) adsorption capacity of 204.4 mg/g (Gao et al., 2017), and sulfhydryl MCM-41 (SH-MCM-41) showed maximum adsorption capacities of 560.56, 467.60, and 540.68 mg/g for La(III), Gd(III), and Yb(III), respectively (Li et al., 2019).

Mesoporous silica foam (MSF) and fibrous nano-silica KCC-1 were grafted with poly(amidoamine) dendrimers (PAMAM) using 3-aminopropyltriethoxysilane (APTES) and glutaraldehyde (GTA) as cross-linking spacers (Lee et al., 2019). The adsorption capacity of MSF-based adsorbent PAMAM@GTA-NH_2-MSF was higher than KCC-1-based adsorbent PAMAM@GTA-NH_2-KCC-1 with values of 86, 104, 110, and 132 mg/g for La(III), Ce(III), Nd(III), and Gd(III), respectively, for MSF-based adsorbent. Both adsorbents showed selectivity for Gd(III) from a solution containing Na(I), Ca(II), Mg(II), Al(III), and Fe(III) with Gd(III) recoveries of 75.8% with PAMAM@GTA-NH_2-MSF and 67.2%

with PAMAM@GTA-NH_2-KCC-1. The recovery of Gd(III) increased to 86% with further functionalization of PAMAM@GTA-NH_2-MSF with N-(phosphonomethyl) iminodiacetic (PMIDA) (Lee et al., 2019). N-methyl-N-phenyl-1,10-phenanthroline-2-carboxamide (MePhPTA) functionalized mesoporous silica adsorbed Eu(III) and Sm(III) with maximum adsorption capacities of 125.63 and 124.38 mg/g, respectively, at pH 4 (Awual et al., 2013). Ordered mesoporous silica functionalized with 4-tert-octyl-4-((phenyl)diazenyl)phenol (TOPP) reached maximum Yb(III) adsorption of 139.19 mg/g at pH 5 and was selective against Na(I), K(I), Ca(II), Mg(II), Zn(II), and Al(III) (Rahman et al., 2020). Hollow mesoporous silica nanosphere (HMSNs) functionalized with amino-phosphonic acid had an adsorption capacity of 387.3 mg/g for Gd(III) at pH 5 (Yin et al., 2020).

9.6.2. Carbon-Based Supports

Carbon-based adsorbents and supports have various forms, such as mesoporous carbon, activated carbon, graphene oxide, and carbon nanotubes. Carbon-based supports can be an alternative to silica supports to overcome the low stability of silica-based support in acidic conditions (Hu et al., 2018). However, the surface functionalization of carbon-based supports can be more challenging than silica-based supports (Hu et al., 2018).

Ordered Mesoporous Carbons

Ordered mesoporous carbons (OMCs) can work as better solid support under acidic conditions because of higher chemical resistance than mesoporous silica while possessing comparable porosity and surface area (Hu et al., 2018). However, the synthesis and functionalization of OMCs are more challenging than OMS. Ordered mesoporous carbon synthesis can be done by hard- or soft-template methods.

In the hard-template method, an ordered mesoporous solid (mostly ordered mesoporous silica) is used as a mold, and its pores are impregnated with a selected carbon precursor (e.g., sucrose), which is then polymerized and carbonized (Eftekhari & Fan, 2017). In the end, the template is dissolved and removed using strong acid (e.g., HF) or base (e.g., NaOH) (Chuenchom et al., 2012). The porosity and structure of OMC can be tuned by selecting an appropriate hard template.

On the other hand, in the soft-template method, OMC is directly synthesized by self-assembling a block of copolymer surfactant and a carbon precursor. In this method, the template is removed by a thermal process. The pH, temperature, and gel composition can be varied to control the porosity and morphology of the mesoporous carbon (Ma et al., 2013). In comparison with the hard-template method, OMCs synthesized through soft templates have a continuous framework with thick pore walls, hence have more stability against thermal and oxidation treatment during functionalization (Eftekhari & Fan, 2017).

In adsorption studies involving OMCs, Parsons-Moss et al. (2014) investigated Eu(III) adsorption using a carbon–silica nanocomposite (C–CS) type OMC functionalized with –COOH and reported adsorption capacity (q_m) of 138 mg/g at pH 4.0. Perreault et al. (2017) functionalized CMK-8, a 3D cubic ordered mesoporous carbon, with different DGA-based ligands. Among these CMK-8-based adsorbents, the K_d values for different Ln(III) in the presence of Al(III) and Fe(III) were highest for diglycolylester-grafted CMK-8 (CMK-8-DGO) followed by oxidized CMK-8 (CMK-8-O) and chloropropyl-diglycolylamide-grafted CMK-8 (CMK-8-PGA) (legends 6–8, Figure 9.2). Bertelsen et al. (2019) modified two OMC materials, MC-*l*-MSN and CMK-3 type OMC, with bis-(2-ethylhexyl) phosphoric acid (HDEHP) and investigated batch adsorption of Eu(III) as well as column separation of Nd(III) and Eu(III) using HDEHP-OMC-*l*-MSN. The HDEHP-OMC showed q_m of 0.35 mmol (or 53.2 mg) Eu(III)/g. Mesoporous carbon was functionalized with a single-stranded oligo consisting of 100 thymine units, and it achieved adsorption of 9.57, 38.27, and 52.15 mg/g, for Lu(III), Dy(III), and La(III), respectively. The adsorbed amounts of REEs increased with an increase in the metallic radius from Lu to La (Gismondi et al., 2022; Unsworth et al., 2020).

Activated Carbon/Graphene Oxide/Carbon Nanotube

Activated carbon: Activated carbon (AC) is an environmentally friendly and low-cost adsorbent with a high surface area but has little affinity for REE(III) in native form; thus, functionalization with ligands that have affinity for REEs is required for its application in REE adsorption (Asadollahzadeh et al., 2020). The ligands used for AC functionalization are Schiff's base derived from 3-diethylenetriamine and 3,4-dihydroxybenzaldehyde (AC-DETADHBA) (Marwani et al., 2017), Carboxylic acid (Kilian et al., 2017; Marwani et al., 2017), and EDTA (Babu et al., 2018). The EDTA functionalization of AC (AC-EDTA) increased adsorption capacity to 71.4 Nd(III) mg/g from 19.1 Nd(III) mg/g at pH 5 (Babu et al., 2018). The adsorbent AC-EDTA showed higher selectivity toward HREEs. Carboxyl-functionalized AC (AC-COOH) showed an adsorption capacity of 89.50 mg/g for La(III) at pH 6.0, whereas AC-DETADHBA showed a higher adsorption capacity of 144.80 mg/g for La(III) at the same pH (Marwani et al., 2017). In a different study, AC-COOH showed very low adsorption of 2.1 mg/g for Sc(III) at pH 2.0

(Kilian et al., 2017), suggesting that the pH plays a vital role in AC-COOH adsorbents, likely due to the weakly acidic nature of –COOH group.

Graphene oxide: Graphene oxide (GO) contains hydroxyl, epoxide, carbonyl, and carboxyl groups, essential sites for functionalization with ligands. Moreover, due to the presence of these groups, GO can adsorb REEs without functionalization. A colloidal graphene oxide suspension had Gd(III) adsorption capacity of 286.86 mg/g at pH 5.9 (Chen, W. et al., 2014). Graphene oxide nanosheets (GONS) obtained a maximum adsorption of 175.44 mg/g for Eu(III) at pH 6 (Sun et al., 2012). The maximum Eu(III) adsorption decreased slightly to 167.16 mg/g at pH 2 (Sun et al., 2012). In another study with GO nanosheet, the maximum adsorption capacities were in the order of Gd(III) > Nd(III) > Y(III) > La(III) with values 225.5, 188.6, 135.7, and 85.7 mg/g at pH 6, respectively (Ashour et al., 2017). Unmodified graphene oxide showed an adsorption capacity of 39.7 mg/g for Sc(III) at pH 4, which was better than –COOH-functionalized AC and chelating resin Chelex 100 (contains iminodiacetic acid) but lower than –COOH-modified MWCNT (Kilian et al., 2017). Higher adsorption of 89.7 mg/g and 70.2 mg/g for Eu(III) using GO and magnetic GO (by modifying with Fe_3O_4) was achieved at pH 7 (Li et al., 2015).

GO was used to produce amine-functionalized mesoporous graphene (AMG) followed by functionalization with carbamoyl phosphine oxide moiety (CMPO, triethylphosphonoacetate), and it reached an adsorption capacity of 26.9 mg/g for La(III) (Kim et al., 2019). Fe_3O_4- and MnO_2-modified GO nanocomposites, Fe_3O_4/MnO_2/rGO (FMG), provided high maximum adsorption capacities of 1016 and 981 mg/g for La(III) and Ce(III), respectively, at pH 7 (Liu et al., 2020), and is among the adsorbents with the highest capacity for REE(III) (Table 9.1). Another adsorbent, polyaniline-functionalized GO, achieved 250.74 mg/g Eu(III) adsorption (Sun et al., 2013). Polyurethane sponge was used as a support polymer for titanium phosphate with graphene oxide (GO@TiP-sponge) and showed a maximum capacity of 576.17 mg/g for Dy(III) (Peng et al., 2020). In comparison, GO@TiP had a maximum Dy(III) adsorption capacity of 316.75 mg/g (Peng et al., 2020). GO-tris(4-aminophenyl)amine composite had a Langmuir adsorption capacity of 46.35 mg/g for Yb(III) at pH 6 (Zhao et al., 2021). Another GO-based adsorbent, graphene oxide/poly (N-isopropyl acrylamide-maleic acid) [GO/P(NIPAM-MA)] cryogel, adsorbs La(III) with equilibrium adsorption capacity of 33.1 mg/g at pH 5 with separation factors of 2.86, 7.57, 8.00, 6.69, and 1.45 for La over Cu, Co, Ni, Nd, and Yb, respectively (Yang et al., 2020). GO functionalized with 3-[2-(2-aminoethylamino) ethylamino] propyl-trimethoxysilane (GO-APTS) showed adsorption capacities of 110.0, 93.4, 103.2, 83.7, 97.2, 48.3, and 92.8 mg/g for Ho(III), Er(III), Eu(III), Lu(III), Tm(III), Y(III), and Yb(III), respectively (Bao et al., 2022). The adsorbent was selective for Ho(III) over alkali or alkaline earth metal ions with SFs of above 600. A further modified adsorbent combining GO-ATPS with Fe_3O_4 resulted in lower capacities of 72.1, 67.1, 65.2, 65.0, 73.2, 36.3, and 68.7 mg/g for Ho(III), Er(III), Eu(III), Lu(III), Tm(III), Y(III), and Yb(III), respectively, due to lower APTS fraction in the composite (Bao et al., 2022).

Carbon nanotubes: CNTs are tubes of graphite sheets and can be divided into two types: single-walled carbon nanotubes (SWCNTs) and MWCNTs with a set of concentric nanotubes. The examples of CNT-based adsorbent studied for REE separation include oxidized MWCNTs (Alguacil et al., 2020; Behdani et al., 2013; Fan et al., 2009; Koochaki-Mohammadpour et al., 2014; Li et al., 2015), CNT modified with –COOH (Kilian et al., 2017), MWCNT modified with tannic acid (Tong et al., 2011), and MWCNT/Fe_3O_4 composites (Chen et al., 2009; Fan et al., 2009). The CNT modified with –COOH showed higher adsorption (37.9 mg/g) for Sc(III) than AC-COOH (2.1 mg/g) at pH 2.0 (Kilian et al., 2017). The adsorption of Sc(III) onto CNT-COOH increased to 42.5 mg/g at a pH of 4.0. Oxidized MWCNTs showed good adsorption (80–100%) for target REE(III) in the presence of competitive ions, and the adsorption increases with pH since, at higher pH, the surface is negatively charged, leading to higher sorption of positively charged REE(III) cations (Cardoso et al., 2019). CNT/GO hybrid hydrosols had Gd(III) adsorption capacity of 534.76 mg/g at pH 5.9 (Guo et al., 2018). CNT modified with furfuryl amine and further functionalized with poly(acrylic acid) reached a maximum equilibrium Eu(III) adsorption capacity of 130.8 mg/g as opposed to 46.67 mg/g for pristine CNT at pH 7 (Guo et al., 2020).

9.6.3 Other Supports/Adsorbents

Magnetic Nanomaterials

Researchers have explored adsorbents obtained after coating Fe_3O_4 nanoparticles with SiO_2 or TiO_2 shell and then the functionalization of the surface of the coated layer. The advantage of magnetic nano-adsorbents is the convenience of collecting/separating adsorbents by using magnets after adsorption, avoiding the issue of the isolation of nanoparticles, which tend to form stable sols in batch reactors. The coating with SiO_2 or TiO_2 provides higher surface-OH density, which results in higher surface-ligand density and easier functionalization. The coating also protects the Fe_3O_4 nanoparticle from oxidation.

Polido Legaria, Samouhos, et al. (2017) used EDTA, DTPA, and triethylenetetraminehexaacetic acid (TTHA) functionalized SiO_2 nano-adsorbents, both magnetic and non-magnetic, and reported uptake up to 300 mg REE(III)/g. REE uptake order was Dy(III) > Nd(III) > La(III) for EDTA-functionalized, Nd(III) > La(III) > Dy(III) for DTPA-functionalized, and La(III) > Nd(III) > Dy(III) for TTHA-functionalized adsorbents. Kostenko et al. (2019) synthesized Fe_3O_4 MNPs with and without a coating of silica and functionalized them with aminomethylenephosphonic (MNPs/SiO_2-AMPA and MNPs/AMPA). The adsorption capacities for Eu(III) were 69 and 77 mg/g for MNPs/SiO_2-AMPA and MNPs/AMPA at pH 7.0, respectively. Magnetic Fe_2O_3 coated with silica and functionalized with APTMS (γ-Fe_2O_3-NH_4OH@SiO_2(APTMS)) achieved maximum Dy(III) adsorption of 23.3 mg at pH 7 (Kegl et al., 2019). Magnetic nanoparticle functionalized with poly(aminoethylene N-methyl 1-formic acid, 1-phosphonic acid) (PAEMFP) reached 25 mg/g Nd(III) adsorption capacity at pH 6 (Miraoui et al., 2016). Other research involving coated magnetic nano-adsorbents for REEs adsorption includes silica-coated magnetic particles grafted with 2-ethylhexyl phosphonic acid mono-2-ethylhexyl (P507) using 3-chloropropyltryethosysilane as silane-coupling agent (Wu et al., 2013) and magnetic Fe_3O_4 coated with SiO_2 and TiO_2 and modified with N-[(3-trimethoxysilyl) propyl] EDTA (TMS-EDTA) (Dupont et al., 2014).

Some studies have directly functionalized the magnetic nanoparticles (Chen et al., 2012; Gaete et al., 2021; Ngomsik et al., 2012; Yang et al., 2012). Citric-acid-functionalized maghemite nanoparticles had q_m of 74.4 mg/g and 71 mg/g for Eu(III) and La(III), respectively (Ngomsik et al., 2012). Magnetic nanoparticle Fe_3O_4 with humic acid coating (Fe_3O_4@HA MNPs) synthesized through chemical coprecipitation was able to remove ~99% of Eu(III) at pH 8.5 (Yang et al., 2012). In another study, bentonite was used as a support to prepare di-(2-thylhexly)phosphoric acid-immobilized magnetic GMZ bentonite and was used to adsorb Eu(III) (capacity of 48.02 mg/g at pH 3.05) (Chen et al., 2012). Another adsorbent, phosphonic-acid-functionalized magnetite nanoparticle (PA-MNP), showed the capacities of 18.4, 17.6, and 23.9 mg/g for La(III), Pr(III), and Sm(III), respectively, at pH 4 (Gaete et al., 2021).

Metal-Organic Frameworks

MOFs have ordered crystalline porous structures composed of a three-dimensional network of metal or metal oxide clusters held together by multidentate organic ligands (i.e., linker), for example, aromatic polycarboxylates. MOFs have received considerable attention as support for grafting various REE-selective ligands. The desirability of MOFs as support lies in their high surface area, high porosity, adjustable pore size, and tunable surface functionality (Baumann et al., 2019).

The MOF, HKUST-1, showed an adsorption capacity of 353 mg/g for Ce(III) at pH 6 (Zhao et al., 2019). MOF ZIF-8 showed a higher adsorption capacity (385 mg/g) for La(III) than another MOF ZIF-90 (q_m = 168 mg/g for La(III)) that had a different linker (Jiang et al., 2016). MOF Zeolitic imidazolate frameworks-8 nanoparticle (ZIF-8 NP) has adsorption capacities of 28.8 mg/g, 281.1 mg/g, and 430.4 mg/g for La(III), Sm(III), and Dy(III), respectively, at pH 7 (Abdel-Magied et al., 2019). MOF ZIF-8 functionalized with –COOH showed an adsorption capacity of 175 mg/g for Nd(III), significantly higher than unfunctionalized ZIF-8 (57 mg Nd(III)/g)) (Ahmed et al., 2021). Another MOFsUiO-66-$(COOH)_2$ functionalized with polyacrylonitrile had maximum adsorption capacities of 214.1 mg/g and 191.9 mg/g for Tb(III) and Eu(III), respectively, at pH 6 (Hua et al., 2019). Similar MOF UiO-66 dual functionalized with –COOH and $-NH_2$ obtained an equilibrium adsorption capacity of 79 mg/g for Gd(III), significantly higher than pristine UiO-66 with 16 mg Gd(III)/g (Ahmed, Lee, et al., 2019). A MOF, ZnGA, synthesized by the reaction of zinc acetate and glutaric and then functionalized with polymer based on p-chlorocresol and piperazine had Y(III) adsorption capacity of 377.02 mg/g at pH 6 (Mahmoud et al., 2019).

In several different studies, MIL-101 was functionalized with different functional groups and used for REE adsorption (de Decker et al., 2016; Kavun et al., 2021; Lee et al., 2018; Lou et al., 2019; Ryu et al., 2021). Among MIL-101 functionalized with $-NH_2$, ethylenediamine (ED), diethylenetriamine (DETA), and N-(phosphonomethyl)iminodiacetic acid (PMIDA), the REE-adsorption capacity increased in the order of MIL-101 < MIL-101-NH_2 < MIL-101-ED < MIL-101-DETA < MIL-101-PMIDA with the highest being 90.0 mg Gd(III)/g onto MIL-101-PMIDA at pH 4.5 (Lee et al., 2018). Acrylic-acid-functionalized MIL-101 had adsorption capacities of 90.21, 104.59, 58.29, and 74.94 mg/g for Sc(III), Nd(III), Gd(III), and Er(III), respectively, at pH 5.5, and was selective for Sc(III) in the presence of other REEs and competitive cations (Lou et al., 2019). In a different study involving MIL-101 functionalization with tributyl phosphate (TBP), HDEHP, and bis(2,4,4-trimethylpentyl) phosphinic acid (Cyanex 272), the adsorption capacity increased in the order of MIL-101-TBP < MIL-101-Cyanex 272 < MIL-HDEHP with a maximum adsorption capacity of 57.47 mg/g at pH 5.5 for HDEHP-functionalized adsorbent (Kavun et al., 2021). The MIL-101-HDEHP showed selectivity factors of 22.8 and 7.7 for Er(III) over Nd(III) and Gd(III) in Er(III)–Nd(III)–Gd(III) mix solution with all three adsorbents showing high selectivity (>90%) for Er(III) over Co(II), Ni(II), Cu(II), and Zn(II).

Metal-Based Supports

A metal-based support mesoporous zirconium titanate (ZrTi-0.33) was functionalized with methylphosphonic acid (MeP), amino tris-methylenephosphonic acid (ATMP), phosphono-imido-dicarboxylic acid (PIDC), 4-amino,1-hydroxy,1,1-bis-phosphonic acid butane (HABDP), and 1-hydroxylethylene-1,1-bis-phosphonic acid (HEDP) bis-phosphonic acids (1,4-diphosphonic acid butane (BuDP) and 1,4-diphosphonic acid benzene(BenDP)), and used for $^{153}Gd(III)$ adsorption (Griffith et al., 2010). ATMP-based adsorbent achieved the highest K_d (>10,000 in 10^{-3} M and 10^{-5} M HNO_3).

Similar support, mesoporous TiO_2 particle, was functionalized with dimethylphosphato-ethyltriethoxysilane and reached Ce(III) adsorption capacity of 92.6 mg/g while completely separating Ce(III) from Sr(II) and Cs(I) (Moloney et al., 2014). Mg–Fe hydrotalcite modified with Cyanex 272 (bis(2,4,4-trimethylpentyl) phosphinic acid) showed separation factor (SF) >6.3 for La(III)/Nd(III) (Gasser & Aly, 2013) and >3.6 for Ce(III)/Eu(III) (Gasser et al., 2017).

Composite Supports

A composite support of nanosilica (~12 nm particle size) and AC (0.8 mm pellets) was modified with 1-(2-pyridylazo) 2-naphthol (PAN) (Ramasamy, Puhakka, Repo, et al., 2018). The adsorbent obtained adsorption capacities of 103.5, 112.7, and 84.1 mg/g for La(III), Sc(III), and Y(III), respectively. Another composite support consisting of nanosilica-grafted carbon nanotubes (both SWCNT and MWCNT) functionalized with 1-(2-pyridylazo)-2-naphthol (PAN) was selective for Sc(III) in mixed REE solution (Ramasamy et al., 2019). However, the adsorption capacities for Sc(III) were lower than La(III) and Y(III) in mono-element solution with adsorption capacities of 12.68, 80.68, and 48.34 mg/g for Sc(III), La(III), and Y(III), respectively, in the SWCNT and 103.2, 32.92, and 68.78 mg/g, respectively, with the MWCNT (Ramasamy et al., 2020). MWCNTs embedded with Fe_3O_4 nanoparticles were functionalized with carbon disulfide and achieved 23.23 mg/g La(III) adsorption (Huang et al., 2021). A nanoporous graphene and zinc–trimesic acid (Zn–BTC) MOF composite showed selectivity for Ce(III) with SFs 12,081.04, 20.05, and 328.80 for Ce(III) over Lu(III), La(III), and Pr(III), respectively (Wu et al., 2021). The SFs between adjacent lanthanide Nd/Pr, Sm/Eu, Gd/Tb, and Tb/Dy reached 9.80, 3.11, 2.20, and 1.68. The maximum capacity of the adsorbent was about 300 mg REEs/g. Another composite consisting of nanoporous graphene oxide and zinc–terephthalic acid (Zn-BDC) MOF reached a maximum capacity of 344.48 mg REEs/g with high SFs for Sc, that is, Sc/Tm ≈ 529.57, Sc/Er ≈ 461.91, and Sc/Y ≈ 445.70 (Chen et al., 2022). It was also selective for Tm and Er with SFs for Tm/Eu ≈ 4.55, Tm/Pr ≈ 4.20, Tm/Nd ≈ 3.96, and Er/Eu ≈ 3.9. A composite obtained via the introduction of a MOF MIL-101 shell over a magnetite (Fe_3O_4) core and functionalized with $-SO_3$ and diethylenetriamine (DETA) extracted REEs from aqueous and brine solution with efficiency up to 99.99% (Elsaidi et al., 2018).

A silica/polymer (SiO_2–P) composite support synthesized by immobilizing styrene–divinylbenzene copolymer (SDB) in porous silica (SiO_2) was functionalized with 2,6-bis(5,6,7,8-tetrahydro-5,8,9,9-tetramethyl-5,8-methano-1,2,4-benzotriazin-3-yl)pyridine (Me_2-CA-BTP) and was used for Ln(III) adsorption. Similar SiO_2–P supports were functionalized with a mixed trialkyl phosphine oxide (TRPO) (where R_1, R_2, and R_3 are different alkyl groups) (Yu et al., 2018) and functionalized with HDEHP for REE(III) adsorption (Zhang et al., 2019).

An organic–inorganic hybrid hydrogel, polyethylenimine–acrylamide/SiO_2, showed selectivity factors of 105–450 for REEs over other metals (Wang, Wilfong, et al., 2017). A porous carboxymethyl cellulose hydrogel containing polyacrylic acid had a monolayer adsorption capacity (q_m) of 381.72 mg/g for La(III) at pH 5.58 and 320.47 mg/g for Ce(III) at pH 5.83 (Roosen et al., 2016; Zhu et al., 2016). Chitosan containing acrylic acid-co-styrene sulfonate/smectite clay hybrid granular hydrogel (CTS-g-(AA-co-SS)/ISC) had adsorption capacities of 232.97 mg/g for Gd(III) and 185.47 mg/g for Ce(III) at pH 5 (Wang, F., Wang, W., et al., 2017). A magnetic $MnFe_2O_4@SiO_2$–chitosan composite achieved one of the highest adsorption capacities of 1,030 mg/g and 1,020 mg/g for La(III) and Ce(III), respectively, (Table 9.1) (Liu et al., 2021).

Similar to chitosan-containing hydrogel, many composite supports are based on composites of biological materials with inorganic materials or polymers (Iftekhar, Ramasamy, et al., 2018). REE adsorption has been studied with silica–chitosan hybrid beads functionalized with EDTA, DTPA (Roosen et al., 2014), PAN, and acetylacetone (Ramasamy, Puhakka, Iftekhar, et al., 2018), Zn/Al-layered double hydroxide (LDH) intercalated cellulose (CL) (Iftekhar et al., 2017b), cellulose–silica nanocomposite (Iftekhar et al., 2017a), and gum Arabic (GA)–polyacrylamide–silica nanocomposite (Iftekhar, Srivastava, et al., 2018). Among these adsorbents, silica–chitosan–PAN showed the highest adsorption capacities of 199.8, 198.8, and 123.4 mg/g for La(III), Sc(III), and Y(III), respectively, at pH 5. The LDH-CL adsorbent had q_m values of 102.25, 92.51, and 96.25 mg/g for Y(III), La(III), and Ce(III), respectively, at pH 7 (Iftekhar et al., 2017b). The cellulose–silica nanocomposite and the GA–polyacrylamide–silica nanocomposite reached q_m of 29.48 mg/g and 7.90 mg/g for La(III), respectively, and 23.76 mg/g and 11.05 mg/g for

Sc(III), respectively, at pH 6 (Iftekhar et al., 2017a; Iftekhar, Srivastava, et al., 2018). The EDTA- and PDTA-functionalized silica–chitosan had adsorption capacity (q_m) of 0.27 mmol/g (or 38.9 mg/g) for Nd(III) at pH 6 (Roosen et al., 2014), and equilibrium capacity of 6.74 mg/g and 8.99 mg/g for Sc(III) at pH 2.

9.7. Molecular-/Ion-Imprinted Polymers

Molecular-/ion-imprinted polymers (MIPs/IIPs) are sorbents with specific recognition sites for a desired molecule or ion (Chen et al., 2016; Ensing & De Boer, 1999; Fu et al., 2015; Wackerlig & Schirhagl, 2016). This is achieved by copolymerizing functional monomers and cross-linkers in the presence of the target molecule/ion that acts as a template molecule. Then, the template ion/molecule is removed chemically, leaving behind a binding site specific to the template (Chen et al., 2016; Fu et al., 2015).

The IIPs show high selectivity for the target REEs. Biju et al. (2003) achieved high selectivity for Dy(III) over Y(III), Nd(III), Lu(III), and La(III) with selectivity factors (SFs) of around 66, 82, 175, and 116, respectively, by using Dy(III)-ion-imprinted polymer based on 5,7-dichloroquinoline-8-ol and 4-vinyl pyridine. Similarly, Er(III) (Kala et al., 2004, 2005; Rao et al., 2006) and Nd(III) (Guo et al., 2009) ion-imprinted polymers based on 5,7-dichloroquinoline-8-ol and 4-vinyl pyridine were prepared and tested for selectivity over different lanthanides. In all of these investigations, the selectivity of the imprinted ion was significantly increased (Guo et al., 2009; Kala et al., 2004, 2005; Rao et al., 2006). Similarly, several IIPs (Pr(III)-IIP, Nd(III)-IIP, Sm(III)-IIP, Eu(III)-IIP, and Gd(III)-IIPs) were synthesized with maximum sorption capacities of 125.3, 126.5, 127.6, 128.2, and 129.1 mg/g for Pr(III), Nd(III), Sm(III), Eu(III), and Gd(III), respectively (Yusoff et al., 2017). A phosphonic-based La(III)-ion-imprinted polymer showed an adsorption capacity of 62.8 mg/g for La(III) and a selectivity factor of 54.57 for La(III) over Cu(II) (Ni et al., 2021).

Dolak et al. (2015) used Nd(III)-IIP to selectively extract Nd(III) in the presence of Ce(III), La(III), and Eu(III). The selectivity coefficients for Nd(III) over Ce(III), La(III), Eu(III) were 234.3, 129.6, and 248.4, respectively; however, the maximum adsorption capacity was only 15.03 mg/g. In different investigations, Gd(III)-ion-imprinted polymers bearing N,N″-bisacetamidostyrene diethylenetriaminetriacetic acid (DTPA–bisamide) (Garcia et al., 1998; Vigneau et al., 2001) and N-acetamidostyrene ethylenetriaminetriacetic acid (EDTA–amide) (Vigneau et al., 2001) exhibited at least threefold increase in selectivity than analogous non-imprinted polymers. The authors noted that DTPA (nine-coordinate chelating agent)-based

Gd(III)-IIP showed higher Gd/La selectivity (maximum value of approximately 8.3) for Gd(III) in a binary solution of Gd(III) and La(III) than EDTA (five-coordinate chelating agent)-based Gd(III)-IIP (around 3.0). The same DTPA-based Gd(III)-IIP was used to separate Gd(III) and other lanthanides and actinides (Vigneau et al., 2002). Other studies on imprinted polymer include La(III)-IIP (Nik Mustapa et al., 2016), Ce(III)-IIP (Keçili et al., 2018; Nik Mustapa et al., 2016), Y(III)-IIPs (Zulfikar et al., 2017), Nd(III)-IIP (Moussa et al., 2017), ionic-imprinted mesoporous bilayer films (IIBFs) for Dy(III) and Nd(III) adsorption (Zheng et al., 2018), Gd (III)-imprinted membranes (GIMs) (Cui et al., 2019), and Gd(III)-imprinted divinylbenzene and methacrylic acid copolymer (Bunina et al., 2021). The IIPs show a stronger affinity for the imprinted REE cation compared to other REEs.

Surface ion-imprinted polymers are derivatives of IIP technology and include the formation of IIP on a surface. A surface-IIP was prepared for L(III) using MCM-41 as support and APTES as the functional monomer (Qin et al., 2022). The surface-IIP showed an adsorption capacity of 272.2 mg/g. Other examples of surface-IIP are Ce(III)-imprinted polymer with attapulgite support (Pan et al., 2010), Ce(III)-imprinted-polymer-functionalized potassium tetratitanate whisker (Zhang et al., 2010), Ce(III)-IIP functionalized on SBA-15 (Meng et al., 2014), and Gd(III)-imprinted magnetic-ordered mesoporous carbon (Patra et al., 2017).

The limitation in the application of IIP stems from the large number of inaccessible adsorption sites since these sites are embedded inside the polymer (Canfarotta et al., 2016; Dong et al., 2021; Huang et al., 2019). Considering that IIP preparations are usually a sophisticated and chemically expensive process, the presence of inaccessible sites is a big hindrance for widespread industrial application. Additional challenges arise due to low mass transfer rates (Aravind & Mathew, 2018; Dong et al., 2021; Xu & Guo, 2012). These limitations can be improved with surface-ion-imprinted polymers (Dong et al., 2021), but more research effort is required in this direction.

9.8. Elution Solutions

9.8.1. Mineral Acids

In the adsorption-based separation process, the use of mineral acids for the desorption of adsorbed ions and regeneration of the adsorbent is commonplace. However, in the case of REEs chromatographic separation using strongly acidic cation-exchange resin, the use of mineral acids such as HCl, H_2SO_4, HNO_3, and HBr as eluents results in very low separation among REEs (Korkisch et al., 1967; Nelson, 1965). However, these acids can be used to separate REEs from other monovalent and divalent ions using strongly acidic cation resin (Fritz & Garralda, 1963; Page et al., 2019; Strelow, 1960; Strelow et al., 1965). Between various strong mineral acids, the distribution coefficients (K_d) for different REE ions with strongly acidic cation-exchange resin are higher for HCl (Strelow, 1960) than H_2SO_4 and HNO_3 (Strelow et al., 1965) (i.e., in 0.5 M HCl, K_d = 2,480, 2,460, and 1,460; in 0.5 N H_2SO4, K_d = 1,860, 1,800, and 1,380; and in 0.5 N HNO_3, K_d = 1,870, 1,840, and 1,020 for La(III), Ce(III), and Y(III), respectively).

In the case of weakly acidic cation-exchange resins, strong mineral acids can be used for desorption. At very low pH, the excess of hydrogen ions results in the desorption of REE ions from the resin surface through ion exchange. Thus, strong mineral acids are used for the regeneration of resin in many cases. For example, HCl was used with various carboxyl-based weakly acidic cation exchangers D152 (0.5 M HCl) (Xiong et al., 2008), D155 (0.25 M HCl + 0.5 M NaCl) (Xiong, 2008), D113 (4 M HCl) (Xiong et al., 2009), and D151 (0.5 M HCl) (Yao, 2010). In the case of strongly basic anion exchangers, the elution of the adsorbed anionic REE–ligand complex is carried out with HNO_3, H_2SO_4, NH_4SCN, and CH_3COOH solutions (Faris & Warton, 1962; Hubicki & Olszak, 1998, 2002; Koodynska & Hubicki, 2012).

Similarly, chelating resins can be regenerated using mineral acids. In a study with multiple chelating resins, Page et al. (2017) achieved high elution (>97%) for iminodiacetic (IDA) resins with 0.1–1 M HCl, the highest being 99.9% with 1 M HCl. For sulfonic/phosphonic (SP) and aminophosphonic (AP) resins, a higher strength of HCl (2–4 M) achieved 58–80% elution. In another study, 2 M HCl and 2 M HNO_3 were used for desorption of La(III), Ce(III), and Nd(III) ions from chelating resin Dowex M 4195; however, the highest desorption was only 43.9% for Ce(III) achieved with 2 M HNO_3 (Kołodyńska et al., 2019). Other examples of mineral acid as eluent for chelating resins are 2 M HCl for iminodiacetic acid resin (IDAAR) (Xiong et al., 2006); 1 M HCl, 6 M HCl, 1 M HNO_3, and 6 M HNO_3 for Tulsion CH-90 resin (Dutta et al., 2008).

Mineral acid, HNO_3, was used in the desorption step for REEs adsorbed onto alumina and hematite (Kosmulski, 1997b; Marmier et al., 1997). The elution of REEs from amorphous ZrP with 1 M mineral acids decreased in the order of H_2SO_4 > HNO_3 > HCl > H_3PO_4, with the highest being 85.1% and the lowest of 73.2% (Xu et al., 2018). In the case of inorganic TiP, 0.5 M HNO_3 + 0.5 M H_3PO_4 eluted 98.3% Sc(III) (Wenzhong Zhang, Koivula, et al., 2017). Other inorganic nanomaterials have been desorbed using 0.001 M HCl (Zhao et al., 2019), 0.3 M HCl (Marwani et al., 2018), 0.2 M HNO_3 (Ghobadi et al., 2017, 2018), and 0.05–0.8 M HNO_3 (Tu & Johnston, 2018). 0.2 M HNO_3 desorbed >80% REEs from $Co_{0.2}Mn_{0.8}Fe_2O_4$ and $Co_{0.8}Mn_{0.2}Fe_2O_4$ (Ghobadi et al., 2017) and >90% REEs from $MnFe_2O4$ and $MnFe_2O_4$-GO (Ghobadi et al., 2018). With $CuFe_2O_4$ nanoparticles, 100% REE desorption was obtained with 0.05–0.8 M HNO_3 (Tu & Johnston, 2018).

Mineral acid HCl is very frequently used as an eluent in solvent-impregnated resins. Examples of the use of different concentrations of HCl as eluent for SIRs include the use of 0.15 M HCl (Lee et al., 2010a, 2010b), 0.2 M HCl (Park et al., 2005), 0.5 M HCl (Yin et al., 2020), 1 M HCl (Kumar et al., 2011), and 2 M HCl (Aardaneh et al., 2008). Kumar et al. (2011) used HCl, HNO_3, H_2SO_4, and $(NH_4)_2CO_3$ and reported 99.7%, 99.1%, 100%, and 74.6% elution at 1 M concentration. HCl with strength >0.1 M achieved almost complete desorption of La(III) from P227-impregnated XAD-7HP (Yang et al., 2020). Nitric acid solutions are also effective for elution from different SIRs (Helaly et al., 2012; Louis & Duyckaerts, 1984, 1985). İnan et al. (2018) used 0.01 M HNO_3 to desorb REEs from Cyanex-impregnated Amberlite XAD-7. 0.1 M HNO_3 was used as a stripping agent for SIL obtained through ([A336] [CA-100]) impregnation into XAD-7 (Sun et al., 2009). Another SIL loaded with Sc(III) was stripped using 1 M HCl and 1 M HNO_3 with efficiencies >90% and 100% with 1 M H_2SO_4 (Avdibegović et al., 2017).

Similarly, for surface-functionalized adsorbents and ion-imprinted polymers, mineral acids are mostly used for the desorption of REEs. The strength of the acid solution used for desorption has varied from 0.1 M HCl to 2.0 M HCl (e.g., 0.1 M HCl (Ahmed, Lee, et al., 2019; Kostenko et al., 2019; Lee et al., 2018; Ryu et al., 2021), 0.25 M HCl (Hua et al., 2019), 0.3 M HCl (Lou et al., 2019), 0.5 M HCl (Chen, H. et al., 2014; Ravi, Lee, et al., 2018), 1.0 M HCl (Babu et al., 2018; Gasser et al., 2017; Moussa et al., 2017), 2 M HCl (Li et al., 2019; Qin et al., 2022; Yusoff et al., 2017), and 1.0–2.0 M HCl (Ogata et al., 2015a). Other mineral acids such as HNO_3 with the strength of 0.01–6.55 M (Ashour et al., 2017; Bao et al., 2022; Bertelsen et al., 2019; Callura et al., 2018; Chen, W. et al., 2014; Noack et al., 2014; Rahman et al., 2020; Ravi, Zhang, et al., 2018; Zhao et al., 2021) and H_2SO_4 (e.g., 0.1–1 M H_2SO_4 (Ogata et al., 2015a), 0.1 M H_2SO_4 (Gaete et al., 2021)) have been used as well. Mineral acids usually provide 90–100% desorption for functionalized adsorbent and ion-imprinted polymers. The desorption mechanism with mineral acid is the ion-exchange reaction between protons and the adsorbed ions (Ogata et al., 2015a).

9.8.2. Organic Acids and Complexing Ligands

Along with mineral acids, various organic acids and complexing ligand solutions have been used for the desorption of REEs from different adsorbents. In the case of strongly acidic cation-exchange resins, the organic acid and complexing ligands play a vital part as eluent, since they can increase the chromatographic separation between REEs in the elution step due to their difference in affinity for different REEs (Boyd, 1978; Chen et al., 2017; Spedding et al., 1947, 1954; Strelow & Victor, 1990). Examples of organic acids used as eluents in REE separation include acetic, malonic, phthalic, and other carboxyl acids, hydroxy acids, and phosphonic and amino phosphonic acids (Koodynska & Hubicki, 2012).

However, the best results for REE desorption and separation from strong acidic cation-exchange resins were obtained using complexing ligand as eluent, primarily eluents based on polyamino-polycarboxylic acid such as EDTA (Smith & Hoffman, 1956; Spedding et al., 1954; Spedding & Powell, 1954); diethylenetriaminepentaacetic acid (DTPA) (Chuveleva et al., 1974, 1995; Hale & Hammer, 1972; Kharitonov et al., 2009b; Kogan & Ratner, 1971), nitrilotriacetic acid (Kharitonov et al., 2009a; Martynenko et al., 1968, 1972; Wheelwright, 1969), hydroxy-ethylethylenediaminetriacetic acid (Hagiwara, 1969; James et al., 1961; Strelow & Victor, 1990), cyclohexane-1,2-diaminetetraacetic acid (Gschneidner & Eyring, 1982), and iminodiacetic acid (Martynenko et al., 1972).

The advantage of using polyamino-polycarboxylic acid as eluent is that they form strong complexes with REEs, reducing the consumption of ligands and resulting in concentrated elutes. Also, the stability constant of different REEs with these complexes differs more than their ion-exchange constants, which results in a higher degree of separation (Chen et al., 2017; Ehrlich & Lisichkin, 2017; Koodynska & Hubicki, 2012). The most widely used polyamino–polycarboxylic acid is EDTA. EDTA is readily available, inexpensive, and easy to regenerate. Additionally, due to the multidentate nature of EDTA, the REEs form 1:1 complexes with EDTA, which reduces metal zone tailing.

EDTA has been used as a stripping agent for SLEs and functionalized adsorbents (Gupta & Sengupta, 2017; Li et al., 2018; Liu et al., 2009; Zhao et al., 2016). EDTA and oxalic acid were used for desorption with amide-functionalized multiwalled carbon nanotubes (Gupta & Sengupta, 2017). Other organic compounds used for desorption of REEs include acetonitrile (Abdel-Magied et al., 2019; Jiang et al., 2016; Zheng et al., 2015), methanol (Abdel-Magied et al., 2019), and ammonium oxalate (Florek et al., 2014, 2015; Giret et al., 2018; Hu et al., 2017, 2019; Juère et al., 2016; Perreault et al., 2017).

9.9. Application to Real Samples

Examples of using solid–liquid separation at an industrial level are scarce. Diphonix was used to extract REES from the nitric acid solution from apatite processing on a pilot scale at Akron Public Corporation (Ehrlich & Lisichkin, 2017). However, there are lab studies in literature where various adsorbents have been used to separate REEs from various natural samples such as bauxite mud (Hu et al., 2019; Ochsenkühn-Petropulu et al., 1995; Roosen et al., 2016; Zhou et al., 2008); Silicate and niobium mining deposits (Hu et al., 2017); apatite, orthite, and slag (Kostenko et al., 2019; Ogata et al., 2016); bastnäsite (Dolak et al., 2015), acidic mine drainage (Hermassi et al., 2021; Ramasamy et al., 2019; Ramasamy, Puhakka, Iftekhar, et al., 2018; Ramasamy, Puhakka, Repo, et al., 2018); spiked seawater (Callura et al., 2018; Noack et al., 2016); spiked river/groundwater and sewage water (Marwani et al., 2018; Yantasee et al., 2009); geothermal brines (Liu et al., 2021), acidic streams, and dialysate (Yantasee et al., 2009); REE mineral leachate (Giret et al., 2018); industrial wastewater (Li et al., 2013); leached solution of spent Ni-MH cells (Gasser & Aly, 2013; Kołodyńska et al., 2019); thin-film phosphors leached solution (Schaeffer et al., 2017); phosphoric acid plant effluent (Al-Thyabat & Zhang, 2015); zinc mine ore (Fonseka et al., 2021); and coal fly ash (Brown & Balkus, 2021; Hovey et al., 2021; Mondal et al., 2019).

These adsorbents showed varying degrees of success for REE extraction (Figure 9.4). REE extraction from red mud used ion-exchanger Dowex 50W-X8 in combination with the solvent-extraction method (Ochsenkühn-Petropulu et al., 1995). The whole separation process resulted in an average 93% recovery of Sc(III). High La(III) and Nd(III) recovery (98%) from spent Ni-MH cells was achieved using layered double hydroxide (or hydrotalcite) functionalized with Cyanex-272 (Fe-Mg-LDH-Cyanex-272) (Gasser & Aly, 2013). In another study, 95% of REEs were separated at pH 4.2 from the waste sulfate solutions of uranium leaching solution using carboxyl adsorbent CYBBER LX 280, after an initial step involving $Fe(OH)_3$ precipitation (Lokshin et al., 2013). Bidentate phthaloyl diamide (PA) functionalized KIT-6 showed high selectivity for Lu(III) from a leached solution of silicate and niobium mining deposit samples (legends 6–11, Figure 9.4) (Hu et al., 2017). The adsorbent tetradentate PDDA-functionalized KIT-6 showed a higher affinity for HREEs in the case of KIT-6-1,2–PDDA (legend 14, Figure 9.4) and LREEs with KIT-6-1,3–PDDA (legend 15, Figure 9.4), while extracting REEs from red bauxite residue (Hu et al., 2019). Silica gels functionalized with phosphonoacetic acid (PAA), N,N-bis(phosphonomethyl) gylcine (BPG), diethylenetriaminepentaacetic acid (DTPA), and diethylenetriaminepentaacetic

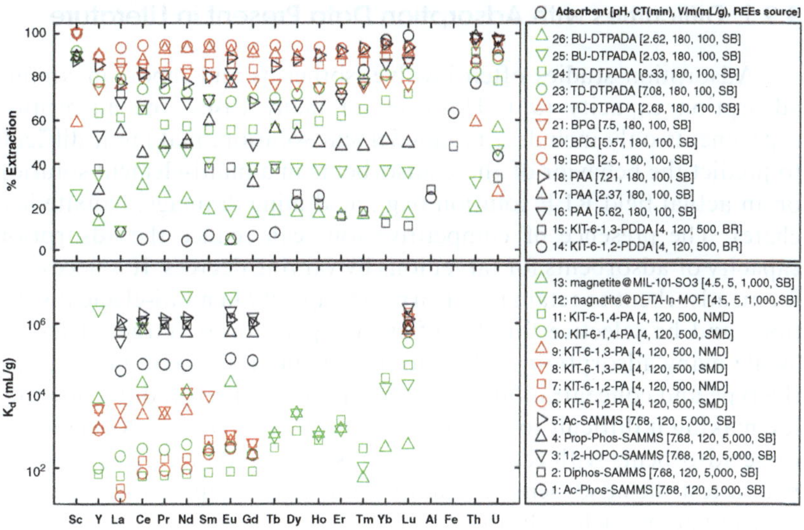

Figure 9.4 Distribution coefficient (K_d) for various adsorbents in actual REE solution. The legends include adsorbents with the experimental condition (pH, contact time (CT), *V/m* ratio (volume of solution/mass of adsorbent and source of the REE solution)). Sources of REEs: SRW: spiked river water, SB: spiked brine, SMD: silicate mining deposit, NMD: niobium mining deposit, BR: bauxite residue. Source for data: adsorbent 1–5 (Yantasee et al., 2009), 6–11 (Hu et al., 2017), 12–13 (Elsaidi et al., 2018), 14–15 (Hu et al., 2019), and 16–26 (reported pH is equilibrium pH) (Callura et al., 2018).

dianhydride (DTPADA) were able to adsorb REEs from spiked brine samples with high efficiency (legends 16–26, Figure 9.4) (Callura et al., 2018; Noack et al., 2016). Nd(III)-ion-imprinted polymer extracted 100% Nd(III) from bastnäsite leachate solution (Dolak et al., 2015). In fly ash leachate solution, bis(ethylhexyl)amido diethylenetriaminepentaacetic acid-functionalized silica showed selectivity for REEs over Fe and Al that were present in much higher concentrations than REEs (approximately 700-fold to 90,000-fold excess) (Hovey et al., 2021). Eu(III) was selectively recovered (88% Eu(III) recovery, <10% recovery of competitive ions) from zinc mine ore leachate solution adjusted to pH 5.5 using PMIDA-functionalized MOF (Cr-MIL-PMIDA) (Fonseka et al., 2021). The single-element Eu(III) adsorption at pH 5.5 was 90–91%; thus, the presence of a high concentration of competing ions did not affect the Eu(III) adsorption. The Langmuir maximum capacity (q_m) for Eu(III) on Cr-MIL-PMIDA was 69.14 mg/g.

9.9.1. Difficulties with Adsorption Data Present in Literature

Adsorption studies often involve isotherms and report maximum adsorption capacity (q_m). However, in many cases, the adsorption experiments only include a mono-element solution, making it difficult to predict the behavior of these adsorbents in a multi-element solution or an actual field REE solution (e.g., acid mine drainage, bauxite leachate). The presence of competitive ions can reduce the adsorption capacity of adsorbents for target ions by varying degrees. It is necessary for an adsorbent to retain its adsorption capacity in multi-element solutions and be selective for the efficient separation of target elements. While many adsorption studies do include multi-element adsorptions, the type and concentration of competitive ions differ in different studies, making it difficult to compare the mechanism and performance of these adsorbents with each other. Even so, it should be a general practice to test the adsorbent in a multi-element solution with competitive ions and actual field solutions.

9.10. Synopsis

The adsorption-based REE separation has been explored for a long time, and the technology has evolved. For separating REEs from other cations with a different charge, cation exchangers may be recommended. However, for the selective separation of specific REEs, ligand-functionalized adsorbents promoting complexation are more suitable.

There remain strikingly large gaps in two directions with adsorption-based REE separation: (1) lack of industrial-level applications and setups, and (2) lack of fundamental level modeling and understanding of surface ligand behavior with the REEs. Although there have been many REE separation studies at lab scales, there is a lack of examples of conversion to industrial-level applications.

The challenges in SPE with functionalized adsorbents are the selection of selective ligands and predicting the behavior of these ligands once they are grafted onto a specific solid surface. No quantitative relationship has been developed between free-solution thermodynamics data and surface complexation constants for the functionalized adsorbents. Future research can employ quantum chemistry calculations to select ligands and predict the behavior after grafting. Understanding and modeling the ligand grafted on the solid surface and recognizing the difference in metal–ligand interaction on the surface against the metal–ligand interaction in the free solution can help better develop functionalized adsorbents.

Acknowledgments

This work was supported by the College of Earth and Minerals Sciences and the John and Willie Leone Family Department of Energy and Mineral Engineering of Pennsylvania State University.

Appendix A. Abbreviations

AC	Activated carbon
AER	Anion-exchange resin
APTES	3-aminopropyl triethoxysilane
APTS	3-[2-(2-aminoethylamino) ethylamino] propyl-trimethoxysilane
APTMS	3-aminopropyl trimethoxysilane
ATMP	Amino tris(methylene phosphonic acid)
ATS	Aluminum silicotitanates
BHPA	N-Benzoyl-N-phenylhydroxylamine
BPG	Bis(phosphonomethyl) glycine
CER	Cation-exchange resin
CN	Coordination number
CNT	Carbon nanotube
COK	Centre for Research Chemistry and Catalysis
Cyanex 272	Bis(2,4,4-trimethylpentyl)-phosphinic acid
DETA	Diethylenetriamine
DGA	Diglycolamide
DMF	N,N-dimethylformamide
DOODA	3,6-dioxaoctanediamidopropyl
DTPA	Diethylenetriaminepentaacetic acid
DTPADA	Diethylenetriaminepentaacetic dianhydride
EDTA	Ethylenediaminetetraacetic acid
FDGA	Furan-2,4-diamidopropyltriethoxysilane
GO	Graphene oxide
GONS	Graphene oxide nanosheet
HDEHP	Bis(2-ethylhexyl) hydrogen phosphate
HREE	Heavy rare earth element
IER	Ion-exchange resin
IIP	Ion-imprinted polymer
KIT	Korean Advance Institute of Science and Technology
LREE	Light rare earth element
MCM	Mobil Composition of Matter
MIP	Molecular imprinted polymer
MOF	Metal-organic framework

MSF	Mesoporous silica foam
MWCNT	Multi-walled carbon nanotube
OMC	Ordered mesoporous carbon
OMS	Ordered mesoporous silica
PA	Phthaloyl diamide
PAA	Phosphonoacetic acid
PAN	1-(2-pyridylazo)-2-naphthol
PC88A	2-ethylhexyl phosphonic acid mono-2-ethylhexyl ester
PDDA	Phenylenedioxy diamide
PES	Polyethersulfone
PMIDA	N-(phosphonomethyl)iminodiacetic acid
PVA	Polyvinyl alcohol
REE	Rare earth element
SBA	Santa Barbara Amorphous
SIL	Supported ionic liquid
SIR	Solvent impregnated resin
SLE	Supported liquid extraction
SLLE	Supported liquid–liquid extraction
SPE	Solid-phase extraction
SWCNT	Single-walled carbon nanotube
TEHDGA	N,N,N′N′-tetrakis-2-ethylhexyldiglycolamide
TMO	Transition metal oxide
TODGA	N,N,N′,N′-tetraoctyl diglycolamide
TTHA	Triethylenetetraminehexaacetic acid
ZrP	Zirconium organophosphonates

References

Aardaneh, K., Saal, D., Swarts, G., & Dewindt, S. C. (2008). TBP and TBP impregnated Amberlite XAD-4 resin for radiochemical separation of 88Y from Sr and Al. *Journal of Radioanalytical and Nuclear Chemistry*, *275*(3), 665–669. https://doi.org/10.1007/s10967-007-7074-6

Abdel-Magied, A. F., Abdelhamid, H. N., Ashour, R. M., Zou, X., & Forsberg, K. (2019). Hierarchical porous zeolitic imidazolate frameworks nanoparticles for efficient adsorption of rare-earth elements. *Microporous and Mesoporous Materials*, *278*(September 2018), 175–184. https://doi.org/10.1016/j.micromeso.2018.11.022

Abu Elgoud, E. M., Ismail, Z. H., Ahmad, M. I., El-Nadi, Y. A., Abdelwahab, S. M., & Aly, H. F. (2019). Sorption of Lanthanum(III) and Neodymium(III) from concentrated phosphoric acid by strongly acidic cation exchange resin (SQS-6). *Russian Journal of Applied Chemistry*, *92*(11), 1581–1592. https://doi.org/10.1134/S1070427219110156

Ahmed, I., Adhikary, K. K., Lee, Y. R., Ho Row, K., Kang, K. K., & Ahn, W. S. (2019). Ionic liquid entrapped UiO-66: Efficient adsorbent for Gd3+ capture from water. *Chemical Engineering Journal*, *370*, 792–799. https://doi.org/10.1016/j.cej.2019.03.265

Ahmed, I., Bhattacharjee, S., Lee, C. S., Kang, K. K., Ahn, J. W., & Ahn, W. S. (2021). Aqueous Nd3+ capture using a carboxyl-functionalized porous carbon derived from ZIF-8. *Journal of Colloid and Interface Science*, *594*, 702–712. https://doi.org/10.1016/J.JCIS.2021.03.036

Ahmed, I., Lee, Y.-R., Yu, K., Bhattacharjee, S., & Ahn, W.-S. (2019). Gd 3+ adsorption over carboxylic- and amino-group dual-functionalized UiO-66 [research-article]. *Industrial & Engineering Chemistry Research*, *58*(6), 2324–2332. https://doi.org/10.1021/acs.iecr.8b05220

Aja, S. U. (1998). Sorption of the rare earth element, Nd, onto kaolinite at 25 °C. *Clays and Clay Minerals*, *46*(1), 103–109. https://doi.org/10.1346/CCMN.1998.0460112

Al-Thyabat, S., & Zhang, P. (2015). In-line extraction of REE from Dihydrate (DH) and HemiDihydrate (HDH) wet processes. *Hydrometallurgy*, *153*, 30–37. https://doi.org/10.1016/j.hydromet.2015.01.010

Alexandratos, S. D., & Hussain, L. A. (1995). Bifunctionality as a means of enhancing complexation kinetics in selective ion exchange resins. *Industrial and Engineering Chemistry Research*, *34*(1), 251–254. https://doi.org/10.1021/ie00040a026

Alexandratos, S. D., & Hussain, L. A. (1998). Synthesis of α-, β-, and γ-ketophosphonate polymer-supported reagents: The role of intra-ligand cooperation in the complexation of metal ions. *Macromolecules*, *31*(10), 3235–3238. https://doi.org/10.1021/ma971587d

Alexandratos, S. D., & Natesan, S. (1999). Ion-selective polymer-supported reagents: The principle of bifunctionality. *European Polymer Journal*, *35*(3), 431–436. https://doi.org/10.1016/S0014-3057(98)00142-6

Alexandratos, S. D., & Smith, S. D. (2004a). High stability solvent impregnated resins: Metal ion complexation as a function of time. *Solvent Extraction and Ion Exchange*, *22*(4), 713–720. https://doi.org/10.1081/SEI-120038701

Alexandratos, S. D., & Smith, S. D. (2004b). Intraligand cooperation in metal-ion binding by immobilized ligands: The effect of bifunctionality. *Journal of Applied Polymer Science*, *91*(1), 463–468. https://doi.org/10.1002/app.13131

Alexandratos, S. D., & Zhu, X. (2005). Bifunctional coordinating polymers: Auxiliary groups as a means of tuning the ionic affinity of immobilized phosphate ligands. *Macromolecules*, *38*(14), 5981–5986. https://doi.org/10.1021/ma050057b

Alexandratos, S. D., & Zhu, X. (2008). Polyols as scaffolds in the development of ion-selective polymer-supported reagents: The effect of auxiliary groups on the mechanism of metal ion complexation. *Inorganic Chemistry*, *47*(7), 2831–2836. https://doi.org/10.1021/ic702263x

Alexandratos, S. D., & Zhu, X. (2015). The role of polarizability in determining metal ion affinities in polymer-supported reagents: monoprotic phosphates and the effect of hydrogen bonding. *New Journal of Chemistry*, *39*(7), 5366–5373. https://doi.org/10.1039/C5NJ00387C

Alguacil, F. J., García-Díaz, I., Escudero Baquero, E., Rodríguez Largo, O., & López, F. A. (2020). On the adsorption of cerium(III) using multiwalled carbon nanotubes. *Metals*, *10*(8), 1057. https://doi.org/10.3390/met10081057

Amarasekara, A. S., Owereh, O. S., & Aghara, S. K. (2009). Synthesis of 4-acylpyrazolone Schiff base ligand grafted silica and selectivity in adsorption of lanthanides from aqueous solutions. *Journal of Rare Earths*, *27*(5), 870–874. https://doi.org/10.1016/S1002-0721(08)60352-X

Aravind, A., & Mathew, B. (2018). Electrochemical sensor based on nanostructured ion imprinted polymer for the sensing and extraction of Cr(III) ions from industrial wastewater. *Polymer International*, *67*(12), 1595–1604. https://doi.org/10.1002/PI.5683

Arnold, R., & Hing, L. B. S. (1967). Selectivity of carboxylic ion-exchange resin for lanthanide ions. *Journal of the Chemical Society A: Inorganic, Physical, Theoretical*, *2*, 306. https://doi.org/10.1039/j19670000306

Artiushenko, O., Ávila, E. P., Nazarkovsky, M., & Zaitsev, V. (2020). Reusable hydroxamate immobilized silica adsorbent for dispersive solid phase extraction and separation of rare earth metal ions. *Separation and Purification Technology*, *231*(August 2019), 115934. https://doi.org/10.1016/j.seppur.2019.115934

Artiushenko, O., Kostenko, L., & Zaitsev, V. (2020). Influence of competitive eluting agents on REEs recovery from silica gel adsorbent with immobilized aminodiphosphonic acid. *Journal of Environmental Chemical Engineering*, *8*(4), 103883. https://doi.org/10.1016/j.jece.2020.103883

Asadollahzadeh, M., Torkaman, R., & Torab-Mostaedi, M. (2020). Extraction and separation of rare earth elements by adsorption approaches: current status and future trends. *Separation & Purification Reviews*, *50*(4), 417–444. https://doi.org/10.1080/15422119.2020.1792930

Ashour, R. M., Abdelhamid, H. N., Abdel-Magied, A. F., Abdel-Khalek, A. A., Ali, M. M., Uheida, A., et al. (2017). Rare earth ions adsorption onto graphene oxide nanosheets. *Solvent Extraction and Ion Exchange*, *35*(2), 91–103. https://doi.org/10.1080/07366299.2017.1287509

Attallah, M. F., Elgazzar, A. H., Borai, E. H., & El-Tabl, A. S. (2016). Preparation and characterization of aluminum silicotitanate: ion exchange behavior for some lanthanides and iron. *Journal of Chemical Technology & Biotechnology*, *91*(8), 2243–2252. https://doi.org/10.1002/jctb.4810

Attia, M. A., Moussa, S. I., Sheha, R. R., Someda, H. H., & Saad, E. A. (2019). Hydroxyapatite/NiFe 2 O 4 superparamagnetic composite: Facile synthesis and adsorption of rare elements. *Applied Radiation and Isotopes*, *145*(December 2018), 85–94. https://doi.org/10.1016/j.apradiso.2018.12.003

Avdibegović, D., Regadío, M., & Binnemans, K. (2017). Recovery of scandium(<scp>iii</scp>) from diluted aqueous solutions by a supported ionic liquid phase (SILP). *RSC Advances*, *7*(78), 49664–49674. https://doi.org/10.1039/C7RA07957E

Awual, M. R., Kobayashi, T., Miyazaki, Y., Motokawa, R., Shiwaku, H., Suzuki, S., et al. (2013). Selective lanthanide sorption and mechanism using novel hybrid Lewis base (N-methyl-N-phenyl-1,10-phenanthroline-2-carboxamide) ligand modified adsorbent. *Journal of Hazardous Materials, 252–253*, 313–320. https://doi.org/10.1016/j.jhazmat.2013.03.020

Babu, C. M., Binnemans, K., & Roosen, J. (2018). Ethylenediaminetriacetic acid-functionalized activated carbon for the adsorption of rare earths from aqueous solutions. *Industrial and Engineering Chemistry Research*, *57*(5), 1487–1497. https://doi.org/10.1021/acs.iecr.7b04274

Bao, S., Tang, Y., Zhang, Y., & Liang, L. (2016). Recovery and separation of metal ions from aqueous solutions by solvent-impregnated resins. *Chemical Engineering & Technology*, *39*(8), 1377–1392. https://doi.org/10.1002/ceat.201500324

Bao, S., Wang, Y., Wei, Z., Yang, W., & Yu, Y. (2022). Highly efficient recovery of heavy rare earth elements by using an amino-functionalized magnetic graphene oxide with acid and base resistance. *Journal of Hazardous Materials*, *424*, 127370. https://doi.org/10.1016/j.jhazmat.2021.127370

Barrak, H., Ahmedi, R., Chevallier, P., M'nif, A., Laroche, G., & Hamzaoui, A. H. (2019). Highly efficient extraction of rare earth elements and others ions from green phosphoric acid medium using TMSEDTA@GO@Fe3O4 core-shell. *Separation and Purification Technology*, *222*, 145–151. https://doi.org/10.1016/j.seppur.2019.04.016

Baumann, A. E., Burns, D. A., Liu, B., & Thoi, V. S. (2019). Metal-organic framework functionalization and design strategies for advanced electrochemical energy storage devices. *Communications Chemistry*, *2*(1), 1–14. Springer Nature. https://doi.org/10.1038/s42004-019-0184-6

Beck, J. S., Vartuli, J. C., Roth, W. J., Leonowicz, M. E., Kresge, C. T., Schmitt, K. D., et al. (1992). A new family of mesoporous molecular sieves prepared with liquid crystal templates. *Journal of the American Chemical Society*, *114*(27), 10834–10843. https://doi.org/10.1021/ja00053a020

Behdani, F. N., Rafsanjani, A. T., Torab-Mostaedi, M., & Mohammadpour, S. M. A. K. (2013). Adsorption ability of oxidized multiwalled carbon nanotubes towards aqueous Ce(III) and Sm(III). *Korean Journal of Chemical Engineering*, *30*(2), 448–455. https://doi.org/10.1007/s11814-012-0126-9

Bertelsen, E. R., Deodhar, G., Kluherz, K. T., Davidson, M., Adams, M. L., Trewyn, B. G., & Shafer, J. C. (2019). Microcolumn lanthanide separation using bis-(2-ethylhexyl) phosphoric acid functionalized ordered mesoporous carbon materials. *Journal of Chromatography A*, *1595*, 248–256. https://doi.org/10.1016/j.chroma.2019.02.057

Bezzina, J. P., Ogden, M. D., Moon, E. M., & Soldenhoff, K. L. (2018). REE behavior and sorption on weak acid resins from buffered media. *Journal of Industrial and Engineering Chemistry*, *59*, 440–455. https://doi.org/10.1016/j.jiec.2017.11.005

Bhatnagar, A., & Sillanpää, M. (2010). Utilization of agro-industrial and municipal waste materials as potential adsorbents for water treatment—A review. *Chemical Engineering Journal*, *157*(2–3), 277–296. Elsevier. https://doi.org/10.1016/j.cej.2010.01.007

Biju, V., Gladis, J. M., & Rao, T. P. (2003). Ion imprinted polymer particles: synthesis, characterization and dysprosium ion uptake properties suitable for analytical applications. *Analytica Chimica Acta*, *478*(1), 43–51. https://doi.org/10.1016/S0003-2670(02)01416-2

Botelho Junior, A. B., Pinheiro, É. F., Espinosa, D. C. R., Tenório, J. A. S., & Baltazar, M. dos P. G. (2021). Adsorption of lanthanum and cerium on chelating ion exchange resins: kinetic and thermodynamic studies. *Separation Science and Technology*, *57*(1), 60–69. doi: 10.1080/01496395.2021.188472

Boyd, G. E. (1978). Thermodynamic property changes in lanthanide(III) cation exchange reactions with poly(styrenesulfonic acid) type cation exchangers. *Journal of Physical Chemistry*, *82*(25), 2704–2709. https://doi.org/10.1021/j100514a015

Brown, A. T., & Balkus, K. J. (2021). Critical rare earth element recovery from coal ash using microsphere flower carbon. *ACS Applied Materials and Interfaces*, *13*(41), 48492–48499. https://doi.org/10.1021/ACSAMI.1C09298/SUPPL_FILE/AM1C09298_SI_001.PDF

Bunina, Z., Bryleva, K., & Belikov, K. (2021). Synthesis and adsorption properties of gadolinium-imprinted divinylbenzene-based copolymers. *ACS Omega*, *6*(4), 3336–3344. https://doi.org/10.1021/ACSOMEGA.0C05812/SUPPL_FILE/AO0C05812_SI_001.PDF

Callura, J. C. (2018). *Ligand-functionalized adsorbents for the extraction and recovery of rare earth elements* (Issue December). Carnegie Mellon University.

Callura, J. C., Perkins, K. M., Baltrus, J. P., Washburn, N. R., Dzombak, D. A., & Karamalidis, A. K. (2019). Adsorption kinetics, thermodynamics, and isotherm studies for functionalized lanthanide-chelating resins. *Journal of Colloid and Interface Science*, *557*, 465–477. https://doi.org/10.1016/j.jcis.2019.08.097

Callura, J. C., Perkins, K. M., Noack, C. W., Washburn, N. R., Dzombak, D. A., & Karamalidis, A. K. (2018). Selective adsorption of rare earth elements onto functionalized silica particles. *Green Chemistry*, *20*(7), 1515–1526. https://doi.org/10.1039/C8GC00051D

Callura, J. C., Shi, Q., Dzombak, D. A., & Karamalidis, A. K. (2021). Selective recovery of rare earth elements with ligand-functionalized polymers in fixed-bed adsorption columns. *Separation and Purification Technology*, *265*, 118472. https://doi.org/10.1016/J.SEPPUR.2021.118472

Canfarotta, F., Poma, A., Guerreiro, A., & Piletsky, S. (2016). Solid-phase synthesis of molecularly imprinted nanoparticles. *Nature Protocols*, *11*(3), 443–455. https://doi.org/10.1038/nprot.2016.030

Cardoso, C. E. D., Almeida, J. C., Lopes, C. B., Trindade, T., Vale, C., & Pereira, E. (2019). Recovery of rare earth elements by carbon-based nanomaterials—A review. *Nanomaterials*, *9*(6), 814. https://doi.org/10.3390/nano9060814

Chen, B., He, M., Zhang, H., Jiang, Z., & Hu, B. (2017). Chromatographic techniques for rare earth elements analysis. *Physical Sciences Reviews*, *2*(4), 20160057. https://doi.org/10.1515/psr-2016-0057

Chen, C. L., Wang, X. K., & Nagatsu, M. (2009). Europium adsorption on multiwall carbon nanotube/iron oxide magnetic composite in the presence of polyacrylic acid. *Environmental Science and Technology*, *43*(7), 2362–2367. https://doi.org/10.1021/es803018a

Chen, H., Shao, D., Li, J., & Wang, X. (2014). The uptake of radionuclides from aqueous solution by poly(amidoxime) modified reduced graphene oxide.

Chemical Engineering Journal, *254*, 623–634. https://doi.org/10.1016/j.cej.2014.05.091

Chen, L., Wang, X., Lu, W., Wu, X., & Li, J. (2016). Molecular imprinting: Perspectives and applications. *Chemical Society Reviews*, *45*(8), 2137–2211. https://doi.org/10.1039/c6cs00061d

Chen, W., Wang, L., Zhuo, M., Liu, Y., Wang, Y., & Li, Y. (2014). Facile and highly efficient removal of trace Gd(III) by adsorption of colloidal graphene oxide suspensions sealed in dialysis bag. *Journal of Hazardous Materials*, *279*, 546–553. https://doi.org/10.1016/j.jhazmat.2014.06.075

Chen, Y., Zhu, B., Wu, D., Wang, Q., Yang, Y., Ye, W., & Guo, J. (2012). Eu(III) adsorption using di(2-thylhexly) phosphoric acid-immobilized magnetic GMZ bentonite. *Chemical Engineering Journal*, *181–182*, 387–396. https://doi.org/10.1016/j.cej.2011.11.100

Chen, Z., Li, Z., Chen, J., Tan, H., Wu, J., & Qiu, H. (2022). Selective adsorption of rare earth elements by Zn-BDC MOF/graphene oxide nanocomposites synthesized via in situ interlayer-confined strategy. *Industrial & Engineering Chemistry Research*, *61*(4), 1841–1849. https://doi.org/10.1021/ACS.IECR.1C04180/SUPPL_FILE/IE1C04180_SI_001.PDF

Chuenchom, L., Kraehnert, R., & Smarsly, B. M. (2012). Recent progress in soft-templating of porous carbon materials. *Soft Matter*, *8*(42), 10801–10812. https://doi.org/10.1039/c2sm07448f

Chuveleva, E. A., Kharitonov, O. V, & Firsova, L. A. (1995). Effect of diethylenetriamine of pentaacetic acid on the chromatographic-separation of rare-earth elements on Ku-2 cation-exchange resin. *Zhurnal Fizicheskoi Khimii*, *69*(7), 1322–1326.

Chuveleva, E. A., Nazarov, P. P., & Chmutov, K. V. (1974). Chromatographic-separation of curium, americium and rare-earth elements. *Zhurnal Fizicheskoi Khimii*, *48*(12), 3078–3081.

Coppin, F., Berger, G., Bauer, A., Castet, S., & Loubet, M. (2002). Sorption of lanthanides on smectite and kaolinite. *Chemical Geology*, *182*(1), 57–68. https://doi.org/10.1016/S0009-2541(01)00283-2

Cotton, S. A., & Harrowfield, J. M. (2012). Lanthanides: Coordination chemistry. In *Encyclopedia of Inorganic and Bioinorganic Chemistry*. John Wiley & Sons https://doi.org/10.1002/9781119951438.eibc2062

Croissant, J. G., Fatieiev, Y., & Khashab, N. M. (2017). Degradability and clearance of silicon, organosilica, silsesquioxane, silica mixed oxide, and mesoporous silica nanoparticles. In *Advanced Materials*, Vol. *29*, Issue 9, p. 1604634 https://doi.org/10.1002/adma.201604634

Crundwell, F. K. (2017). On the mechanism of the dissolution of quartz and silica in aqueous solutions. *ACS Omega*, *2*(3), 1116–1127. https://doi.org/10.1021/acsomega.7b00019

Cui, K., Gao, B., Tai, M., & Su, B. (2019). A facile bionic strategy towards Gd (III)-imprinted membranes via interlaced stacking of one-dimensional/two-dimensional nanocomposite materials. *Journal of the Taiwan Institute of Chemical Engineers*, *95*, 652–659. https://doi.org/10.1016/j.jtice.2018.09.036

Dardenne, K., Schäfer, T., Lindqvist-Reis, P., Denecke, M. A., Plaschke, M., Rothe, J., & Kim, J. I. (2002). Low temperature XAFS investigation on the Lutetium binding changes during the 2-line ferrihydrite alteration process. *Environmental Science and Technology*, *36*(23), 5092–5099. https://doi.org/10.1021/es025513f

Davranche, M., Pourret, O., Gruau, G., & Dia, A. (2004). Impact of humate complexation on the adsorption of REE onto Fe oxyhydroxide. *Journal of Colloid and Interface Science*, *277*(2), 271–279. https://doi.org/10.1016/j.jcis.2004.04.007

Davranche, M., Pourret, O., Gruau, G., Dia, A., & Le Coz-Bouhnik, M. (2005). Adsorption of REE(III)-humate complexes onto MnO2: Experimental evidence for cerium anomaly and lanthanide tetrad effect suppression. *Geochimica et Cosmochimica Acta*, *69*(20), 4825–4835. https://doi.org/10.1016/j.gca.2005.06.005

de Decker, J., de Clercq, J., Vermeir, P., & van der Voort, P. (2016). Functionalized metal-organic-framework CMPO@MIL-101(Cr) as a stable and selective rare earth adsorbent. *Journal of Materials Science*, *51*(10), 5019–5026. https://doi.org/10.1007/s10853-016-9807-9

Dolak, İ., Keçili, R., Hür, D., Ersöz, A., & Say, R. (2015). Ion-imprinted polymers for selective recognition of neodymium(III) in environmental samples. *Industrial & Engineering Chemistry Research*, *54*(19), 5328–5335. https://doi.org/10.1021/acs.iecr.5b00212

Dong, C., Shi, H., Han, Y., Yang, Y., Wang, R., & Men, J. (2021). Molecularly imprinted polymers by the surface imprinting technique. *European Polymer Journal*, *145*, 110231. https://doi.org/10.1016/j.eurpolymj.2020.110231

Dubey, S. S., & Grandhi, S. (2016). Sorption studies of yttrium (III) ions on nano maghemite. *Journal of Environmental Chemical Engineering*, *4*(4), 4719–4730. https://doi.org/10.1016/j.jece.2016.11.006

Dubey, S. S., & Grandhi, S. (2019). Sorption studies of yttrium(III) ions on surfaces of nano-thorium(IV) oxide and nano-zirconium (IV) oxide. *International Journal of Environmental Science and Technology*, *16*(1), 59–70. https://doi.org/10.1007/s13762-017-1589-3

Dupont, D., Brullot, W., Bloemen, M., Verbiest, T., & Binnemans, K. (2014). Selective uptake of rare earths from aqueous solutions by EDTA-functionalized magnetic and nonmagnetic nanoparticles. *ACS Applied Materials & Interfaces*, *6*(7), 4980–4988. https://doi.org/10.1021/am406027y

Dutta, S., Mohapatra, P. K., Dhekane, G. D., Das, A. K., & Manchanda, V. K. (2008). Solid phase extraction of europium and uranium using Tulsion CH-90 resin. *Desalination*, *232*(1–3), 216–224. https://doi.org/10.1016/j.desal.2007.10.038

Eftekhari, A., & Fan, Z. (2017). Ordered mesoporous carbon and its applications for electrochemical energy storage and conversion. *Materials Chemistry Frontiers*, *1*(6), 1001–1027. https://doi.org/10.1039/c6qm00298f

Egawa, H., Yamabe, K., & Jyo, A. (1994). Studies on selective adsorption resins. XXXIII. Behavior of macroreticular chelating resins containing phosphinic and/or phosphonic acid groups in the adsorption of trivalent lanthanides. *Journal of Applied Polymer Science*, *52*(8), 1153–1164. https://doi.org/10.1002/app.1994.070520815

Ehrlich, G. V., & Lisichkin, G. V. (2017). Sorption in the chemistry of rare earth elements. *Russian Journal of General Chemistry*, *87*(6), 1220–1245. https://doi.org/10.1134/S1070363217060196

El Mourabit, S., Guillot, M., Toquer, G., Cambedouzou, J., Goettmann, F., & Grandjean, A. (2012). Stability of mesoporous silica under acidic conditions. *RSC Advances*, *2*(29), 10916–10924. https://doi.org/10.1039/c2ra21569a

El-Nahhal, I. M., & El-Ashgar, N. M. (2007). A review on polysiloxane-immobilized ligand systems: Synthesis, characterization and applications. *Journal of Organometallic Chemistry*, *692*(14), 2861–2886. https://doi.org/10.1016/j.jorganchem.2007.03.009

Elsaidi, S. K., Sinnwell, M. A., Devaraj, A., Droubay, T. C., Nie, Z., Murugesan, V., et al. (2018). Extraction of rare earth elements using magnetite@MOF composites. *Journal of Materials Chemistry A*, *6*(38), 18438–18443. https://doi.org/10.1039/C8TA04750B

Elsofany, E. (2008). Removal of lanthanum and gadolinium from nitrate medium using Aliquat-336 impregnated onto Amberlite XAD-4. *Journal of Hazardous Materials*, *153*(3), 948–954. https://doi.org/10.1016/j.jhazmat.2007.09.046

Ensing, K., & De Boer, T. (1999). Tailor-made materials for tailor-made applications: Application of molecular imprints in chemical analysis. *TrAC–Trends in Analytical Chemistry*, *18*(3), 138–145. https://doi.org/10.1016/S0165-9936(98)00103-4

Fan, Q. H., Shao, D. D., Hu, J., Chen, C. L., Wu, W. S., & Wang, X. K. (2009). Adsorption of humic acid and Eu (III) to multi-walled carbon nanotubes: EFFECT of pH, ionic strength and counterion effect. In *Radiochimica Acta* (Vol. *97*, Issue 3, pp. 141–148). https://doi.org/10.1524/ract.2009.1586

Faris, J. P., & Warton, J. W. (1962). Anion exchange resin separation of the rare earths, yttrium, and scandium in nitric acid–methanol mixtures. *Analytical Chemistry*, *34*(9), 1077–1080. https://doi.org/10.1021/ac60189a013

Farley, K. J., Dzombak, D. A., & Morel, F. M. M. (1985). A surface precipitation model for the sorption of cations on metal oxides. *Journal of Colloid And Interface Science*, *106*(1), 226–242. https://doi.org/10.1016/0021-9797(85)90400-X

Feng, X., Onel, O., Council-Troche, M., Noble, A., Yoon, R.-H., & Morris, J. R. (2021). A study of rare earth ion-adsorption clays: The speciation of rare earth elements on kaolinite at basic pH. *Applied Clay Science*, *201*, 105920. https://doi.org/10.1016/j.clay.2020.105920

Florek, J., Chalifour, F., Bilodeau, F., Larivière, D., & Kleitz, F. (2014). Nanostructured hybrid materials for the selective recovery and enrichment of rare earth elements. *Advanced Functional Materials*, *24*(18), 2668–2676. https://doi.org/10.1002/adfm.201303602

Florek, J., Giret, S., Juère, E., Larivière, D., & Kleitz, F. (2016). Functionalization of mesoporous materials for lanthanide and actinide extraction. *Dalton Transactions*, *45*(38), 14832–14854. https://doi.org/10.1039/C6DT00474A

Florek, J., Larivière, D., Kählig, H., Fiorilli, S. L., Onida, B., Fontaine, F. G., & Kleitz, F. (2020). Understanding selectivity of mesoporous silica-grafted diglycolamide-type ligands in the solid-phase extraction of rare earths. *ACS Applied Materials and Interfaces*, *12*(51), 57003–57016. https://doi.org/10.1021/acsami.0c16282

Florek, J., Mushtaq, A., Larivière, D., Cantin, G., Fontaine, F. G., & Kleitz, F. (2015). Selective recovery of rare earth elements using chelating ligands grafted on mesoporous surfaces. *RSC Advances*, *5*(126), 103782–103789. https://doi.org/10.1039/c5ra21027e

Fonseka, C., Ryu, S., Choo, Y., Mullett, M., Thiruvenkatachari, R., Naidu, G., & Vigneswaran, S. (2021). Selective recovery of rare earth elements from mine ore by Cr-MIL metal–organic frameworks. *ACS Sustainable Chemistry and Engineering*, *9*(50), 16896–16904. https://doi.org/10.1021/ACSSUSCHEMENG.1C04775/SUPPL_FILE/SC1C04775_SI_001.PDF

Friend, M. T., Parker, T. G., Mastren, T., Mocko, V., Brugh, M., Birnbaum, E. R., & Fassbender, M. E. (2020). Extraction chromatography of 225Ac and lanthanides on N,N-dioctyldiglycolamic acid /1-butyl-3-methylimidazolium bis(trifluoromethylsulfonyl)imide solvent impregnated resin. *Journal of Chromatography A*, *1624*, 461219. https://doi.org/10.1016/j.chroma.2020.461219

Fritz, J. S., & Garralda, B. B. (1963). Cation-exchange separation of bivalent metal ions from rare earths. *Talanta*, *10*(1), 91–95. https://doi.org/10.1016/0039-9140(63)80209-X

Fryxell, G. E., Chouyyok, W., & Rutledge, R. D. (2011). Design and synthesis of chelating diamide sorbents for the separation of lanthanides. *Inorganic Chemistry Communications*, *14*(6), 971–974. https://doi.org/10.1016/j.inoche.2011.03.045

Fu, J., Chen, L., Li, J., & Zhang, Z. (2015). Current status and challenges of ion imprinting. *Journal of Materials Chemistry A*, *3*(26), 13598–13627. https://doi.org/10.1039/c5ta02421h

Gabor, A., Davidescu, C. M., Negrea, A., Ciopec, M., Muntean, C., Duteanu, N., & Negrea, P. (2016). Sorption properties of Amberlite XAD 7 functionalized with sodium β-glycerophosphate. *Pure and Applied Chemistry*, *88*(12), 1143–1154. https://doi.org/10.1515/pac-2016-0806

Gaete, J., Molina, L., Valenzuela, F., & Basualto, C. (2021). Recovery of lanthanum, praseodymium and samarium by adsorption using magnetic nanoparticles functionalized with a phosphonic group. *Hydrometallurgy*, *203*, 105698. https://doi.org/10.1016/J.HYDROMET.2021.105698

Gao, Q., Xie, J. F., Shao, Y. T., Chen, C., Han, B., Xia, K. S., & Zhou, C. G. (2017). Ultrafast and high-capacity adsorption of Gd(III) onto inorganic phosphorous acid modified mesoporous SBA-15. *Chemical Engineering Journal*, *313*, 197–206. https://doi.org/10.1016/j.cej.2016.12.068

Garcia, R., Pinel, C., Madic, C., & Lemaire, M. (1998). Ionic imprinting effect in gadolinium/lanthanum separation. *Tetrahedron Letters*, *39*(47), 8651–8654. https://doi.org/10.1016/S0040-4039(98)01970-4

Garg, B. S., Sharma, R. K., Bhojak, N., & Mittal, S. (1999). Chelating resins and their applications in the analysis of trace metal ions. *Microchemical Journal*, *61*(2), 94–114. https://doi.org/10.1006/mchj.1998.1681

Gasser, M. S., & Aly, M. I. (2013). Separation and recovery of rare earth elements from spent nickel-metal-hydride batteries using synthetic adsorbent. *International Journal of Mineral Processing*, *121*, 31–38. https://doi.org/10.1016/j.minpro.2013.02.012

Gasser, M. S., El Sherif, E., & Abdel Rahman, R. O. (2017). Modification of Mg-Fe hydrotalcite using Cyanex 272 for lanthanides separation. *Chemical Engineering Journal*, *316*, 758–769. https://doi.org/10.1016/j.cej.2017.01.129

Ghobadi, M., Gharabaghi, M., Abdollahi, H., Boroumand, Z., & Moradian, M. (2018). MnFe2O4-graphene oxide magnetic nanoparticles as a high-performance adsorbent for rare earth elements: Synthesis, isotherms, kinetics, thermodynamics and desorption. *Journal of Hazardous Materials*, *351*, 308–316. https://doi.org/10.1016/j.jhazmat.2018.03.011

Ghobadi, M., Gharabaghi, M., Abdollahi, H., & Shafiee Kisomi, A. (2017). A simple and low-cost route to recycle rare earth elements (La, Ce) from aqueous solution using magnetic nanoparticles of Co x Mn 1−x Fe 2 O 4 (x = 0.2 and 0.8): synthesis, isotherms, kinetics, thermodynamics and desorption. *New Journal of Chemistry*, *41*(20), 11906–11914. https://doi.org/10.1039/C7NJ02125A

Giret, S., Hu, Y., Masoumifard, N., Boulanger, J. F., Estelle, J., Kleitz, F., & Larivière, D. (2018). Selective separation and preconcentration of scandium with mesoporous silica. *ACS Applied Materials and Interfaces*, *10*(1), 448–457. https://doi.org/10.1021/acsami.7b13336

Gismondi, P., Kuzmin, A., Unsworth, C., Rangan, S., Khalid, S., & Saha, D. (2022). Understanding the adsorption of rare-earth elements in oligo-grafted mesoporous carbon. *Langmuir*, *38*(1), 203–210. https://doi.org/10.1021/ACS.LANGMUIR.1C02403/SUPPL_FILE/LA1C02403_SI_001.PDF

Gok, C., Seyhan, S., Merdivan, M., & Yurdakoc, M. (2007). Separation and preconcentration of La3+, Ce3+ and Y3+ using calix[4]resorcinarene impregnated on polymeric support. *Microchimica Acta*, *157*(1–2), 13–19. https://doi.org/10.1007/s00604-006-0646-2

Griffith, C. S., Reyes, M. D. L., Scales, N., Hanna, J. V., & Luca, V. (2010). Hybrid inorganic–organic adsorbents Part 1: Synthesis and characterization of mesoporous zirconium titanate frameworks containing coordinating organic functionalities. *ACS Applied Materials & Interfaces*, *2*(12), 3436–3446. https://doi.org/10.1021/am100891u

Gschneidner, K. A., & Eyring, L. R. (1982). Preface. In *Handbook on the physics and chemistry of rare earths* (Vol. 5, p. 5). Elsevier. https://doi.org/10.1016/S0168-1273(82)05001-6

Guo, J., Cai, J., & Su, Q. (2009). Ion imprinted polymer particles of neodymium: synthesis, characterization and selective recognition. *Journal of Rare Earths*, *27*(1), 22–27. https://doi.org/10.1016/S1002-0721(08)60183-0

Guo, L., Liu, Y., Dou, J., Huang, Q., Lei, Y., Chen, J., et al. (2020). Highly efficient removal of Eu3+ ions using carbon nanotubes-based polymer composites synthesized from the combination of Diels-Alder and multicomponent reactions. *Journal of Molecular Liquids*, *308*, 112964. https://doi.org/10.1016/j.molliq.2020.112964

Guo, L., Xu, Y., Zhuo, M., Liu, L., Xu, Q., Wang, L., et al. (2018). Highly efficient removal of Gd(III) using hybrid hydrosols of carbon nanotubes/graphene oxide in dialysis bags and synergistic enhancement effect. *Chemical Engineering Journal*, *348*, 535–545. https://doi.org/10.1016/j.cej.2018.04.212

Gupta, N. K., & Sengupta, A. (2017). Understanding the sorption behavior of trivalent lanthanides on amide functionalized multi walled carbon nanotubes. *Hydrometallurgy*, *171*, 8–15. https://doi.org/10.1016/j.hydromet.2017.03.016

Hagiwara, Z. (1969). Elution of heavier rare earths with H.E.D.T.A. eluant at a high temperature. *Journal of Inorganic and Nuclear Chemistry*, *31*(9), 2933–2949. https://doi.org/10.1016/0022-1902(69)80213-7

Hale, W. H., & Hammer, C. A. (1972). Cation exchange elution sequence with DTPA. *Ion Exchange Membranes*, *1*(2), 81–85.

Harris, D. C., & Lucy, C. A. (2015). *Quantitative Chemical Analysis* (9th ed.). W. H. Freeman. https://books.google.com/books?id=PJhaMQAACAAJ

Helaly, O. S., Abd El-Ghany, M. S., Moustafa, M. I., Abuzaid, A. H., Abd El-Monem, N. M., & Ismail, I. M. (2012). Extraction of cerium(IV) using tributyl phosphate impregnated resin from nitric acid medium. *Transactions of Nonferrous Metals Society of China (English Edition)*, *22*(1), 206–214. https://doi.org/10.1016/S1003-6326(11)61162-X

Hérès, X., Blet, V., Di Natale, P., Ouaattou, A., Mazouz, H., Dhiba, D., & Cuer, F. (2018). Selective extraction of rare earth elements from phosphoric acid by ion exchange resins. *Metals*, *8*(9), 682. https://doi.org/10.3390/met8090682

Hermassi, M., Granados, M., Valderrama, C., Ayora, C., & Cortina, J. L. (2021). Recovery of Rare Earth Elements from acidic mine waters by integration of a selective chelating ion-exchanger and a solvent impregnated resin. *Journal of Environmental Chemical Engineering*, *9*(5), 105906. https://doi.org/10.1016/J.JECE.2021.105906

Hidayah, N. N., & Abidin, S. Z. (2017). The evolution of mineral processing in extraction of rare earth elements using solid-liquid extraction over liquid-liquid extraction: A review. *In Minerals Engineering*, Vol. *112*, pp. 103–113. https://doi.org/10.1016/j.mineng.2017.07.014

Hoffmann, F., Cornelius, M., Morell, J., & Fröba, M. (2006). Silica-based mesoporous organic–inorganic hybrid materials. *Angewandte Chemie International Edition*, *45*(20), 3216–3251. https://doi.org/10.1002/anie.200503075

Hovey, J. L., Dardona, M., Allen, M. J., & Dittrich, T. M. (2021). Sorption of rare-earth elements onto a ligand-associated media for pH-dependent extraction and recovery of critical materials. *Separation and Purification Technology*, *258*, 118061. https://doi.org/10.1016/j.seppur.2020.118061

Hu, Y., Drouin, E., Larivière, D., Kleitz, F., & Fontaine, F. G. (2017). Highly efficient and selective recovery of rare earth elements using mesoporous silica functionalized by preorganized chelating ligands. *ACS Applied Materials and Interfaces*, *9*(44), 38584–38593. https://doi.org/10.1021/acsami.7b12589

Hu, Y., Florek, J., Larivière, D., Fontaine, F.-G., & Kleitz, F. (2018). Recent advances in the separation of rare earth elements using mesoporous hybrid materials. *The Chemical Record*, *18*(7–8), 1261–1276. https://doi.org/10.1002/tcr.201800012

Hu, Y., Misal Castro, L. C., Drouin, E., Florek, J., Kählig, H., Larivière, D., et al. (2019). Size-selective separation of rare earth elements using functionalized mesoporous silica materials. *ACS Applied Materials & Interfaces*, *11*(26), 23681–23691. https://doi.org/10.1021/acsami.9b04183

Hua, W., Zhang, T., Wang, M., Zhu, Y., & Wang, X. (2019). Hierarchically structural PAN/UiO-66-(COOH)2 nanofibrous membranes for effective recovery of

Terbium(III) and Europium(III) ions and their photoluminescence performances. *Chemical Engineering Journal*, *370*, 729–741. https://doi.org/10.1016/j.cej.2019.03.255

Huang, L., Liu, L., Huang, W., Zhao, B., Shen, Z., Bao, Y., & Znad, H. (2021). Recovery of lanthanum cations by functionalized magnetic multi-walled carbon nanotube bundles. *RSC Advances*, *11*(8), 4751–4759. https://doi.org/10.1039/D0RA09902C

Huang, R., Shao, N., Hou, L., & Zhu, X. (2019). Fabrication of an efficient surface ion-imprinted polymer based on sandwich-like graphene oxide composite materials for fast and selective removal of lead ions. *Colloids and Surfaces A: Physicochemical and Engineering Aspects*, *566*, 218–228. https://doi.org/10.1016/j.colsurfa.2019.01.011

Hubicka, H., & Drobek, D. (1997). Anion-exchange method for separation of ytterbium from holmium and erbium. *Hydrometallurgy*, *47*(1), 127–136. https://doi.org/10.1016/S0304-386X(97)00040-6

Hubicka, H., & Drobek, D. (1998). Studies on separation of iminodiacetate complexes of lanthanum (III) from neodymium (III) and praseodymium (III) on anion-exchangers. *Hydrometallurgy*, *50*(1), 51–60. https://doi.org/10.1016/s0304-386x(98)00045-0

Hubicka, H., & Drobek, D. (1999). Separation of Y(III) complexes from Dy(III), Ho(III) and Er(III) complexes with iminodiacetic acid on the anion-exchangers type 1 and type 2. *Hydrometallurgy*, *53*(1), 89–100. https://doi.org/10.1016/S0304-386X(99)00035-3

Hubicka, H., & Hubicki, Z. (1986). Separation op rare earth—Polyalzdtocarboxylic acids complexes ok various types of anion-exchangers. *Solvent Extraction and Ion Exchange*, *4*(2), 383–399. https://doi.org/10.1080/07366298608917872

Hubicka, H., & Kołodyńska, D. (2004). Separation of rare-earth element complexes with trans-1,2-diaminocyclohexane-N,N,N′,N′-tetraacetic acid on polyacrylate anion exchangers. *Hydrometallurgy*, *71*(3–4), 343–350. https://doi.org/10.1016/S0304-386X(03)00086-0

Hubicka, H., & Kołodyńska, D. (2008). Application of monodispersive anion exchangers in sorption and separation of Y3+ from Nd3+ and Sm3+ complexes with dcta. *Journal of Rare Earths*, *26*(5), 619–625. https://doi.org/10.1016/S1002-0721(08)60149-0

Hubicki, Z., & Olszak, M. (1998). Studies of the sorption of rare earth element nitrate complexes in the C2H5OH-HNO3 system on the strongly basic anion exchanger Wofatit SBW. *Adsorption Science and Technology*, *16*(6), 487–492. https://doi.org/10.1177/026361749801600606

Hubicki, Z., & Olszak, M. (2002). Studies on separation of rare earth elements on various types of anion-exchangers in the C3H7OH-7 M HNO3 systems. *Journal of Chromatography A*, *955*(2), 257–262. https://doi.org/10.1016/S0021-9673(02)00212-1

Huo, Q., Margolese, D. I., & Stucky, G. D. (1996). Surfactant control of phases in the synthesis of mesoporous silica-based materials. *Chemistry of Materials*, *8*(5), 1147–1160. https://doi.org/10.1021/cm960137h

Ide, M., El-Roz, M., De Canck, E., Vicente, A., Planckaert, T., Bogaerts, T., et al. (2013). Quantification of silanol sites for the most common mesoporous ordered silicas and organosilicas: Total versus accessible silanols. *Physical Chemistry Chemical Physics*, *15*(2), 642–650. https://doi.org/10.1039/c2cp42811c

Iftekhar, S., Ramasamy, D. L., Srivastava, V., Asif, M. B., & Sillanpää, M. (2018). Understanding the factors affecting the adsorption of Lanthanum using different adsorbents: A critical review. In *Chemosphere* (Vol. *204*, pp. 413–430). https://doi.org/10.1016/j.chemosphere.2018.04.053

Iftekhar, S., Srivastava, V., Casas, A., & Sillanpää, M. (2018). Synthesis of novel GA-g-PAM/SiO2 nanocomposite for the recovery of rare earth elements (REE) ions from aqueous solution. *Journal of Cleaner Production*, *170*, 251–259. https://doi.org/10.1016/j.jclepro.2017.09.166

Iftekhar, S., Srivastava, V., & Sillanpää, M. (2017a). Enrichment of lanthanides in aqueous system by cellulose based silica nanocomposite. *Chemical Engineering Journal*, *320*, 151–159. https://doi.org/10.1016/j.cej.2017.03.051

Iftekhar, S., Srivastava, V., & Sillanpää, M. (2017b). Synthesis and application of LDH intercalated cellulose nanocomposite for separation of rare earth elements (REEs). *Chemical Engineering Journal*, *309*, 130–139. https://doi.org/10.1016/j.cej.2016.10.028

Ihara, T., Jyo, A., & Yamabe, K. (2001). Metal ion selectivity of macroreticular chelating cation exchange resins with phosphonic acid groups attached to phenyl groups of a styrene-divinylbenzene copolymer matrix. *Separation Science and Technology*, *36*(15), 3511–3528. https://doi.org/10.1081/SS-100107917

İnan, S., Tel, H., Sert, S., Çetinkaya, B., Sengül, S., Özkan, B., & Altaş, Y. (2018). Extraction and separation studies of rare earth elements using Cyanex 272 impregnated Amberlite XAD-7 resin. *Hydrometallurgy*, *181*(April), 156–163. https://doi.org/10.1016/j.hydromet.2018.09.005

Jackson, M. L. (1954). Ion exchangers in analytical chemistry. *Soil Science Society of America Journal*, *18*(1), 99. https://doi.org/10.2136/sssaj1954.03615995001800010025x

James, D. B., Powell, J. E., & Spedding, F. H. (1961). Cation-exchange elution sequences—I Divalent and rare-earth cations with EDTA, hedta and citrate. *Journal of Inorganic and Nuclear Chemistry*, *19*(1–2), 133–141. https://doi.org/10.1016/0022-1902(61)80055-9

Jia, Q., Wang, Z. H., Li, D. Q., & Niu, C. J. (2004). Adsorption of heavy rare earth(III) with extraction resin containing bis(2,4,4-trimethylpentyl) monothiophosphinic acid. *Journal of Alloys and Compounds*, *374*(1–2), 434–437. https://doi.org/10.1016/j.jallcom.2003.11.056

Jiang, L., Zhang, W., Luo, C., Cheng, D., & Zhu, J. (2016). Adsorption toward trivalent rare earth element from aqueous solution by zeolitic imidazolate frameworks. *Industrial & Engineering Chemistry Research*, *55*(22), 6365–6372. https://doi.org/10.1021/acs.iecr.6b00422

Johannesson, K. H., Lyons, W. B., Stetzenbach, K. J., & Byrne, R. H. (1995). The solubility control of rare earth elements in natural terrestrial waters and the significance of PO43- and CO32- in limiting dissolved rare earth concentrations: A review of recent information. *Aquatic Geochemistry*, *1*(2), 157–173. https://doi.org/10.1007/BF00702889

Juère, E., Florek, J., Larivière, D., Kim, K., & Kleitz, F. (2016). Support effects in rare earth element separation using diglycolamide-functionalized mesoporous silica. *New Journal of Chemistry*, *40*(5), 4325–4334. https://doi.org/10.1039/c5nj03147h

Jyo, A., Yamabe, K., & Egawa, H. (1997). Metal ion selectivity of a macroreticular styrene-divinylbenzene copolymer-based methylenephosphonic acid resin. *Separation Science and Technology*, *32*(6), 1099–1105. https://doi.org/10.1080/01496399708000948

Kabay, N., Cortina, J. L., Trochimczuk, A., & Streat, M. (2010). Solvent-impregnated resins (SIRs)—Methods of preparation and their applications. *Reactive and Functional Polymers*, *70*(8), 484–496. https://doi.org/10.1016/j.reactfunctpolym.2010.01.005

Kala, R., Biju, V. M., & Rao, T. P. (2005). Synthesis, characterization, and analytical applications of erbium(III) ion imprinted polymer particles prepared via γ-irradiation with different functional and crosslinking monomers. *Analytica Chimica Acta*, *549*(1–2), 51–58. https://doi.org/10.1016/j.aca.2005.06.024

Kala, R., Mary Gladis, J., & Prasada Rao, T. (2004). Preconcentrative separation of erbium from Y, Dy, Ho, Tb and Tm by using ion imprinted polymer particles via solid phase extraction. *Analytica Chimica Acta*, *518*(1–2), 143–150. https://doi.org/10.1016/j.aca.2004.05.029

Kaneko, T., Hikosaka, R., Nagata, F., Inagaki, M., & Kato, K. (2019). Effective adsorption of dysprosium ions on amino and carboxyl functionalized mesoporous silica sheets. *Journal of Asian Ceramic Societies*, *7*(2), 213–220. https://doi.org/10.1080/21870764.2019.1606139

Kaneko, T., Nagata, F., Kugimiya, S., & Kato, K. (2018). Optimization of carboxyl-functionalized mesoporous silica for the selective adsorption of dysprosium. *Journal of Environmental Chemical Engineering*, *6*(5), 5090–5098. https://doi.org/10.1016/j.jece.2018.09.018

Karraker, R. H. (1961). *Stability constants of some rare-earth-metal chelates. 1961*, 105. https://lib.dr.iastate.edu/cgi/viewcontent.cgi?article=2968&context=rtd

Kavun, V., van der Veen, M. A., & Repo, E. (2021). Selective recovery and separation of rare earth elements by organophosphorus modified MIL-101(Cr). *Microporous and Mesoporous Materials*, *312*, 110747. https://doi.org/10.1016/j.micromeso.2020.110747

Kazantsev, E. I., Fisenko, V. V., & Mal'tsev, G. I. (1974). Complexing of rare earth ions with carboxylic cationite KB-2x7. *Izvestiya Vysshikh Uchebnykh Zavedenii, Khimiya i Khimicheskaya Tekhnologiya*, *17*(9), 1304–1306. https://inis.iaea.org/search/search.aspx?orig_q=RN:6174755

Keçili, R., Dolak, İ., Ziyadanoğulları, B., Ersöz, A., & Say, R. (2018). Ion imprinted cryogel-based supermacroporous traps for selective separation of cerium(III) in real samples. *Journal of Rare Earths*, *36*(8), 857–862. https://doi.org/10.1016/j.jre.2018.02.008

Kegl, T., Ban, I., Lobnik, A., & Košak, A. (2019). Synthesis and characterization of novel Γ-Fe2O3-NH4OH@SiO2(APTMS) nanoparticles for dysprosium adsorption. *Journal of Hazardous Materials*, *378*, 120764. https://doi.org/10.1016/j.jhazmat.2019.120764

Kharitonov, O. V., Firsova, L. A., & Chuveleva, E. A. (2009a). The accumulation of the competing displacer ion in the retaining ion zone in the separation of rare-earth metals by complex formation displacement chromatography. *Russian Journal of Physical Chemistry A*, *83*(7), 1217–1219. https://doi.org/10.1134/S0036024409070309

Kharitonov, O. V., Firsova, L. A., & Chuveleva, E. A. (2009b). The influence of a competing eluting ion content on the efficiency of separation of rare-earth metals in displacement complex forming chromatography. *Russian Journal of Physical Chemistry A*, *83*(6), 1030–1031. https://doi.org/10.1134/S0036024409060284

Khawassek, Y. M., Eliwa, A. A., Haggag, E. S. A., Omar, S. A., & Abdel-Wahab, S. M. (2019). Adsorption of rare earth elements by strong acid cation exchange resin thermodynamics, characteristics and kinetics. *SN Applied Sciences*, *1*(1), 51. https://doi.org/10.1007/s42452-018-0051-6

Kilian, K., Pyrzyńska, K., & Pęgier, M. (2017). Comparative study of Sc(III) sorption onto carbon-based materials. *Solvent Extraction and Ion Exchange*, *35*(6), 450–459. https://doi.org/10.1080/07366299.2017.1354580

Kim, H., Lee, J., & Jung, H. (2019). Study on the carbamoyl phosphine oxide moiety functionalized mesoporous graphene for the removal of rare earth elements. *Journal of Porous Materials*, *26*(4), 931–939. https://doi.org/10.1007/s10934-018-0691-3

Kogan, L., & Ratner, R. (1971). The separation of yttrium from the lanthanides by ion exchange with diethylenetriaminepentaacetic acid as eluant. I. Separation of yttrium-erbium mixtures. *Journal of Chromatography A*, *62*(3), 449–457. https://doi.org/10.1016/S0021-9673(00)91397-9

Kołodyńska, D., Hubicki, Z., & Fila, D. (2019). Recovery of rare earth elements from acidic solutions using macroporous ion exchangers. *Separation Science and Technology*, *54*(13), 2059–2076. https://doi.org/10.1080/01496395.2019.1604753

Koochaki-Mohammadpour, S. M. A., Torab-Mostaedi, M., Talebizadeh-Rafsanjani, A., & Naderi-Behdani, F. (2014). Adsorption isotherm, kinetic, thermodynamic, and desorption studies of lanthanum and dysprosium on oxidized multiwalled carbon nanotubes. *Journal of Dispersion Science and Technology*, *35*(2), 244–254. https://doi.org/10.1080/01932691.2013.785361

Koodynska, D., & Hubicki, Z. (2012). Investigation of sorption and separation of lanthanides on the ion exchangers of various types. In *Ion Exchange Technologies: Vol. I* (Issue tourism). InTech. https://doi.org/10.5772/50857

Korkisch, J., Feik, F., & Ahluwalia, S. S. (1967). Cation-exchange behaviour of several elements in nitric acid-organic solvent media. *Talanta*, *14*(9), 1069–1081. https://doi.org/10.1016/0039-9140(67)80145-0

Kosmulski, M. (1997a). Standard enthalpies of adsorption of di- and trivalent cations on alumina. *Journal of Colloid and Interface Science*, *192*(1), 215–227. https://doi.org/10.1006/jcis.1997.4994

Kosmulski, M. (1997b). Adsorption of Trivalent Cations on Silica. *Journal of Colloid and Interface Science*, *195*(2), 395–403. https://doi.org/10.1006/jcis.1997.5155

Kostenko, L., Kobylinska, N., Khainakov, S., & Granda, S. G. (2019). Magnetite nanoparticles with aminomethylenephosphonic groups: Synthesis, characterization and uptake of europium(III) ions from aqueous media. *Microchimica Acta*, *186*(7), 474. https://doi.org/10.1007/s00604-019-3520-8

Kowal-Fouchard, A., Drot, R., Simoni, E., Marmier, N., Fromage, F., & Ehrhardt, J. J. (2004). Structural identification of europium(III) adsorption complexes on montmorillonite. *New Journal of Chemistry*, *28*(7), 864–869. https://doi.org/10.1039/b400306c

Kumar, B. N., Radhika, S., Kantam, M. L., & Reddy, B. R. (2011). Solid-liquid extraction of terbium from phosphoric acid solutions using solvent-impregnated resin containing TOPS 99. *Journal of Chemical Technology & Biotechnology*, *86*(4), 562–569. https://doi.org/10.1002/jctb.2553

Kumar, D., Schumacher, K., Du Fresne von Hohenesche, C., Grün, M., & Unger, K. K. (2001). MCM-41, MCM-48 and related mesoporous adsorbents: Their synthesis and characterisation. *Colloids and Surfaces A: Physicochemical and Engineering Aspects*, *187–188*, 109–116. https://doi.org/10.1016/S0927-7757(01)00638-0

Lee, G. S., Uchikoshi, M., Mimura, K., & Isshiki, M. (2009). Distribution coefficients of La, Ce, Pr, Nd, and Sm on Cyanex 923-, D2EHPA-, and PC88A-impregnated resins. *Separation and Purification Technology*, *67*(1), 79–85. https://doi.org/10.1016/j.seppur.2009.03.033

Lee, G. S., Uchikoshi, M., Mimura, K., & Isshiki, M. (2010a). Preparation and evaluation of high-purity La2O3. *Metallurgical and Materials Transactions B: Process Metallurgy and Materials Processing Science*, *41*(3), 509–519. https://doi.org/10.1007/s11663-010-9348-6

Lee, G. S., Uchikoshi, M., Mimura, K., & Isshiki, M. (2010b). Separation of major impurities Ce, Pr, Nd, Sm, Al, Ca, Fe, and Zn from La using bis(2-ethylhexyl) phosphoric acid (D2EHPA)-impregnated resin in a hydrochloric acid medium. *Separation and Purification Technology*, *71*(2), 186–191. https://doi.org/10.1016/j.seppur.2009.11.020

Lee, Y.-R., Yu, K., Ravi, S., & Ahn, W.-S. (2018). Selective adsorption of rare earth elements over functionalized Cr-MIL-101 [research-article]. *ACS Applied Materials & Interfaces*, *10*(28), 23918–23927. https://doi.org/10.1021/acsami.8b07130

Lee, Y. R., Zhang, S., Yu, K., Choi, J., & Ahn, W. S. (2019). Poly(amidoamine) dendrimer immobilized on mesoporous silica foam (MSF) and fibrous nano-silica KCC-1 for Gd3+ adsorption in water. *Chemical Engineering Journal*, *378*, 122133. https://doi.org/10.1016/j.cej.2019.122133

Li, C., Zhuang, Z., Huang, F., Wu, Z., Hong, Y., & Lin, Z. (2013). Recycling rare earth elements from industrial wastewater with flowerlike nano-Mg(OH)2. *ACS Applied Materials and Interfaces*, *5*(19), 9719–9725. https://doi.org/10.1021/am4027967

Li, D., Zhang, B., & Xuan, F. (2015). The sorption of Eu(III) from aqueous solutions by magnetic graphene oxides: A combined experimental and modeling studies. *Journal of Molecular Liquids*, *211*, 203–209. https://doi.org/10.1016/j.molliq.2015.07.012

Li, J., Gong, A., Li, F., Qiu, L., Zhang, W., Gao, G., et al. (2018). Synthesis and characterization of magnetic mesoporous Fe3O4@mSiO2-DODGA nanoparticles for adsorption of 16 rare earth elements. *RSC Advances*, *8*(68), 39149–39161. https://doi.org/10.1039/c8ra07762b

Li, J., Gong, A., Qiu, L., Zhang, W., Shi, G., Li, X., et al. (2020). Selective extraction and column separation for 16 kinds of rare earth element ions by using N, N-dioctyl diglycolacid grafted silica gel particles as the stationary phase. *Journal of Chromatography A*, *1627*. https://doi.org/10.1016/j.chroma.2020.461393

Li, K., Gao, Q., Yadavalli, G., Shen, X., Lei, H., Han, B., et al. (2015). Selective adsorption of Gd3+ on a magnetically retrievable imprinted chitosan/carbon nanotube composite with high capacity. *ACS Applied Materials and Interfaces*, *7*(38), 21047–21055. https://doi.org/10.1021/acsami.5b07560

Li, X., Lu, T., Wang, Y., & Yang, Y. (2019). Study on the controllable synthesis of SH-MCM-41 mesoporous materials and their adsorption properties of the La3+, Gd3+ and Yb3+. *Chinese Chemical Letters*, *30*(12), 2318–2322. https://doi.org/10.1016/j.cclet.2019.05.056

Liao, C., Nie, H., Jiao, Y., Liang, Y., & Yang, S. (2010). Study on the diffusion kinetics of adsorption of heavy rare earth with Cyanex272-P507 impregnated resin. *Journal of Rare Earths*, *28*(Suppl. 1), 120–124. https://doi.org/10.1016/S1002-0721(10)60290-6

Liu, J., Martin, P. F., & Peter McGrail, B. (2021). Rare-earth element extraction from geothermal brine using magnetic core-shell nanoparticles-techno-economic analysis. *Geothermics*, *89*, 101938. https://doi.org/10.1016/j.geothermics.2020.101938

Liu, Y., Zhu, L., Sun, X., Chen, J., & Luo, F. (2009). Silica materials doped with bifunctional ionic liquid extractant for yttrium extraction. *Industrial and Engineering Chemistry Research*, *48*(15), 7308–7313. https://doi.org/10.1021/ie900468c

Liu, Z., Chen, G., Li, X., & Lu, X. (2021). Removal of rare earth elements by MnFe2O4 based mesoporous adsorbents: Synthesis, isotherms, kinetics, thermodynamics. *Journal of Alloys and Compounds*, *856*, 158185. https://doi.org/10.1016/J.JALLCOM.2020.158185

Liu, Z., Gao, Z., Xu, L., & Hu, F. (2020). Efficient and rapid adsorption of rare earth elements from water by magnetic Fe3O4/MnO2 decorated reduced graphene oxide. *Journal of Molecular Liquids*, *313*, 113510. https://doi.org/10.1016/j.molliq.2020.113510

Liu, Z., Feng, Y., & Li, H. (2021). Application of titanium phosphate prepared from acidic titanium dioxide wastewater to remove cerium (III) in aqueous solution. *Colloids and Surfaces A: Physicochemical and Engineering Aspects*, *630*, 127613. https://doi.org/10.1016/J.COLSURFA.2021.127613

Liu, Z., Liu, Y., & Gong, A. (2019). Preparation of diglycolamide polymer modified silica and its application as adsorbent for rare earth ions. *Designed Monomers and Polymers*, *22*(1), 1–7. https://doi.org/10.1080/15685551.2018.1564425

Lokshin, E. P., Ivanenko, V. I., Tareeva, O. A., & Korneikov, R. I. (2013). Sorption of rare earth elements of waste solution of leaching uranium. *Russian Journal of Applied Chemistry*, *86*(3), 450–452. https://doi.org/10.1134/S1070427213030269

Losev, V., Buyko, O., Metelitsa, S., Borodina, E., Kuzmin, N., & Shimanskiy, A. (2020). Novel silica-based adsorbent layer-by-layer modified with polyhexamethylene guanidine and Arsenazo reagents for solid-phase extraction of lanthanides from lignites and products of their processing. *Separation Science and Technology*, *56*(9), 1510–1519,*)*. https://doi.org/10.1080/01496395.2020.1785500

Lou, Z., Xiao, X., Huang, M., Wang, Y., Xing, Z., & Xiong, Y. (2019). Acrylic acid-functionalized metal–organic frameworks for Sc(III) selective adsorption [research-article]. *ACS Applied Materials & Interfaces*, *11*(12), 11772–11781. https://doi.org/10.1021/acsami.9b00476

Louis, R. E., & Duyckaerts, G. (1984). Some parameters affecting the extraction chromatographic performance of TBP impregnated macroporous XAD-4 columns for Am(III)-Eu(III) separations. *Journal of Radioanalytical and Nuclear Chemistry Articles*, *81*(2), 305–315. https://doi.org/10.1007/BF02135383

Louis, R. E., & Duyckaerts, G. (1985). Some parameters affecting column distribution ratios of Am(III) and Eu(III) and column resolution of TBP impregnated macroporous XAD-4 polymers. *Journal of Radioanalytical and Nuclear Chemistry Articles*, *90*(1), 105–112. https://doi.org/10.1007/BF02037325

Ma, T. Y., Liu, L., & Yuan, Z. Y. (2013). Direct synthesis of ordered mesoporous carbons. *Chemical Society Reviews*, *42*(9), 3977–4003. https://doi.org/10.1039/c2cs35301f

Mackey, J. L. (1960). *A study of the rare-earth chelate stability constants of some aminopolyacetic acids*. PhD Thesis, Iowa state University. https://doi.org/10.31274/rtd-180813-498

Mahmoud, M. E., Mohamed, A. K., Amira, M. F., & Seleim, S. M. (2019). Novel nanostructured metal–organic framework-bonded silica amine and polymer: Facile synthesis, kinetics, isotherms, and thermodynamics evaluation for adsorption of yttrium(III) ions [research-article]. *Journal of Chemical & Engineering Data*, *64*(12), 6060–6070. https://doi.org/10.1021/acs.jced.9b00918

Marmier, N., Delisée, A., & Fromage, F. (1999). Surface complexation modeling of Yb(III) and Cs(I) sorption on silica. *Journal of Colloid and Interface Science*, *212*(2), 228–233. https://doi.org/10.1006/jcis.1999.6086

Marmier, N., Dumonceau, J., & Fromage, F. (1997). Surface complexation modeling of Yb(III) sorption and desorption on hematite and alumina. *Journal of Contaminant Hydrology*, *26*(1–4), 159–167. https://doi.org/10.1016/S0169-7722(96)00065-4

Marmier, N., & Fromage, F. (1999). Comparing electrostatic and nonelectrostatic surface complexation modeling of the sorption of lanthanum on hematite. *Journal of Colloid and Interface Science*, *212*(2), 252–263. https://doi.org/10.1006/jcis.1998.6039

Martell, A. E., & Hancock, R. D. (1996). *Metal Complexes in Aqueous Solutions*. New York: Springer. https://doi.org/10.1007/978-1-4899-1486-6

Martynenko, L. I., Kupriyanova, G. N., & Prutkova, N. M. (1972). Non-ion exchange sorption of monoiminodiacetates of rare-earth elements on cationite. *Zhurnal Neorganicheskoj Khimii*, *17*(1), 214–217. http://inis.iaea.org/Search/search.aspx?orig_q=RN:3033722

Martynenko, L. I., Mitrofanova, N. D., & Spitsyn, V. I. (1968). Mechanism of the ion-exchange separation of mixtures of rare earth elements during elution with solutions of nitrilotriacetic acid. *Bulletin of the Academy of Sciences of the USSR Division of Chemical Science*, *17*(9), 1854–1861. https://doi.org/10.1007/BF00904970

Marwani, H. M., Albishri, H. M., Jalal, T. A., & Soliman, E. M. (2017). Study of isotherm and kinetic models of lanthanum adsorption on activated carbon loaded with recently synthesized Schiff's base. *Arabian Journal of Chemistry*, *10*, S1032–S1040. https://doi.org/10.1016/j.arabjc.2013.01.008

Marwani, H. M., & Alsafrani, A. E. (2013). New solid phase extractor based on ionic liquid functionalized silica gel surface for selective separation and determination of lanthanum. *Journal of Analytical Science and Technology*, *4*(1), 1–10. https://doi.org/10.1186/2093-3371-4-13

Marwani, H. M., Bakhsh, E. M., Khan, S. B., Danish, E. Y., & Asiri, A. M. (2018). Cerium oxide-cadmium oxide nanomaterial as efficient extractant for yttrium ions. *Journal of Molecular Liquids*, *269*, 252–259. https://doi.org/10.1016/j.molliq.2018.08.046

Masi, A. N., & Olsina, R. A. (1993). Preconcentration and determination of Ce, La and Pr by X-ray fluorescence analysis, using Amberlite XAD resins loaded with 8-Quinolinol and 2-(2-(5 chloropyridylazo)-5-dimethylamino)-phenol. *Talanta*, *40*(6), 931–934. https://doi.org/10.1016/0039-9140(93)80054-U

Mathur, J. N., & Khopkar, P. K. (1985). Ion exchange behaviour of chelating resin Dowex A-1 with actinides and lanthanides. *Solvent Extraction and Ion Exchange*, *3*(5), 753–762. https://doi.org/10.1080/07366298508918538

Mehmood, A., Ghafar, H., Yaqoob, S., Gohar, U. F., & Ahmad, B. (2017). Mesoporous silica nanoparticles: A review. *Journal of Developing Drugs*, *06*(02). https://doi.org/10.4172/2329-6631.1000174

Meng, M., Meng, X., Liu, Y., Liu, Z., Han, J., Wang, Y., et al. (2014). An ion-imprinted functionalized SBA-15 adsorbent synthesized by surface imprinting technique via reversible addition-fragmentation chain transfer polymerization for selective removal of Ce(III) from aqueous solution. *Journal of Hazardous Materials*, *278*, 134–143. https://doi.org/10.1016/j.jhazmat.2014.06.002

Metwally, S. S., & Rizk, H. E. (2014). Preparation and characterization of nano-sized iron–titanium mixed oxide for removal of some lanthanides from aqueous solution. *Separation Science and Technology*, *49*(15), 2426–2436. https://doi.org/10.1080/01496395.2014.926457

Meynen, V., Cool, P., & Vansant, E. F. (2009). Verified syntheses of mesoporous materials. *Microporous and Mesoporous Materials*, *125*(3), 170–223. https://doi.org/10.1016/j.micromeso.2009.03.046

Miklishanskii, A. Z., Yakovlev, Y. V., Dogadkin, N. N., Leikin, Y. A., & Davankov, A. B. (1968). Investigation of the sorption of the rare-earth elements and uranium by phosphorus-containing cation-exchange resins. *Bulletin of the Academy of Sciences of the USSR Division of Chemical Science*, *17*(7), 1373–1376. https://doi.org/10.1007/BF00907827

Minczewski, J, Chwastowska, J., Dybczynski, R., & Masson, M. R. (1982). *Separation and preconcentration methods in inorganic trace analysis*. E. Horwood; Halsted Press.

Minczewski, J., & Dybczyński, R. (1962). Separation of rare earths on anion exchange resins II. Anion exchange behaviour of the rare earth complexes with ethylenediaminetetraacetic acid. *Journal of Chromatography A*, *7*(C), 98–111. https://doi.org/10.1016/s0021-9673(01)86386-x

Miraoui, A., Didi, M. A., & Villemin, D. (2016). Neodymium(III) removal by functionalized magnetic nanoparticles. *Journal of Radioanalytical and Nuclear Chemistry*, *307*(2), 963–971. https://doi.org/10.1007/s10967-015-4267-2

Mohamed, W. R., Metwally, S. S., Ibrahim, H. A., El-Sherief, E. A., Mekhamer, H. S., Moustafa, I. M. I., & Mabrouk, E. M. (2017). Impregnation of task-specific ionic liquid into a solid support for removal of neodymium and gadolinium ions from aqueous solution. *Journal of Molecular Liquids*, *236*, 9–17. https://doi.org/10.1016/j.molliq.2017.04.013

Mohammedi, H., Miloudi, H., Tayeb, A., Bertagnolli, C., & Boos, A. (2019). Study on the extraction of lanthanides by a mesoporous MCM-41 silica impregnated with Cyanex 272. *Separation and Purification Technology*, *209*(February 2018), 359–367. https://doi.org/10.1016/j.seppur.2018.07.035

Molina, L., Gaete, J., Alfaro, I., Ide, V., Valenzuela, F., Parada, J., & Basualto, C. (2019). Synthesis and characterization of magnetite nanoparticles functionalized with organophosphorus compounds and its application as an adsorbent for La (III), Nd (III) and Pr (III) ions from aqueous solutions. *Journal of Molecular Liquids*, *275*, 178–191. https://doi.org/10.1016/j.molliq.2018.11.074

Moloney, M. P., Causse, J., Loubat, C., & Grandjean, A. (2014). Sodium "activation" of silano-phosphonate modified mesoporous TiO 2 leading to improved rare-earth element extraction. *European Journal of Inorganic Chemistry*, *2014*(13), 2268–2277. https://doi.org/10.1002/ejic.201400027

Mondal, S., Ghar, A., Satpati, A. K., Sinharoy, P., Singh, D. K., Sharma, J. N., et al. (2019). Recovery of rare earth elements from coal fly ash using TEHDGA impregnated resin. *Hydrometallurgy*, *185*, 93–101. https://doi.org/10.1016/j.hydromet.2019.02.005

Mosai, A. K., Chimuka, L., Cukrowska, E. M., Kotzé, I. A., & Tutu, H. (2019). The Recovery of Rare Earth Elements (REEs) from aqueous solutions using natural zeolite and bentonite. *Water, Air, and Soil Pollution*, *230*(8). https://doi.org/10.1007/s11270-019-4236-4

Moussa, M., Ndiaye, M. M., Pinta, T., Pichon, V., Vercouter, T., & Delaunay, N. (2017). Selective solid phase extraction of lanthanides from tap and river waters with ion imprinted polymers. *Analytica Chimica Acta*, *963*, 44–52. https://doi.org/10.1016/j.aca.2017.02.012

Muraviev, D. (1998). Surface impregnated sulfonate ion exchangers: Preparation, properties and application. *Solvent Extraction and Ion Exchange*, *16*(1), 381–457. https://doi.org/10.1080/07366299808934533

Muraviev, D., Ghantous, L., & Valiente, M. (1998). Stabilization of solvent-impregnated resin capacities by different techniques. *Reactive and Functional Polymers*, *38*(2–3), 259–268. https://doi.org/10.1016/s1381-5148(98)00075-3

Narayan, R., Nayak, U. Y., Raichur, A. M., & Garg, S. (2018). Mesoporous silica nanoparticles: A comprehensive review on synthesis and recent advances. *Pharmaceutics*, *10*(3), 1–49. https://doi.org/10.3390/pharmaceutics10030118

Naser, A. A., El-deen, G. E. S., Bhran, A. A., Metwally, S. S., & El-Kamash, A. M. (2015). Elaboration of impregnated composite for sorption of europium and neodymium ions from aqueous solutions. *Journal of Industrial and Engineering Chemistry*, *32*, 264–272. https://doi.org/10.1016/j.jiec.2015.08.024

Nash, K. L. (1993). A review of the basic chemistry and recent developments in trivalent f-elements separations. *Solvent Extraction and Ion Exchange*, *11*(4), 729–768. https://doi.org/10.1080/07366299308918184

Nelson, F. (1965). Ion-exchange procedures. *Journal of Chromatography A*, *20*, 378–383. https://doi.org/10.1016/s0021-9673(01)97426-6

Ngomsik, A. F., Bee, A., Talbot, D., & Cote, G. (2012). Magnetic solid-liquid extraction of Eu(III), La(III), Ni(II) and Co(II) with maghemite nanoparticles. *Separation and Purification Technology*, *86*, 1–8. https://doi.org/10.1016/j.seppur.2011.10.013

Ni, C., Liu, Q., Ren, Z., Hu, H., Sun, B., Liu, C., et al. (2021). Selective removal and recovery of La(III) using a phosphonic-based ion imprinted polymer: Adsorption performance, regeneration, and mechanism. *Journal of Environmental Chemical Engineering*, *9*(6), 106701. https://doi.org/10.1016/J.JECE.2021.106701

Nik Mustapa, N. R., Malek, N. F. A., Yusoff, M. M., & Rahman, M. L. (2016). Ion imprinted polymers for selective recognition and separation of lanthanum and cerium ions from other lanthanide. *Separation Science and Technology (Philadelphia)*, *51*(17), 2762–2771. https://doi.org/10.1080/01496395.2016.1225091

Nishihama, S., Kohata, K., & Yoshizuka, K. (2013). Separation of lanthanum and cerium using a coated solvent-impregnated resin. *Separation and Purification Technology*, *118*, 511–518. https://doi.org/10.1016/j.seppur.2013.07.047

Niu, F., Xie, Z., Fu, C., Xu, H., Liu, D., Zhang, X., et al. (2021). Adsorption–desorption of La3+, Eu3+, and Y3+ by Mg(OH)2-pretreated TP207 resin. *Journal of Operations Management*, *73*(1), 32–38. https://doi.org/10.1007/s11837-020-04472-2

Noack, C. W., Dzombak, D. A., & Karamalidis, A. K. (2014). Rare earth element distributions and trends in natural waters with a focus on groundwater. *Environmental Science and Technology*, *48*(8), 4317–4326. https://doi.org/10.1021/es4053895

Noack, C. W., Perkins, K. M., Callura, J. C., Washburn, N. R., Dzombak, D. A., & Karamalidis, A. K. (2016). Effects of ligand chemistry and geometry on rare earth element partitioning from saline solutions to functionalized adsorbents. *ACS Sustainable Chemistry & Engineering*, *4*(11), 6115–6124. https://doi.org/10.1021/acssuschemeng.6b01549

Ochsenkühn-Petropulu, M., Lyberopulu, T., & Parissakis, G. (1995). Selective separation and determination of scandium from yttrium and lanthanides in red mud by a combined ion exchange/solvent extraction method. *Analytica Chimica Acta, 315*(1–2), 231–237. https://doi.org/10.1016/0003-2670(95)00309-N

Ogata, T., Narita, H., & Tanaka, M. (2014). Immobilization of diglycol amic acid on silica gel for selective recovery of rare earth elements. *Chemistry Letters, 43*(9), 1414–1416. https://doi.org/10.1246/cl.140446

Ogata, T., Narita, H., & Tanaka, M. (2015a). Adsorption behavior of rare earth elements on silica gel modified with diglycol amic acid. *Hydrometallurgy, 152*, 178–182. https://doi.org/10.1016/j.hydromet.2015.01.005

Ogata, T., Narita, H., & Tanaka, M. (2015b). Rapid and selective recovery of heavy rare earths by using an adsorbent with diglycol amic acid group. *Hydrometallurgy, 155*, 105–109. https://doi.org/10.1016/j.hydromet.2015.04.015

Ogata, T., Narita, H., Tanaka, M., Hoshino, M., Kon, Y., & Watanabe, Y. (2016). Selective recovery of heavy rare earth elements from apatite with an adsorbent bearing immobilized tridentate amido ligands. *Separation and Purification Technology, 159*, 157–160. https://doi.org/10.1016/j.seppur.2016.01.008

Ohta, A., & Kawabe, I. (2001). REE(III) adsorption onto Mn dioxide (δ-MnO2) and Fe oxyhydroxide: Ce(III) oxidation by δ-MnO2. *Geochimica et Cosmochimica Acta, 65*(5), 695–703. https://doi.org/10.1016/S0016-7037(00)00578-0

Page, M. J., Quinn, J. E., & Soldenhoff, K. H. (2019). The impact of sulfate ions on the ion exchange of rare earth elements. *Hydrometallurgy, 186*, 12–20. https://doi.org/10.1016/j.hydromet.2019.03.003

Page, M. J., Soldenhoff, K., & Ogden, M. D. (2017). Comparative study of the application of chelating resins for rare earth recovery. *Hydrometallurgy, 169*, 275–281. https://doi.org/10.1016/j.hydromet.2017.02.006

Pallavicini, P., Dacarro, G., Diaz-Fernandez, Y. A., & Taglietti, A. (2014). Coordination chemistry of surface-grafted ligands for antibacterial materials. *Coordination Chemistry Reviews 275*, 37–53. https://doi.org/10.1016/j.ccr.2014.04.013

Pan, J., Zou, X., Li, C., Liu, Y., Yan, Y., & Han, J. (2010). Synthesis and applications of Ce(III)-imprinted polymer based on attapulgite as the sacrificial support material for selective separation of cerium(III) ions. *Microchimica Acta, 171*(1), 151–160. https://doi.org/10.1007/s00604-010-0416-z

Park, J., Han, C., Lee, J., Kim, S., Kim, J., & Wee, J. (2005). Synthesis of extraction resin containing 2-ethylhexyl phosphonic acid mono-2-ethylhexyl ester and its performance for separation of rare earths (Gd, Tb). *Separation and Purification Technology, 43*(2), 111–116. https://doi.org/10.1016/j.seppur.2004.10.007

Parsons-Moss, T., Wang, J., Jones, S., May, E., Olive, D., Dai, Z., et al. (2014). Sorption interactions of plutonium and europium with ordered mesoporous carbon. *Journal of Materials Chemistry A, 2*(29), 11209–11221. https://doi.org/10.1039/c4ta01740d

Patra, S., Roy, E., Madhuri, R., & Sharma, P. K. (2017). Removal and recycling of precious rare earth element from wastewater samples using imprinted magnetic ordered mesoporous carbon [research-article]. *ACS Sustainable Chemistry and Engineering, 5*(8), 6910–6923. https://doi.org/10.1021/acssuschemeng. 7b01124

Pearson, R. G. (1963). Hard and soft acids and bases. *Journal of the American Chemical Society, 85*(22), 3533–3539. https://doi.org/10.1021/ja00905a001

Peng, X., Mo, S., Li, R., Li, J., Tian, C., Liu, W., & Wang, Y. (2020). Effective removal of the rare earth element dysprosium from wastewater with polyurethane sponge-supported graphene oxide–titanium phosphate. *Environmental Chemistry Letters*, *1–10*. https://doi.org/10.1007/s10311-020-01073-y

Perreault, L. L., Giret, S., Gagnon, M., Florek, J., Larivière, D., & Kleitz, F. (2017). Functionalization of mesoporous carbon materials for selective separation of lanthanides under acidic conditions. *ACS Applied Materials and Interfaces*, *9*(13), 12003–12012. https://doi.org/10.1021/acsami.6b16650

Piasecki, W., & Sverjensky, D. A. (2008). Speciation of adsorbed yttrium and rare earth elements on oxide surfaces. *Geochimica et Cosmochimica Acta*, *72*(16), 3964–3979. https://doi.org/10.1016/j.gca.2008.05.049

Polido Legaria, E., Rocha, J., Tai, C. W., Kessler, V. G., & Seisenbaeva, G. A. (2017). Unusual seeding mechanism for enhanced performance in solid-phase magnetic extraction of rare earth elements. *Scientific Reports*, *7*(January), 1–13. https://doi.org/10.1038/srep43740

Polido Legaria, E., Samouhos, M., Kessler, V. G., & Seisenbaeva, G. A. (2017). Toward Molecular recognition of REEs: Comparative analysis of hybrid nanoadsorbents with the different complexonate ligands EDTA, DTPA, and TTHA. *Inorganic Chemistry*, *56*(22), 13938–13948. https://doi.org/10.1021/acs.inorgchem.7b02056

Qin, Y., Wang, X., Shi, M., Huang, Y., Liu, X., & Li, X. (2022). Preparation and adsorption properties of lanthanide ion surface-imprinted polymer based on reaming MCM-41. *Journal of Inorganic and Organometallic Polymers and Materials*, *32*(1), 161–168. https://doi.org/10.1007/S10904-021-02108-9/FIGURES/7

Quinn, K. A., Byrne, R. H., & Schijf, J. (2006). Sorption of yttrium and rare earth elements by amorphous ferric hydroxide: Influence of solution complexation with carbonate. *Geochimica et Cosmochimica Acta*, *70*(16), 4151–4165. https://doi.org/10.1016/j.gca.2006.06.014

Quinn, K. A., Byrne, R. H., & Schijf, J. (2007). Sorption of yttrium and rare earth elements by amorphous ferric hydroxide: Influence of temperature. *Environmental Science & Technology*, *41*(2), 541–546. https://doi.org/10.1021/es0618191

Rahman, M. M., Awual, M. R., & Asiri, A. M. (2020). Preparation and evaluation of composite hybrid nanomaterials for rare-earth elements separation and recovery. *Separation and Purification Technology*, *253*, 117515. https://doi.org/10.1016/j.seppur.2020.117515

Ramasamy, D. L., Khan, S., Repo, E., & Sillanpää, M. (2017). Synthesis of mesoporous and microporous amine and non-amine functionalized silica gels for the application of rare earth elements (REE) recovery from the waste water-understanding the role of pH, temperature, calcination and mechanism in Light REE and Hea. *Chemical Engineering Journal*, *322*, 56–65. https://doi.org/10.1016/j.cej.2017.03.152

Ramasamy, D. L., Puhakka, V., Doshi, B., Iftekhar, S., & Sillanpää, M. (2019). Fabrication of carbon nanotubes reinforced silica composites with improved rare earth elements adsorption performance. *Chemical Engineering Journal*, *365*(January), 291–304. https://doi.org/10.1016/j.cej.2019.02.057

Ramasamy, D. L., Puhakka, V., Iftekhar, S., Wojtuś, A., Repo, E., Ben Hammouda, S., et al. (2018). N- and O- ligand doped mesoporous silica-chitosan hybrid beads for the efficient, sustainable and selective recovery of rare earth elements (REE) from acid mine drainage (AMD): Understanding the significance of physical modification and conditioning of th. *Journal of Hazardous Materials, 348*(October 2017), 84–91. https://doi.org/10.1016/j.jhazmat.2018.01.030

Ramasamy, D. L., Puhakka, V., Repo, E., Ben Hammouda, S., & Sillanpää, M. (2018). Two-stage selective recovery process of scandium from the group of rare earth elements in aqueous systems using activated carbon and silica composites: Dual applications by tailoring the ligand grafting approach. *Chemical Engineering Journal, 341*(February), 351–360. https://doi.org/10.1016/j.cej.2018.02.024

Ramasamy, D. L., Puhakka, V., Repo, E., Khan, S., & Sillanpää, M. (2017). Coordination and silica surface chemistry of lanthanides (III), scandium (III) and yttrium (III) sorption on 1-(2-pyridylazo)-2-napththol (PAN) and acetylacetone (acac) immobilized gels. *Chemical Engineering Journal, 324*, 104–112. https://doi.org/10.1016/j.cej.2017.05.025

Ramasamy, D. L., Puhakka, V., Repo, E., & Sillanpää, M. (2018). Selective separation of scandium from iron, aluminium and gold rich wastewater using various amino and non-amino functionalized silica gels—A comparative study. *Journal of Cleaner Production, 170*(January), 890–901. https://doi.org/10.1016/j.jclepro.2017.09.199

Ramasamy, D. L., Repo, E., & Sillanpää, M. (2020). Selective recovery of rare-earth elements from diluted aqueous streams using N- and O-coordination ligand–grafted organic–inorganic hybrid composites. *Advanced Water Treatment, 565–664*. https://doi.org/10.1016/B978-0-12-819216-0.00008-4

Ramasamy, D. L., Repo, E., Srivastava, V., & Sillanpää, M. (2017). Chemically immobilized and physically adsorbed PAN/acetylacetone modified mesoporous silica for the recovery of rare earth elements from the waste water-comparative and optimization study. *Water Research, 114*, 264–276. https://doi.org/10.1016/j.watres.2017.02.045

Rao, T. P., Kala, R., & Daniel, S. (2006). Metal ion-imprinted polymers—Novel materials for selective recognition of inorganics. *Analytica Chimica Acta, 578*(2), 105–116. https://doi.org/10.1016/j.aca.2006.06.065

Ravi, S., Lee, Y. R., Yu, K., Ahn, J. W., & Ahn, W. S. (2018). Benzene triamido-tetraphosphonic acid immobilized on mesoporous silica for adsorption of Nd3+ ions in aqueous solution. *Microporous and Mesoporous Materials, 258*, 62–71. https://doi.org/10.1016/j.micromeso.2017.09.006

Ravi, S., Zhang, S., Lee, Y. R., Kang, K. K., Kim, J. M., Ahn, J. W., & Ahn, W. S. (2018). EDTA-functionalized KCC-1 and KIT-6 mesoporous silicas for Nd3+ ion recovery from aqueous solutions. *Journal of Industrial and Engineering Chemistry, 67*(June), 210–218. https://doi.org/10.1016/j.jiec.2018.06.031

Ridley, M. K., Machesky, M. L., Wesolowski, D. J., & Palmer, D. A. (2005). Surface complexation of neodymium at the rutile-water interface: A potentiometric and modeling study in NaCl media to 250°C. *Geochimica et Cosmochimica Acta, 69*(1), 63–81. https://doi.org/10.1016/j.gca.2004.06.028

Roosen, J., Spooren, J., & Binnemans, K. (2014). Adsorption performance of functionalized chitosan–silica hybrid materials toward rare earths. *Journal of Materials Chemistry A*, *2*(45), 19415–19426. https://doi.org/10.1039/C4TA04518A

Roosen, J., Van Roosendael, S., Borra, C. R., Van Gerven, T., Mullens, S., & Binnemans, K. (2016). Recovery of scandium from leachates of Greek bauxite residue by adsorption on functionalized chitosan-silica hybrid materials. *Green Chemistry*, *18*(7), 2005–2013. https://doi.org/10.1039/c5gc02225h

Ryu, S., Fonseka, C., Naidu, G., Loganathan, P., Moon, H., Kandasamy, J., & Vigneswaran, S. (2021). Recovery of rare earth elements (Lu, Y) by adsorption using functionalized SBA-15 and MIL-101 (Cr). *Chemosphere*, *281*, 130869. https://doi.org/10.1016/J.CHEMOSPHERE.2021.130869

Sarmadi, N., Gharabaghi, M., Tamadoni Saray, M., Darestani, M., Garman, D., Koshy, P., et al. (2021). Highly mesoporous hybrid transition metal oxide nanowires for enhanced adsorption of rare earth elements from wastewater. *Inorganic Chemistry*, *60*(1), 175–184. https://doi.org/10.1021/acs.inorgchem.0c02762

Schaeffer, N., Grimes, S. M., & Cheeseman, C. R. (2017). Use of extraction chromatography in the recycling of critical metals from thin film leach solutions. *Inorganica Chimica Acta*, *457*, 53–58. https://doi.org/10.1016/j.ica.2016.11.020

Schoeb, V. R. (1965). *The stability constants of rare-earths with some carboxylic acids*. Retrospective Theses and Dissertations. 3376. https://lib.dr.iastate.edu/rtd/3376%0A

Schrobilgen, G. J., & Lang, C. E. (1968). Lanthanide distribution coefficients on Dowex chelating resin A-1. *Journal of Inorganic and Nuclear Chemistry*, *30*(11), 3127–3130. https://doi.org/10.1016/0022-1902(68)80183-6

Sert, Ş., Altaş, Y., Tel, H., Inan, S., Çetinkaya, B., Sengül, S., & Özkan, B. (2021). Investigation of sorption behaviors of La, Pr, Nd, Sm, Eu and Gd on D2EHPA-impregnated XAD7 resin in nitric acid medium. *Separation Science and Technology (Philadelphia)*, *56*(1), 26–35. https://doi.org/10.1080/01496395.2019.1708107

Shinozaki, T., Ogata, T., Kakinuma, R., Narita, H., Tokoro, C., & Tanaka, M. (2018). Preparation of polymeric adsorbents bearing diglycolamic acid ligands for rare earth elements [research-article]. *Industrial & Engineering Chemistry Research*, *57*(33), 11424–11430. https://doi.org/10.1021/acs.iecr.8b01797

Shu, Q., Khayambashi, A., Wang, X., & Wei, Y. (2018). Studies on adsorption of rare earth elements from nitric acid solution with macroporous silica-based bis(2-ethylhexyl)phosphoric acid impregnated polymeric adsorbent. *Adsorption Science & Technology*, *36*(3–4), 1049–1065. https://doi.org/10.1177/0263617417748112

Shu, Q., Khayambashi, A., Zou, Q., Wang, X., Wei, Y., He, L., & Tang, F. (2017). Studies on adsorption and separation characteristics of americium and lanthanides using a silica-based macroporous bi(2-ethylhexyl) phosphoric acid (HDEHP) adsorbent. *Journal of Radioanalytical and Nuclear Chemistry*, *313*(1), 29–37. https://doi.org/10.1007/s10967-017-5293-z

Shu, Z., Xiong, C., Shen, Q., Yao, C., & Gu, Z. (2007). Adsorption behavior and mechanism of D113 resin for lanthanum. *Rare Metals*, *26*(6), 601–606. https://doi.org/10.1016/S1001-0521(08)60013-3

Shumilova, Y. B., Gelis, V. M., Milyutin, V. V., Kharitonov, O. V., & Firsova, L. A. (2012). Separation of rare-earth and transplutonium elements by displacement chromatography on S-957 phosphonic ion exchanger. *Radiochemistry*, *54*(2), 164–167. https://doi.org/10.1134/S1066362212020129

Sivaraman, N., Kumar, R., Subramaniam, S., & Vasudeva Rao, P. R. (2002). Separation of lanthanides using ion-interaction chromatography with HDEHP coated columns. *Journal of Radioanalytical and Nuclear Chemistry*, *252*(3), 491–495. https://doi.org/10.1023/A:1015894418606

Smith, H. L., & Hoffman, D. C. (1956). Ion-exchange separations of the lanthanides and actinides by elution with ammonium alpha-hydroxy-isobutyrate. *Journal of Inorganic and Nuclear Chemistry*, *3*(3–4), 243–247. https://doi.org/10.1016/0022-1902(56)80025-0

Sparfel, D., & Cote, G. (2004). Synthesis and properties of new highly hydrophobic 7-substituted 8-quinolinols. *Solvent Extraction and Ion Exchange*, *22*(1), 1–12. https://doi.org/10.1081/SEI-120027570

Spedding, F. H., & Powell, J. E. (1954). The separation of rare earths by ion exchange. VIII. Quantitative theory of the mechanism involved in elution by dilute citrate solutions. *Journal of the American Chemical Society*, *76*(9), 2550–2557. https://doi.org/10.1021/ja01638a074

Spedding, F. H., Powell, J. E., & Wheelwright, E. J. (1954). The separation of adjacent rare earths with ethylenediamine-tetraacetic acid by elution from an ion-exchange resin. *Journal of the American Chemical Society*, *76*(2), 612–613. https://doi.org/10.1021/ja01631a091

Spedding, F. H., Voigt, A. F., Gladrow, E. M., & Sleight, N. R. (1947). The separation of rare earths by ion exchange.1,2I. Cerium and yttrium. *Journal of the American Chemical Society*, *69*(11), 2777–2781. https://doi.org/10.1021/ja01203a058

Strelow, F. W. E. (1960). An ion exchange selectivity scale of cations based on equilibrium distribution coefficients. *Analytical Chemistry*, *32*(9), 1185–1188. https://doi.org/10.1021/ac60165a042

Strelow, F. W. E., & Bothma, C. J. C. (1964). Separation of scandium from yttrium, lanthanum, and the rare earths by cation exchange chromatography. *Analytical Chemistry*, *36*(7), 1217–1220. https://doi.org/10.1021/ac60213a015

Strelow, F. W. E., Rethemeyer, R., & Bothma, C. J. C. (1965). Ion exchange selectivity scales for cations in nitric acid and sulfuric acid media with a sulfonated polystyrene resin. *Analytical Chemistry*, *37*(1), 106–111. https://doi.org/10.1021/ac60220a027

Strelow, F. W. E., & Victor, A. H. (1990). Separation of yttrium and neodymium from samarium and the heavier lanthanides by cation-exchange chromatography with hydroxyethylenediaminetriacetate in monochloroacetate buffer. *Talanta*, *37*(12), 1155–1161. https://doi.org/10.1016/0039-9140(90)80185-I

Sun, M., Chen, C., Chen, L., & Su, B. (2016). Hierarchically porous materials: Synthesis strategies and emerging applications. *Frontiers of Chemical Science and Engineering*, *10*(3), 301–347. https://doi.org/10.1007/s11705-016-1578-y

Sun, X., Ji, Y., Chen, J., & Ma, J. (2009). Solvent impregnated resin prepared using task-specific ionic liquids for rare earth separation. *Journal of Rare Earths*, *27*(6), 932–936. https://doi.org/10.1016/S1002-0721(08)60365-8

Sun, X., Peng, B., Ji, Y., Chen, J., & Li, D. (2008). The solid-liquid extraction of yttrium from rare earths by solvent (ionic liquid) impreganated resin coupled with complexing method. *Separation and Purification Technology*, *63*(1), 61–68. https://doi.org/10.1016/j.seppur.2008.03.038

Sun, Y., Shao, D., Chen, C., Yang, S., & Wang, X. (2013). Highly efficient enrichment of radionuclides on graphene oxide-supported polyaniline. *Environmental Science and Technology*, *47*(17), 9904–9910. https://doi.org/10.1021/es401174n

Sun, Y., Wang, Q., Chen, C., Tan, X., & Wang, X. (2012). Interaction between Eu(III) and graphene oxide nanosheets investigated by batch and extended X-ray absorption fine structure spectroscopy and by modeling techniques. *Environmental Science and Technology*, *46*(11), 6020–6027. https://doi.org/10.1021/es300720f

Suzuki, Y., Yokoi, S., Katoh, M., Minato, M., & Takizawa, N. (1980). Stability constants of rare-earth complexes with some organic ligands. In G. J. McCarthy, J. J. Rhyne, & H. B. Silber (Eds.), *The rare earths in modern science and technology* (pp. 121–126). New York: Springer. https://doi.org/10.1007/978-1-4613-3054-7_22

Svoboda, K., Kyrš, M., & Vaňura, P. (1997). Synergism in the sorption of europium on chromatographic supports impregnated with dicarbollide acid and bidentate phosphororganic extractant. *Journal of Radioanalytical and Nuclear Chemistry*, *220*(1), 47–54. https://doi.org/10.1007/BF02035346

Tan, X., Fang, M., Li, J., Lu, Y., & Wang, X. (2009). Adsorption of Eu(III) onto TiO2: Effect of pH, concentration, ionic strength and soil fulvic acid. *Journal of Hazardous Materials*, *168*(1), 458–465. https://doi.org/10.1016/j.jhazmat.2009.02.051

Thakkar, J., Wissler, B., Dudenas, N., Yin, X., Vailhe, M., Bricker, J., & Zhang, X. (2019). Recovery of critical rare-earth elements using ETS-10 titanosilicate [Rapid-communication]. *Industrial and Engineering Chemistry Research*, *58*, 11121–11126. https://doi.org/10.1021/acs.iecr.9b02623

Tong, S., Zhao, S., Zhou, W., Li, R., & Jia, Q. (2011). Modification of multi-walled carbon nanotubes with tannic acid for the adsorption of La, *Tb and Lu ions. Microchimica Acta*, *174*(3), 257–264. https://doi.org/10.1007/s00604-011-0622-3

Trewyn, B. G., Slowing, I. I., Giri, S., Chen, H. T., & Lin, V. S. Y. (2007). Synthesis and functionalization of a mesoporous silica nanoparticle based on the sol-gel process and applications in controlled release. *Accounts of Chemical Research*, *40*(9), 846–853. https://doi.org/10.1021/ar600032u

Trochimczuk, A. W., Kabay, N., Arda, M., & Streat, M. (2004). Stabilization of solvent impregnated resins (SIRs) by coating with water soluble polymers and chemical crosslinking. *Reactive and Functional Polymers*, *59*(1), 1–7. https://doi.org/10.1016/j.reactfunctpolym.2003.12.011

Trochimczuk, A. W. (2000). Synthesis of functionalized phenylphosphinic acid resins through Michael reaction and their ion-exchange properties. *Reactive and Functional Polymers*, *44*(1), 9–19. https://doi.org/10.1016/S1381-5148(99)00072-3

Trochimczuk, A. W., & Alexandratos, S. D. (1994). Synthesis of bifunctional ion-exchange resins through the Arbusov reaction: Effect on selectivity and kinetics.

Journal of Applied Polymer Science, *52*(9), 1273–1277. https://doi.org/10.1002/app.1994.070520912

Tu, Y. J., & Johnston, C. T. (2018). Rapid recovery of rare earth elements in industrial wastewater by CuFe2O4 synthesized from Cu sludge. *Journal of Rare Earths*, *36*(5), 513–520. https://doi.org/10.1016/j.jre.2017.11.009

Turanov, A. N., Karandashev, V. K., & Bondarenko, N. A. (2008). Extraction of rare-earth, yttrium, and scandium perchlorates by podands bearing diphenylphosphorylacetamide terminal groups. *Russian Journal of Inorganic Chemistry*, *53*(11), 1801–1808. https://doi.org/10.1134/S0036023608110223

Turanov, A. N., Karandashev, V. K., Sukhinina, N. S., Masalov, V. M., & Emelchenko, G. A. (2016). Adsorption of lanthanides and scandium ions by silica sol-gel material doped with novel bifunctional ionic liquid, trioctylmethylammonium 1-phenyl-3-methyl-4-benzoyl-5-onate. *Journal of Environmental Chemical Engineering*, *4*(4), 3788–3796. https://doi.org/10.1016/j.jece.2016.08.024

Turanov, A. N., Karandashev, V. K., Sukhinina, N. S., Masalov, V. M., Zhokhov, A. A., & Emelchenko, G. A. (2015). A novel sorbent for lanthanide adsorption based on tetraoctyldiglycolamide, modified carbon inverse opals. *RSC Advances*, *5*(1), 529–535. https://doi.org/10.1039/c4ra11999a

Unsworth, C. E., Kuo, C. C., Kuzmin, A., Khalid, S., & Saha, D. (2020). Adsorption of rare earth elements onto DNA-functionalized mesoporous carbon. *ACS Applied Materials and Interfaces*, *12*(38), 43180–43190. https://doi.org/10.1021/ACSAMI.0C09393/SUPPL_FILE/AM0C09393_SI_001.PDF

Vasylyeva, H., Mironyuk, I., Mykytyn, I., & Savka, K. (2021). Equilibrium studies of yttrium adsorption from aqueous solutions by titanium dioxide. *Applied Radiation and Isotopes*, *168*, 109473. https://doi.org/10.1016/j.apradiso.2020.109473

Veliscek-Carolan, J., Hanley, T. L., & Luca, V. (2014). Zirconium organophosphonates as high capacity, selective lanthanide sorbents. *Separation and Purification Technology*, *129*, 150–158. https://doi.org/10.1016/j.seppur.2014.03.028

Vigneau, O., Pinel, C., & Lemaire, M. (2001). Ionic imprinted resins based on EDTA and DTPA derivatives for lanthanides(III) separation. *Analytica Chimica Acta*, *435*(1), 75–82. https://doi.org/10.1016/S0003-2670(00)01279-4

Vigneau, O., Pinel, C., & Lemaire, M. (2002). Solid-liquid separation of lanthanide/lanthanide and lanthanide/actinide using ionic imprinted polymer based on a DTPA derivative. *Chemistry Letters*, *31*(2), 202–203. https://doi.org/10.1246/cl.2002.202

Wackerlig, J., & Schirhagl, R. (2016). Applications of molecularly imprinted polymer nanoparticles and their advances toward industrial use: A review. *Analytical Chemistry*, *88*(1), 250–261. https://doi.org/10.1021/acs.analchem.5b03804

Wakui, Y., Matsunaga, H., & Suzuki, T. M. (1988). Distribution of rare earth elements between (2-ethylhexyl hydrogen 2-ethylhexylphosphonate)-impregnated resin and acid aqueous solution. *Analytical Sciences*, *4*(3), 325–327. https://doi.org/10.2116/analsci.4.325

Wan, Y., & Zhao, D. (2007). On the controllable soft-templating approach to mesoporous silicates. *Chemical Reviews*, *107*(7), 2821–2860. https://doi.org/10.1021/cr068020s

Wang, F., Wang, W., Zhu, Y., & Wang, A. (2017). Evaluation of Ce(III) and Gd(III) adsorption from aqueous solution using CTS-g-(AA-co-SS)/ISC hybrid hydrogel adsorbent. *Journal of Rare Earths*, *35*(7), 697–708. https://doi.org/10.1016/S1002-0721(17)60966-9

Wang, H., & Gao, P. (2007). Adsorption of D113 resin for dysprosium(III). *Journal Wuhan University of Technology, Materials Science Edition*, *22*(4), 653–656. https://doi.org/10.1007/s11595-006-4653-2

Wang, Q., Wilfong, W. C., Kail, B. W., Yu, Y., & Gray, M. L. (2017). Novel polyethylenimine-acrylamide/SiO2 hybrid hydrogel sorbent for rare-earth-element recycling from aqueous sources. *ACS Sustainable Chemistry and Engineering*,*5*(11),10947–10958.https://doi.org/10.1021/acssuschemeng.7b02851

Wang, X., Xu, D., Chen, L., Tan, X., Zhou, X., Ren, A., & Chen, C. (2006). Sorption and complexation of Eu(III) on alumina: Effects of pH, ionic strength, humic acid and chelating resin on kinetic dissociation study. *Applied Radiation and Isotopes*, *64*(4), 414–421. https://doi.org/10.1016/j.apradiso.2005.08.010

Wang, Y., Chen, L., Yan, Y., Chen, J., Dai, J., & Dai, X. (2020). Separation of adjacent heavy rare earth Lutetium (III) and Ytterbium (III) by task-specific ionic liquid Cyphos IL 104 embedded polymer inclusion membrane. *Journal of Membrane Science*, *610*, 118263. https://doi.org/10.1016/j.memsci.2020.118263

Wang, Z., Ma, G., Lu, J., Liao, W., & Li, D. (2002). Separation of heavy rare earth elements with extraction resin containing 1-hexyl-4-ethyloctyl isopropylphosphonic acid. *Hydrometallurgy*, *66*(1–3), 95–99. https://doi.org/10.1016/S0304-386X (02)00109-3

Wang, Z., Ma, G., & Li, D. (1998). Extraction and separation of heavy rare earth(III) with extraction resin containing di(2,4,4-trimethyl pentyl) phosphinic acid (cyanex 272). *Solvent Extraction and Ion Exchange*, *16*(3), 813–828. https://doi.org/10.1080/07366299808934554

Wheelwright, E. J. (1969). A comparison of eluting agents for the ion-exchange purification of promethium. *Journal of Inorganic and Nuclear Chemistry*, *31*(10), 3287–3293. https://doi.org/10.1016/0022-1902(69)80115-6

Wheelwright, E. J., Spedding, F. H., & Schwarzenbach, G. (1953). The stability of the rare earth complexes with ethylenediaminetetraacetic acid. *Journal of the American Chemical Society*, *75*(17), 4196–4201. https://doi.org/10.1021/ja01113a020

Wódkiewicz, L., & Dybczyński, R. (1968). Anion exchange behaviour of the rare earth complexes with trans-1,2-diaminocyclohexane-N,N'-tetraacetic acid. *Journal of Chromatography A*, *32*(C), 394–402. https://doi.org/10.1016/s0021-9673(01)80506-9

Wood, S. A. (1993). The aqueous geochemistry of the rare-earth elements: Critical stability constants for complexes with simple carboxylic acids at 25°C and 1 bar and their application to nuclear waste management. *Engineering Geology*, *34*(3–4), 229–259. https://doi.org/10.1016/0013-7952(93)90092-Q

Wu, D., Sun, Y., & Wang, Q. (2013). Adsorption of lanthanum (III) from aqueous solution using 2-ethylhexyl phosphonic acid mono-2-ethylhexyl ester-grafted magnetic silica nanocomposites. *Journal of Hazardous Materials*, *260*, 409–419. https://doi.org/10.1016/j.jhazmat.2013.05.042

Wu, J., Li, Z., Tan, H., Du, S., Liu, T., Yuan, Y., et al. (2021). Highly selective separation of rare earth elements by Zn-BTC metal-organic framework/nanoporous graphene via in situ green synthesis. *Analytical Chemistry*, *93*(3), 1732–1739. https://doi.org/10.1021/ACS.ANALCHEM.0C04407/SUPPL_FILE/AC0C04407_SI_001.PDF

Xiao, Y., Huang, L., Long, Z., Feng, Z., & Wang, L. (2016). Adsorption ability of rare earth elements on clay minerals and its practical performance. *Journal of Rare Earths*, *34*(5), 543–548. https://doi.org/10.1016/S1002-0721(16)60060-1

Xie, F., Zhang, T. A., Dreisinger, D., & Doyle, F. (2014). A critical review on solvent extraction of rare earths from aqueous solutions. *Minerals Engineering*, *56*, 10–28. https://doi.org/10.1016/j.mineng.2013.10.021

Xiong, C., Chen, X., & Yao, C. (2011). Enhanced adsorption behavior of Nd(III) onto D113-III resin from aqueous solution. *Journal of Rare Earths*, *29*(10), 979–985. https://doi.org/10.1016/S1002-0721(10)60582-0

Xiong, C., Yao, C., & Wang, Y. (2006). Sorption behaviour and mechanism of ytterbium(III) on imino-diacetic acid resin. *Hydrometallurgy*, *82*(3–4), 190–194. https://doi.org/10.1016/j.hydromet.2006.03.012

Xiong, C., Wang, G., & Yao, C. (2011). Adsorption of ytterbium (III) from aqueous solution by SQD–85 resin. *Transactions of Nonferrous Metals Society of China*, *21*(12), 2764–2771. https://doi.org/10.1016/S1003-6326(11)61121-7

Xiong, C. (2008). Study on sorption of D155 resin for gadolinium. *Journal of Rare Earths*, *26*(2), 258–263. https://doi.org/10.1016/S1002-0721(08)60077-0

Xiong, C., Liu, X., & Yao, C. (2008). Effect of pH on sorption for RE(III) and sorption behaviors of Sm(III) by D152 resin. *Journal of Rare Earths*, *26*(6), 851–856. https://doi.org/10.1016/S1002-0721(09)60020-X

Xiong, C., Meng, Y., Yao, C., & Shen, C. (2009). Adsorption of erbium(III) on D113-III resin from aqueous solutions: Batch and column studies. *Journal of Rare Earths*, *27*(6), 923–931. https://doi.org/10.1016/S1002-0721(08)60364-6

Xu, H., & Guo, D. (2012). Synthesis and characterization of an ion-imprinted polymer for selective adsorption of copper ions in aqueous solution. *Adsorption Science and Technology*, *30*(4), 293–306. https://doi.org/10.1260/0263-6174.30.4.293

Xu, J., Koivula, R., Zhang, W., Wiikinkoski, E., Hietala, S., & Harjula, R. (2018). Separation of cobalt, neodymium and dysprosium using amorphous zirconium phosphate. *Hydrometallurgy*, *175*(November 2017), 170–178. https://doi.org/10.1016/j.hydromet.2017.11.010

Yadav, K. K., Dasgupta, K., Singh, D. K., Anitha, M., Varshney, L., & Singh, H. (2015). Solvent impregnated carbon nanotube embedded polymeric composite beads: An environment benign approach for the separation of rare earths. *Separation and Purification Technology*, *143*, 115–124. https://doi.org/10.1016/j.seppur.2015.01.032

Yang, B., Wu, S.-Z., Liu, X.-Y., Yan, Z.-X., Liu, Y.-X., Li, Q.-S., et al. (2020). Solid-phase extraction and separation of heavy rare earths from chloride media using P227-impregnated resins. *Rare Metals*, *1–12*. https://doi.org/10.1007/s12598-020-01549-4

Yang, S., Zong, P., Ren, X., Wang, Q., & Wang, X. (2012). Rapid and highly efficient preconcentration of Eu(III) by core–shell structured Fe3O4 @humic acid magnetic nanoparticles. *ACS Applied Materials & Interfaces*, *4*(12), 6891–6900. https://doi.org/10.1021/am3020372

Yang, X., Wan, Y., Zheng, Y., He, F., Yu, Z., Huang, J., et al. (2019). Surface functional groups of carbon-based adsorbents and their roles in the removal of heavy metals from aqueous solutions: A critical review. *Chemical Engineering Journal*, *366*(February), 608–621. https://doi.org/10.1016/j.cej.2019.02.119

Yang, X., Debeli, D. K., Shan, G., & Pan, P. (2020). Selective adsorption and high recovery of La3+ using graphene oxide/poly (N-isopropyl acrylamide-maleic acid) cryogel. *Chemical Engineering Journal*, *379*, 122335. https://doi.org/10.1016/j.cej.2019.122335

Yantasee, W., Fryxell, G. E., Addleman, R. S., Wiacek, R. J., Koonsiripaiboon, V., Pattamakomsan, K., et al. (2009). Selective removal of lanthanides from natural waters, acidic streams and dialysate. *Journal of Hazardous Materials*, *168*(2–3), 1233–1238. https://doi.org/10.1016/j.jhazmat.2009.03.004

Yao, C. (2010). Adsorption and desorption properties of D151 resin for Ce(III). *Journal of Rare Earths*, *28*(Suppl. 1), 183–188. https://doi.org/10.1016/S1002-0721(10)60324-9

Yin, W., Liu, L., Zhang, H., Tang, S., & Chi, R. (2020). A facile solvent-free and one-step route to prepare amino-phosphonic acid functionalized hollow mesoporous silica nanospheres for efficient Gd(III) removal. *Journal of Cleaner Production*, *243*. https://doi.org/10.1016/j.jclepro.2019.118688

Yu, Q., Ning, S., Zhang, W., Wang, X., & Wei, Y. (2018). Recovery of scandium from sulfuric acid solution with a macro porous TRPO/SiO2-P adsorbent. *Hydrometallurgy*, *181*(May), 74–81. https://doi.org/10.1016/j.hydromet.2018.07.025

Yuan, Y., Liu, J., Zhou, B., Yao, S., Li, H., & Xu, W. (2010). Synthesis of coated solvent impregnated resin for the adsorption of indium (III). *Hydrometallurgy*, *101*(3–4), 148–155. https://doi.org/10.1016/j.hydromet.2009.12.010

Yusoff, M. M., Mostapa, N. R. N., Sarkar, M. S., Biswas, T. K., Rahman, M. L., Arshad, S. E., et al. (2017). Synthesis of ion imprinted polymers for selective recognition and separation of rare earth metals. *Journal of Rare Earths*, *35*(2), 177–186. https://doi.org/10.1016/S1002-0721(17)60897-4

Zhang, W., Avdibegović, D., Koivula, R., Hatanpää, T., Hietala, S., Regadío, M., et al. (2017). Titanium alkylphosphate functionalised mesoporous silica for enhanced uptake of rare-earth ions. *Journal of Materials Chemistry A*, *5*(45), 23805–23814. https://doi.org/10.1039/C7TA08127H

Zhang, W., Hietala, S., Khriachtchev, L., Hatanpää, T., Doshi, B., & Koivula, R. (2018). Intralanthanide separation on layered Titanium(IV) organophosphate materials via a selective transmetalation process. *ACS Applied Materials & Interfaces*, *10*(26), 22083–22093. https://doi.org/10.1021/acsami.8b04480

Zhang, W., Koivula, R., Wiikinkoski, E., Xu, J., Hietala, S., Lehto, J., & Harjula, R. (2017). Efficient and selective recovery of trace scandium by inorganic titanium phosphate ion-exchangers from leachates of waste bauxite residue. *ACS Sustainable Chemistry & Engineering*, *5*(4), 3103–3114. https://doi.org/10.1021/acssuschemeng.6b02870

Zhang, W., Yu, S., Zhang, S., Zhou, J., Ning, S., Wang, X., & Wei, Y. (2019). Separation of scandium from the other rare earth elements with a novel macroporous silica-polymer based adsorbent HDEHP/SiO 2-P. *Hydrometallurgy, 185*(January), 117–124. https://doi.org/10.1016/j.hydromet.2019.01.012

Zhang, X., Li, C., Yan, Y., Pan, J., Xu, P., & Zhao, X. (2010). A Ce3+-imprinted functionalized potassium tetratitanate whisker sorbent prepared by surface molecularly imprinting technique for selective separation and determination of Ce3+. *Microchimica Acta, 169*(3), 289–296. https://doi.org/10.1007/s00604-010-0352-y

Zhang, Z., Fenter, P., Cheng, L., Sturchio, N. C., Bedzyk, M. J., Předota, M., et al. (2004). Ion adsorption at the rutile–water interface: Linking molecular and macroscopic properties. *Langmuir, 20*(12), 4954–4969. https://doi.org/10.1021/la0353834

Zhao, L., Azhar, M. R., Li, X., Duan, X., Sun, H., Wang, S., & Fang, X. (2019). Adsorption of cerium (III) by HKUST-1 metal-organic framework from aqueous solution. *Journal of Colloid and Interface Science, 542*, 421–428. https://doi.org/10.1016/j.jcis.2019.01.117

Zhao, X., Jiang, X., Peng, D., Teng, J., & Yu, J. (2021). Behavior and mechanism of graphene oxide-tris(4-aminophenyl)amine composites in adsorption of rare earth elements. *Journal of Rare Earths, 39*(1), 90–97. https://doi.org/10.1016/j.jre.2020.02.006

Zhao, Z., Sun, X., Dong, Y., & Wang, Y. (2016). Synergistic effect of acid–base coupling bifunctional ionic liquids in impregnated resin for rare earth adsorption. *ACS Sustainable Chemistry & Engineering, 4*(2), 616–624. https://doi.org/10.1021/acssuschemeng.5b01253

Zheng, X., Song, Z., Liu, E., Zhang, Y., & Li, Z. (2020). Preparation of phosphoric acid-functionalized SBA-15 and its high efficient selective adsorption separation of lanthanum ions. *Journal of Chemical and Engineering Data, 65*(2), 746–756. https://doi.org/10.1021/acs.jced.9b00976

Zheng, X., Wang, C., Dai, J., Shi, W., & Yan, Y. (2015). Design of mesoporous silica hybrid materials as sorbents for the selective recovery of rare earth metals. *Journal of Materials Chemistry A, 3*(19), 10334–10335. https://doi.org/10.1039/c4ta06860b

Zheng, X., Zhang, Y., Zhang, F., Li, Z., & Yan, Y. (2018). Dual-template docking oriented ionic imprinted bilayer mesoporous films with efficient recovery of neodymium and dysprosium. *Journal of Hazardous Materials, 353*(March), 496–504. https://doi.org/10.1016/j.jhazmat.2018.04.022

Zheng, Z., & Xiong, C. (2011). Adsorption behavior of ytterbium (III) on gel-type weak acid resin. *Journal of Rare Earths, 29*(5), 407–412. https://doi.org/10.1016/S1002-0721(10)60469-3

Zhou, H., Li, D., Tian, Y., & Chen, Y. (2008). Extraction of scandium from red mud by modified activated carbon and kinetics study. *Rare Metals, 27*(3), 223–227. https://doi.org/10.1016/S1001-0521(08)60119-9

Zhou, J., Liu, H., Liu, D., Yuan, P., Bu, H., Du, P., et al. (2022). Sorption/desorption of Eu(III) on halloysite and kaolinite. *Applied Clay Science, 216*, 106356. https://doi.org/10.1016/J.CLAY.2021.106356

Zhu, L., & Chen, J. (2011). Adsorption of Ce(IV) in nitric acid medium by imidazolium anion exchange resin. *Journal of Rare Earths, 29*(10), 969–973. https://doi.org/10.1016/S1002-0721(10)60580-7

Zhu, L. L., Guo, L., Zhang, Z. J., Chen, J., & Zhang, S. M. (2012). The preparation of supported ionic liquids (SILs) and their application in rare metals separation. *Science China Chemistry*, *55*(8), 1479–1487. https://doi.org/10.1007/s11426-012-4632-8

Zhu, X., & Alexandratos, S. D. (2014). The role of polarizability in determining metal ion affinities in polymer-supported reagents: Phosphorylated ethylene glycol. *Reactive and Functional Polymers*, *81*(1), 77–81. https://doi.org/10.1016/j.reactfunctpolym.2014.05.001

Zhu, X., & Alexandratos, S. D. (2015). Development of a new ion-exchange/coordinating phosphate ligand for the sorption of U(VI) and trivalent ions from phosphoric acid solutions. *Chemical Engineering Science*, *127*, 126–132. https://doi.org/10.1016/j.ces.2015.01.027

Zhu, Y., Wang, W., Zheng, Y., Wang, F., & Wang, A. (2016). Rapid enrichment of rare-earth metals by carboxymethyl cellulose-based open-cellular hydrogel adsorbent from HIPEs template. *Carbohydrate Polymers*, *140*, 51–58. https://doi.org/10.1016/j.carbpol.2015.12.003

Zulfikar, M. A., Zarlina, R., Rusnadi, Handayani, N., Alni, A., & Wahyuningrum, D. (2017). Separation of yttrium from aqueous solution using ionic imprinted polymers. *Russian Journal of Non-Ferrous Metals*, *58*(6), 614–624. https://doi.org/10.3103/S1067821217060189

INDEX

Rare Earth Elements: Sustainable Recovery, Processing, and Purification, Special Publications 79, First Edition. Edited by Athanasios K. Karamalidis and Roderick Eggert.

DOI: 10.1002/9781119515005.index

Printed and bound by CPI Group (UK) Ltd, Croydon, CR0 4YY
17/09/2026
14964242-0001